Penfolds®

The
REWARDS
of PATIENCE

Penfolds®

The
REWARDS
of PATIENCE

SEVENTH EDITION

Andrew Caillard MW

Contents

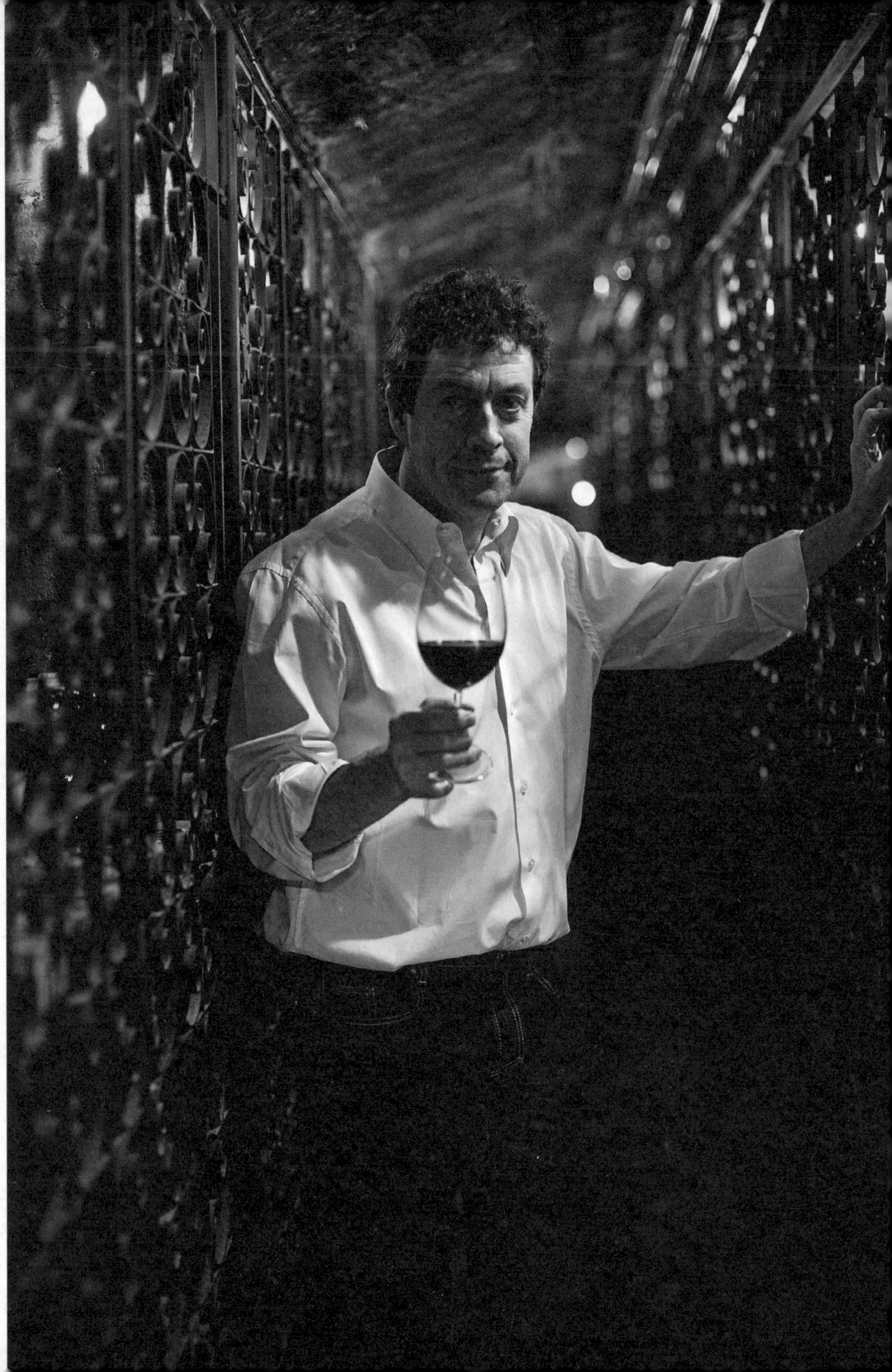

Foreword
Peter Gago

Nothing stands still. Just as the winemaking and culture of Penfolds continues to evolve, so does the story and this endeavour, *Penfolds The Rewards of Patience*. Now into its seventh edition, this book tells the story of 170 years of continuous red, white and fortified winemaking at Penfolds.

Over the years via an intimate involvement with four editions of this project, I have come to recognise that our wines should always remain front and centre of the Penfolds story. Yet much that accompanies the descriptions of that special liquid in a glass, in a bottle or in a cellar needs acknowledgement. The people, the places, the stories and history bring further appreciation of the wines. In this new edition research, interviews and detailed analysis of times past have corrected myths, connected the years and covered previous gaps.

Again, the Museum Cellar has been unlocked. We have shipped hundreds of rare bottles and complete verticals to support extraordinarily probing international media tastings in the US, Europe, China and Australia. As always, independent and forthright critique from the experienced tasters underscores the credibility and authority of *Penfolds The Rewards of Patience* tastings. The tasting notes and opinions in this edition are laced with different accents and descriptors – together with a few laughs and many surprises along the way.

But what a project this has been! We have tasted through an armoury of beguiling wines, many inherited from predecessors rich in character and spirit, and of course those created by the current winemaking team – a team that I wouldn't swap for my bodyweight in '53 Grange!

Revisiting lore, correcting 'truths', attempting to unveil hidden secrets of style, vintage and vineyard is a daunting task, let alone connecting the words and making sense of it all. Where to start, then? Enter Andrew Caillard.

Thank you – we are indebted.

Peter Gago
Penfolds Chief Winemaker
September 2013

Preface
Andrew Caillard, MW

Penfolds The Rewards of Patience, Seventh Edition, is a unique reference guide to the storied history of Penfolds, its wines and vintages. Over a period of sixteen months tastings took place in Beijing (2011), New York and Berlin and at Magill Estate, Adelaide (2012), with an impressive international expert panel of wine writers and journalists. Their tasting notes in this book, compiled by timeline, are a distillation of views about almost every fine table wine made at Penfolds since 1951.

New research material has added an extra richness to the story of Penfolds, the development of Grange and the history of St Henri. Nineteenth-century newspaper reports, aural histories and personal reflections of winemakers past and present have brought insights and previously unknown perspectives.

Indeed, the history of Penfolds is endlessly fascinating. It is a study of Australian ingenuity, colonial vision, modern aspirations and great winemaking. Max Schubert's imagination, Ray Beckwith's science and a loyal support team led to one of the most important collaborations in the history of winemaking. The development of Grange marks the beginning of modern wine.

Loyalty, a feeling of belonging, Penfolds own brand of craftsmanship and its wine styles are values that have been built up over generations. Such attributes, along with its fine wines, have made Penfolds an Australian institution.

For almost thirty years, I have enjoyed a special relationship with the winemakers and people at Penfolds. It has been a great privilege to author this completely revised and updated edition. I hope you enjoy it.

Andrew Caillard, MW

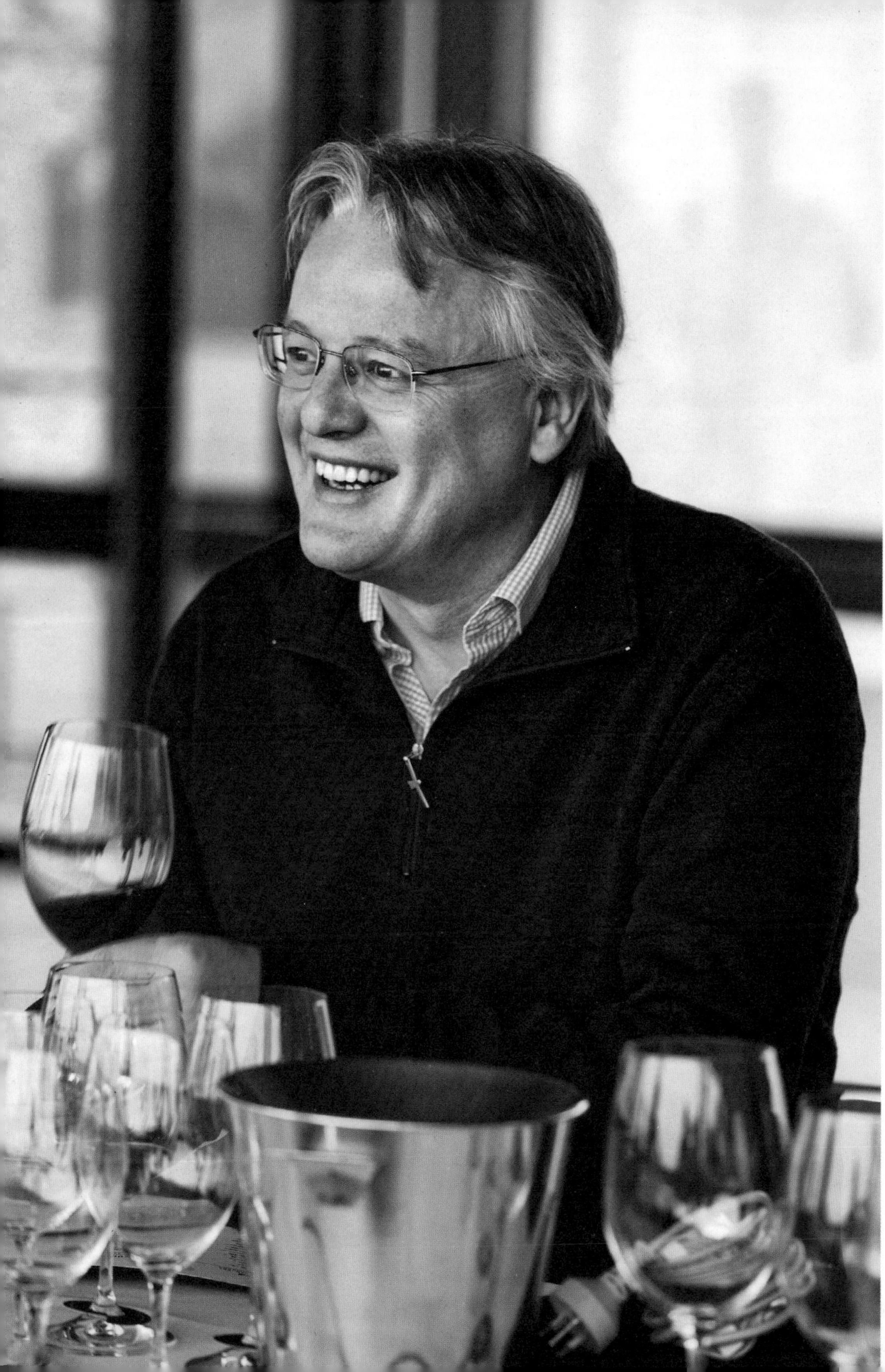

PART ONE

The Penfolds Legend

'These wines, from their age and the great care exercised in their manufacture and after treatment, are pure bright and thoroughly sound, and as a family wine, or for invalids, are unsurpassed.'

MARY PENFOLD

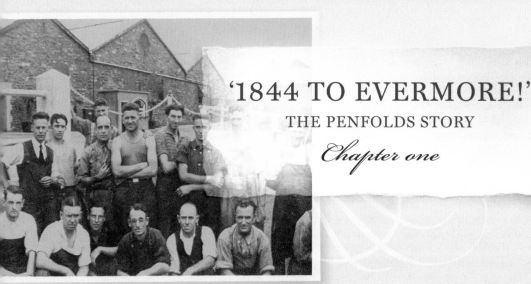

'1844 TO EVERMORE!'
THE PENFOLDS STORY
Chapter one

Penfolds is Australia's oldest iconic wine brand, with a history and heritage that profoundly reflects the country's journey from colonial settlement to the modern era. Established in 1844, just eight years after the foundation of South Australia, Penfolds has played a pivotal role in the evolution of winemaking in Australia – and across the world.

After early beginnings as a provider of iron-rich tonic wines for Adelaide's fledgling society, Penfolds had become Australia's largest producer of fortified wines by the time of Federation in 1901. The acquisition in 1945 of the nineteenth century–planted Kalimna Vineyard, on the western edge of the Barossa Valley, was a precursor to a period of extraordinary ambition and innovation.

In 1844 Marcus Collinson described South Australia as 'emphatically a poor man's country – a land of cheap bread, cheap meat and good wages!' The young British colony, only established in 1836, was enjoying rapid prosperity, with the growing production of wool and wheat and an economy based on a 'self-supporting principle'. In 1845 Adelaide, with a population of just 14 061 people, became a free port 'without charge for pilotage, harbour, wharfage, or other dues'. Its public institutions comprised two banks, a savings institution, a library, two Agricultural Societies, many Masonic and Odd Fellows lodges, a total abstinence society, a mining association and even a vine association. The city also boasted several schools, places of worship and three newspapers.

Within a few miles of Adelaide were 'numerous thriving villages, some of which have several hundreds of inhabitants, who chiefly follow agriculture and horticulture'. Flour mills, breweries, malt-houses, tanneries and brickworks typically dotted the locality, a reflection of the colony's expanding community and economy. South Australia's abundance of mineral wealth was already appreciated, with a handful of lead and copper mines nearby. Iron ore was 'abundant being found on the surface in various parts of the colony'. Already South Australia was exporting its produce and mineral wealth to the United Kingdom and elsewhere.

The settled country around Adelaide comprised 'undulating plains, succeeded by ranges of moderate elevation'. The plains were 'moderately timbered with stately eucalyptus and fragrant acacia'. During the summer months a rich profusion of the most beautiful flowers carpeted the landscape. The hills – covered with grass and numerous shrubs, and supplied with excellent water – were particularly suited to sheep farming and agriculture.

The climate was described by Collinson, 'a fellow South Australian colonist', as salubrious and healthful with 'summer heats moderated by the balmy coolness of the nights and after the lapse of a few days, usually succeeded by refreshing showers'. The hill country was perfectly 'adapted for those of temperate climes' and ideal for the production of stone fruits, citrus, walnuts, almonds and wine grapes.

Mackgill Estate, a vast property on the foothills of the Mount Lofty Ranges, was first acquired by explorer Robert Cock (1801–71) and William Ferguson, both Scotsmen, who travelled together on HMS *Buffalo* en route to the newly founded colony of South Australia in 1836. It was named after David Maitland Makgill Crichton of Rankeillour (1801–51), who was a prominent figure in the establishment of the Free Church of Scotland and Robert Cock's trustee. Soon after farming commenced the enterprise ran out of cash, resulting in a subdivision in 1838 and the establishment of Makgill village. The remaining 524 acre property, producing wheat and barley for local malting houses, was soon advertised for sale.

On 3 October 1844 Adelaide's *Observer* recorded that 'Mr Penfold is the fortunate purchaser of the delightfully situated and truly valuable estate of Makgill, [later named Magill] for the sum of one thousand, two hundred pounds ... comprising 500 acres of the choicest land 200 acres of which are under crops. The site of the residence is worthy of a noble mansion ... its woodlands offer a most agreeable background to this highly picturesque and desirable property.'

Only a few months earlier, on 18 June 1844, Dr Christopher Penfold, his wife Mary, infant daughter Georgina (1844–1911) and two servants had arrived at Port Largs, the disembarkation point near Adelaide, on the 350 ton barque *Taglioni*, with the ambition of building a new life away from the burden of debts left behind in England. At Makgill they established a surgery in the dining room of the Grange Cottage, farmed crops of wheat, barley and oats, and created a kitchen garden, an orchard and vineyard.

We lately saw a remarkably fine bed of asparagus in the garden of Dr Penfold, at Makgill. The plants, one year old, were unusually vigorous. The variety and excellence of this gentleman's collection of English succulents afford equal proof of the goodness of the soil and the judicious care of the cultivator. The Doctor has also a large number of vines and fruit trees, which promise amply to repay his spirited outlay.
—Launceston Chronicle, 1845.

Although the Penfolds first vine cuttings, specially waxed to preserve their moisture, had been purchased in South Africa on the voyage to Australia in 1844, a commercial vineyard required significant bundles of cuttings. These were sourced from William Macarthur (1800–82), whose family had introduced the Merino sheep to Australia and were important colonial pastoralists. William Macarthur and his brother James (1798–1867) had accompanied their father John Macarthur (1767–1834) on a tour of vineyards of France, Switzerland and northern Italy during 1815 and 1816. On their return to New South Wales, vineyards were planted at Camden Park, the Macarthur property, and on the banks of the Nepean River at Penrith. More vine cuttings, sourced from France and Spain by James Busby (1801–71), were later planted at Camden Park.

This vineyard became the mother vineyard of Australia's fledgling wine industry and supplied newly arrived colonial vignerons with cuttings suitable for the southern Australian climate. Christopher and Mary Penfold would have possessed a copy of William Macarthur's *Letters on the culture of the vine, fermentation, and the management of wine in the cellar*, published under his pen-name 'Maro' in Sydney, 1844. It was a key instruction book for vignerons of the time.

While Dr Penfold worked long hours setting up his practice, his wife Mary supervised the running of the house, the garden, the farm, the vineyard and winery. The first wines were prescribed as tonic wines for patients. 'A little wine for thy health's sake' was an age-old prescription, particularly for those suffering from anaemia.

Although the lure of gold was a constant distraction for many settlers during the colonial period, Dr Penfold stayed at Makgill with his family and became a highly respected surgeon and local government member. He was the first mayor of the District Council of Burnside, which was established in 1856 and whose boundaries included Makgill Estate. This property was also known locally as 'Dr Penfolds', as were the wines, illustrating the extent of his reputation in the young colony of South Australia.

By the 1860s, the Penfold family was shipping wheat to England and wine to other Australian colonies, especially Victoria which, since the first gold discoveries in 1851, had experienced an unprecedented influx of migrants and fortune seekers. Records survive today of Mary Penfold's wine production.

Wine in wood:
No. 1 quarter cask white wine – mixture Sweet Water and Frontignac – made March 17/64.
No. 2, 5 gallons red wine – mixture – made March 24/64.
Wicker bottle – all Frontignac – made March 17/64.
April 29th, racked off 10 gallons Muscat from that held – made by A.G. 1862
May 9th, 28 gals. Muscat – made April 30th – turned.
May 9th/1864 exposed to the air 3 days, then covered.

In 1861, Georgina Ann Penfold married Thomas Francis Hyland, an Irishman and an officer in the Victorian Civil Service. He was the Governor of Castlemaine Gaol in central Victoria, one of the colony's model prisons. After the death of Christopher Penfold in 1870, the highly ambitious and devoted son-in-law relinquished his position and worked hard to further expand the growing business. He travelled extensively throughout the Australian colonies and New Zealand. It was really his foresight, administrative

skills and planning that widened the foundations of Penfolds as Australia's first iconic wine brand.

Georgina and Thomas Hyland had seven children, two of whom died at birth. One daughter, Inez Kathleen Hyland (1863–92), was sent to live with Mary Penfold at the Grange Cottage, Makgill, in 1871 when only eight years old. A promising writer, poet and a beloved companion, Inez Hyland is remembered as a colonial literary figure; *In Sunshine and in Shadow*, published in 1893 after her death, is a collection of poems that includes observations of the Grange Vineyard through the seasons.

Joseph Gillard 'the younger' (1846–1927), Mary Penfold's newly appointed cellar manager, played a prominent role in keeping the vineyards and winery in working order. When interviewed by *The South Australian Register* in 1922, the veteran vigneron said, 'When Dr Penfold died in March 1870, Mrs Penfold was undecided what to do. I said if she would carry the business on, I would stick to her and I did, and everyone knows the success of this enterprise.'

Joseph Gillard had arrived in South Australia at five years of age in 1851 with his newly widowed father. His mother – along with his newly born sister – had died of childbirth on the voyage to Adelaide. After leaving school, he worked with his father Joseph Gillard (1820–97) as a blacksmith and vine-grower; his family's Sylvania Vineyard at Norwood was an important colonial wine producer. Joseph Gillard 'the younger' worked with Penfolds for thirty-five years and became an important identity in South Australia's wine industry. He was a foundation member of the Vinegrowers' Association and a local councillor. Nonetheless, it was Mary Penfold who remained firmly in charge at Magill until she became frail with old age.

Mrs Penfold makes but four varieties of wine, namely, sweet and dry red, and sweet and dry white. The work is done under Mrs Penfold's personal direction, not in conformity to any fixed or definite rule, but according to her judgement and taste. There are now in the cellars about 20.000 gallons of wine of an age which is ready for market, but the total stock is close upon 90.000 gallons. Mrs Penfold finds a large market for her wines in Victoria, Tasmania, and New Zealand, and has even sent some to India. There are three large warehouses on the property for making and storing the wine, and some of the vats are of huge dimensions. The largest holds about 5.000 gallons. The prices paid for grapes during the past vintage were very favourable to the growers, who disposed of the produce of their vines to the manufacturers. Tokays and Madeiras fetched about £4 per ton, and frontignacs, verdeilhos and muscats from £4 10/ to £5.
— The South Australian Register, 1874.

In Mary Penfold's wine circular of 1877, Penfolds was selling Mataro, Grenache, Pedro Ximenes [sic], Tokay, Frontignac and Muscadine by the gallon, and by case in quarts or pint bottles:

These wines, from their age and the great care exercised in their manufacture and after treatment, are pure bright and thoroughly sound, and as a family wine, or for invalids, are unsurpassed. To the retail trade, for station use, or where bottling is inconvenient, they offer the advantage that they may be drawn from the cask as required without deteriorating. Trial shipments to India and England have proved them to be unaffected by change of temperature and therefore adapted for export. Great extent of cellarage any [sic] large stocks enable us to ensure buyers the advantage of always receiving good old wines of uniform character and quality, which we will guarantee to keep sound and good.

In 1881, Thomas Hyland drafted a new partnership comprising himself, Mary Penfold and Joseph Gillard, to enable succession and further business growth. By this stage Penfolds was making over one-third of South Australia's wine. Under the management of Joseph Gillard and Thomas Hyland, Penfolds enjoyed increasing prosperity.

Inter-colonial sales of Penfolds wines, including exports to New Zealand, grew significantly during the 1880s, although colonial taxes, complicated border treaties and outdated rules and regulations hampered exports, while transport to Australian markets was not always easy. In 1884, the early days of rail, Thomas Hyland, representing Penfold and Co., took the unusual step of suing the Commissioner of Railways for overcharging on the newly inaugurated Hay to Sydney railway. The main railway between Adelaide and Melbourne was not completed until 1886.

A catalogue of 1889 lists wines from the Grange and Magill vineyards, such as Mataro, Grenache, Constantia, Grange Port, Frontignac, Grange Tawny, Pedro Ximenez, Tokay, Madeira, Grange Sherry and Muscadine. The catalogue adds: 'We have also light Red and White Dinner wines of Claret and Riesling types, suitable for use in clubs.' Penfold's Old Tawny Port, with a guarantee of fifteen years, was awarded a Gold medal at the Paris Exhibition in the same year.

In October 1893, *The South Australian Register* reported that 'Joseph Gillard, Manager of Penfold and Co, had been handed a Silver Salver, by the President of the Royal Agricultural Society as the prize for the best full-bodied red wines at the recent wine show'. This was the Champion Cup, awarded to Penfolds No. 1 Claret. These type of accolades became commonplace over forthcoming years.

During the 1890s exports gathered pace, but not without problems. Salvaged hogsheads of 'the choicest dry and sweet wines', including Penfolds, from the wreck of the *British Empire* were advertised for sale in New Zealand in 1894.

Mary Penfold died in 1896, in Melbourne. Her contribution to the South Australian wine industry was largely overshadowed by her husband's position and popularity. Nonetheless, she played a significant role in establishing Penfolds as the colony's most important wine producer.

* * *

In 1901, a federation of Australian colonies was formed to create the Commonwealth of Australia. At this time the population of Adelaide and environs was 180 825. The newly formed state of South Australia comprised around 363 000 people, while Australia's population was about 3 774 000. The new political union of the colonies resulted in the abolishment of customs and duties, a trade boom and a race to create better transport infrastructure between the Australian states.

Out of affection, and commercial foresight, the Hyland family adopted by deed poll the surname Penfold Hyland in 1905, to preserve continuity of the Penfold name. Thomas and Georgina Hyland's sons Herbert (1873-1940) and Frank Astor Penfold Hyland (1875-1948) were both educated in Australia and England. Frank entered the Penfolds business in 1892. After a stint in Europe studying winemaking methods and buying machinery and equipment, he established cellars in Sydney in 1901, to take advantage of the post-Federation trade boom. Herbert, known to his friends as 'Leslie', took over management of the South Australian cellars and vineyards in 1905.

Penfold's name is known throughout the world, the firm trading in the United Kingdom, India, Japan, China etc. as well as in all the Australian States. Their success at exhibitions has been very great, as they have, in addition to many Australian awards, obtained the gold medal at the Paris Exhibition against the wines of the world; whilst at the London Exhibition the jury of experts appointed by the King pronounced them as the best from Australia.
—The Sydney Morning Herald, 1905.

Over the next thirty years, Penfolds expanded at an extraordinary pace. In New South Wales, it acquired cellars in Sydney, and in 1904 the prized, historic Dalwood Vineyards in the Hunter Valley, established by George Wyndham in 1853, as well as the colonial Minchinbury Vineyards and Winery at Rooty Hill near the Penrith and Parramatta districts of Sydney in 1912, a magnificent property first farmed by Captain William Minchin in 1819.

In South Australia, Penfolds built a state-of-the-art winery at Nuriootpa, which was begun in 1911 and was inconveniently completed after the 1912 vintage; the contracted grower fruit, a guaranteed 1000 tons of grapes at a minimum price of £4 per ton, had to be taken by rail for processing at Magill. Paul 'Alfred' Scholz's meticulous management was 'clearly reflected in the polished brass, brilliant red paint and spotless appointments' of the Nuriootpa winery and distillery. Penfolds also purchased masses of Australian jarrah, for short-term storage, and oak for long-term maturation. After World War I, a consignment of 'shooks' – dismantled 5000 and 10 000 litre oak casks – arrived from France. These were made from Spessart oak and formed a part of Germany's war reparation to France.

In 1911 Penfolds typically produced Old Tawny, V.O. Invalid, No. 1 Rich and RR (Royal Reserve) ports, as well as No. 1, Golden and RR sherries, Grenache, Muscadine, Constantia and Frontignac. All the fortified wines were blended by Alfred Vesey, a Penfolds stalwart of sixty vintages, who worked for Mary Penfold and who trained Max Schubert as his understudy during the late 1930s. A smaller offering of No. 1, No. 0 and RR clarets and burgundy were also marketed.

Messrs Penfold & Co., of the Grange Vineyards, Magill, point out with pardonable pride that the success of Penfold's clarets at this year's Royal Sydney Show, held at Easter time, has never been equalled. Out of six possible first awards Penfolds obtained five firsts and one second.
—The Mail, Adelaide, 1914.

Although the Great War restricted production and manpower, the growing post-war economy and scarcity of imported wine created a pent-up demand for Australian wine. The State governments of New South Wales, Victoria and South Australia were also faced with finding jobs for returning soldiers. The

Murrumbidgee Irrigation Area, centred around the towns of Griffith and Leeton in New South Wales, was earmarked for new soldier settlements: blocks of land that could provide a family living based on horticulture and agriculture. The region, originally explored by John Oxley in 1817, was described by him as 'a country which for bareness and desolation ... has no equal'. Irrigation around the turn of the century transformed the area, now more commonly known as the Riverina district (Riverina is derived from the Spanish *entre rios*, 'between two rivers').

Penfolds was encouraged to build a winery and a distillery here. It also provided vine cuttings, technical support and a minimum price guarantee of £8 per ton of grapes for eight years to contracted growers. The first vintage crop was received in 1921, marking the beginning of irrigated viticulture on a grand scale in Australia, the fruit being used mainly for fortified wine production. As sweet fortified wine was the main Penfolds product, substantial distillation equipment was required to produce the required fortifying spirit in almost all wineries.

During the post–Great War years, Penfolds continued to expand at a rapid rate. New patented equipment including crushing mills, racking devices and 'slagetto' (a type of cement used for paving) tanks were installed at Nuriootpa. In October 1919 *The Mail*, an Adelaide newspaper, wrote an article about Penfolds: after vinification, the wines were 'run off into casks and despatched to Magill by rail from Penfolds station', for maturation. Houses, 'rough-casted on the outside to keep them cool', were offered to workers so that 'the men are satisfied, contented and take a greater interest in their work'.

In 1920, Penfolds built another winery and distillery at Eden Valley township, in the ranges east of Nuriootpa. Initially, wine was vinified at Eden Valley and taken to Nuriootpa for maturing. However, as road transport improved the winery was converted into maturation cellars for sherry. In the same year Penfolds marketed its first light dry white wine, Penfolds Minchinbury Trameah, based on gewürztraminer, known locally as traminer; the 1928 Trameah was regarded as one of the great Australian wines of the 1920s. But fortified wines, particularly tawny port and sherry, remained the dominant wine styles enjoyed by Australians. Nonetheless, sparkling wine production based on the Champagne method was hugely popular in the 1930s and Penfolds Minchinbury Champagne was the leading product of its time. The vineyards and cellars at Minchinbury, at first managed by Leo Buring, were one of the star attractions for leading dignitaries and celebrities of the time.

These Australian wines show a strength which is more comparable with the European wines, and in addition they have a beautiful fullness on the palate which only the sunlight and warmth of an Australian summer can impart.
—An Australian regional newspaper report about Penfolds Blue Label, Royal Reserve and No. 1 clarets, 1925.

In 1934 Penfolds produced a book called *An Empire of Achievement 1844–1934*. It boasts an extraordinary collection of vineyards and wineries in South Australia and New South Wales, as well as cellars in Brisbane, Sydney, Melbourne and Adelaide. An accompanying map, pinpointing every winery and cellar, illustrates the massive distances between each place. It also marks the zenith of Penfolds as a colonial family brand. A new era was about to begin, with worldwide political upheavals and social change. Many of Penfolds vineyards and wineries would succumb to urban encroachment, as Australia's cities expanded. Nonetheless, Penfolds would become a centrepiece of Australia's cultural identity as it rein-vested in new vineyards, new wine regions and the country's future.

During the 1930s Penfolds enjoyed substantial market share for its wines. Almost every newspaper in the country advertised Penfolds and its signage dominated the streets and pubs of Australia's major cities. An observer once remarked that Sydney might as well call itself Penfolds, such was the proliferation of advertising hoardings and merchandising. The Australian public was instructed 'Don't hope for the best, but get it' and 'Be never without Penfolds within'. The sales of Penfolds Royal Reserve Port dominated the Australian market and a typical message for it was 'A Royal Reserve beats a royal flush'. It was around 1947, a few years after Penfolds centenary, that the slogan 'Penfold's to evermore' was first used in advertising. The UK, Canada, Malaya, Singapore and a scattering of Pacific islands were the main export countries for Penfolds 'Sweet Red' – fortified ports. Penfolds also produced an array of table wines. Its clarets, burgundies and sauternes won prizes and medals in Adelaide, Sydney and 'Empire contests' in London and elsewhere.

Over 30 per cent of South Australia's population was unemployed during the Great Depression. To make matters worse, the State Government was proposing new duties and taxes that threatened wineries' profitability. In 1930, Penfolds threatened to close down all its wineries and stand down its workforce. There were also technical problems with wine production: a significant proportion of wine was 'turning'. The causes of sweet wine disease, as it was called, were completely unknown at that time. Up to 25 per cent of some wineries' stock was affected, which all had to be thrown out. It was also a time of generational change. Many of the employees at Penfolds who had worked with Mary Penfold or Joseph Gillard were nearing retirement.

In 1931, Max Schubert, aged sixteen, was employed by the fiery-tempered Austrian-born John Farsch, Penfolds first wine chemist, at Nuriootpa as a junior laboratory assistant. In 1935, after showing some aptitude for winemaking in general, he was transferred to the Magill Cellars to further his education in 'practical winemaking' with Alfred Vesey. In the same year, the brilliant young research scientist Ray Beckwith (1911–2012) was hired by Leslie Penfold Hyland to help tackle emerging technical challenges.

'As Mr Hyland had asked for a decision and a convenient date to start work at Nuriootpa, I wrote and suggested December 13th 1934. He replied and sent a cheque for £4-10-0 for one week's holiday pay with the instruction to start at Nuriootpa on 2nd January 1935.'
—Ray Beckwith's memoirs, circa 2001.

Beckwith recalled that Penfolds, in 1935, did not own any vineyards in the Barossa. All the grapes were purchased from independent growers, many still using 'German wagons and horses' to deliver their crop. The only company vehicle at Nuriootpa was a 'heavy masher dray pulled by a Clydesdale horse'. Wine – in hogsheads – left the winery by rail. However, things were much more advanced in the marketing arm: in 1929 Penfolds had a fleet of fourteen American-manufactured Nash 400 cars in Melbourne for its sales team.

With the support of 'legendary' general manager Alfred Scholz (the creator of Grandfather Port) at Nuriootpa, Ray Beckwith introduced a philosophy of preventative winemaking: improving hygiene standards and ensuring a stable microbiological environment for the wine. His observations, innovations and discoveries, 'kept under strict wraps' by Penfolds management, would profoundly impact the world of fine wine, but not before Penfolds achieved a considerable competitive advantage during the 1950s. Beckwith's importance to Penfolds in a post-war economy was already recognised: during World War II he was listed by the company as *Reserved Occupation*, meaning he was prohibited from enlisting for war service.

Max Schubert, newly appointed as assistant winemaker in 1938, signed up with the 6th Division

Australian Imperial Forces and served in the Middle East and New Guinea before rejoining the firm in 1946. Jeffrey Penfold Hyland, the son of Herbert 'Leslie' Penfold Hyland, rejoined Penfolds as a South Australian manager after war service.

During the war, Penfolds further invested in vineyards in New South Wales and South Australia. Perc McGuigan, 'the first to ever drive a tractor in the Hunter Valley', was employed as Penfolds vineyard manager at Dalwood in 1941. The following year the vineyard was acquired from the old Hunter Valley Distillery (HVD). Although it enjoyed a chequered history under Penfolds ownership, it would inadvertently enter wine history as the nursery vineyard for Australia's first commercial release of chardonnay: 'It was on a moonlit Pokolbin night at HVD in 1967 that Murray Tyrrell leapt the fence to purloin some of the discarded Penfolds chardonnay vine cuttings.'

In 1943, Penfolds acquired the prized Auldana Vineyards and winery, established by Patrick Auld outside Adelaide in 1853. John Davoren, who joined Auldana as its winemaker in 1947, would bottle his first experimental Penfolds St Henri Claret in 1953, celebrating Auldana's centenary and recreating a nineteenth-century tradition, first begun by French winemaker Léon Edmond Mazure (1860–1939).

In 1945, Penfolds purchased the Kalimna Vineyard in the Barossa Valley, the largest vineyard in South Australia at the time. Although it comprised the oldest nineteenth-century plantings of cabernet sauvignon in Australia, it was acquired principally for fortified wine production. Nonetheless, there was at least one bottling of 1948 Penfolds Kalimna Cabernet Sauvignon, mentioned by Max Lake in his *Classic Wines of Australia* (1966).

The Kalimna Vineyard and winery, under the possession of William Salter, was already producing over 400 000 litres of wine at the turn of the century. During the 1930s almost all of the production, then under the ownership of D. & J. Fowler, was exported to the UK. Having experienced wartime trading difficulties and the tragedy of losing a son during war service, the Kalimna Vineyard was offered to Penfolds by a very willing seller. Planted with cabernet sauvignon, shiraz, mataro and cinsaut, it would become the centrepiece of Penfolds post-war vineyard portfolio and the source block for the early Bin 28 and Bin 389 Dry Reds.

The Auldana, Magill and Kalimna vineyards would represent the foundation of Penfolds fine wine business during the heady days of the 1950s and 1960s. Penfolds soon recognised that the wine market would change as soldiers returned from World War II

and new immigrants from Italy, Greece and Eastern Europe settled in Australia. Although fortified wine production would remain important for another twenty years, dry table wine would increasingly feature in the daily lives of multicultural Australia.

After Frank Penfold Hyland died in 1948, his wife Gladys (1886–1974), an important Sydney identity and philanthropist, took over the reins as Chairman and principal shareholder. She relied heavily on her late husband's long-standing secretary Grace Longhurst for guidance. In a business dominated by men, this new arrangement took 'some getting used to', but it began the most exciting period in Penfolds history.

Impressed by the research and development of flor sherry production at Penfolds during the war and early post-war years, Gladys Penfold Hyland sent Max Schubert, the newly appointed senior winemaker at Magill, to Europe to check on stocks lying in London and to investigate sherry production and winemaking in Europe. This fateful visit would lead to an unexpected series of events that would irrevocably change the destiny of the company. Even so, Penfolds was the first wine producer to make and sell Australian flor sherry styles, which were to remain popular among Australian wine consumers until the early 1970s.

'Whatever your preference in Sherry, there's a Penfolds Sherry that will suit your taste exactly.'
—'Five reasons why Penfolds is Australia's most popular name in Sherry', advertising campaign, *The Age*, Melbourne, 1965.

The development of Max Schubert's 'Grange Hermitage' is a modern tale of imagination, a battle against the odds and redemption. It began with a side trip to Bordeaux, where a red wine 'capable of staying alive for a minimum of twenty years', first entered Schubert's mind. In 1950, he began a survey of vineyards, existing grape varieties and oak availability. His first experimental vintage in 1951 began a new way of thinking that would eventually lead to a signature wine style, but not before Grange was discredited and Max Schubert forced to make the wine in secrecy. The experimental Granges are a major body of achievement in the art and science of wine; perhaps the Australian wine industry's equivalent to the chronometer or powered flight. The development of the elegantly styled St Henri Claret, by Auldana's senior winemaker John Davoren, also took place during the 1950s, but most of its early vintages were not commercially released.

'The hidden Grange vintages are symbolic of what Penfolds is about: winemaking and creativity winning out against all the politics and turmoil happening up in the company world ...'
—Stephanie Dutton, Penfolds winemaker.

Jeffrey Penfold Hyland (1911–90), who began work as a cellar hand in the family firm, was assistant general manager of Penfolds in South Australia during the mid-1950s. Max Schubert said, 'without his support Grange would have died a natural, but not peaceful death'. Both having served in the Australian Imperial Forces during World War II, Jeffrey Penfold Hyland and Schubert intuitively gravitated towards each other. Assertive, visionary and confident, they created a remarkable winemaking culture of research and product development, particularly in South Australia.

By the early 1960s, Max Schubert remembered, Penfolds started creating a 'dynasty of wines which may differ in character from year to year, but all bear an unmistakable resemblance and relationship to each other'. Ray Beckwith's ground-breaking wine science and engineering skills allowed Penfolds winemakers to innovate without fear of spoilage.

'The use of pH, phase contrast microscopy, atomic absorption spectroscopy, etc., were all tools that contributed to an understanding and control of the winemaking process.'
—Ray Beckwith's memoirs, circa 2001.

The 1950s and 1960s embraced major advances in winemaking techniques, from yeast technology to fermentation practices (particularly barrel fermentation in American oak) and oak maturation. Prior to the development of Grange, Penfolds had removed new oak flavours from all its new casks and barrels, including its substantial collection of 'old English oak casks' and American oak hogsheads, by treating the timber with an alkaline agent. (In 1952 Penfolds advertising boasted that 'if placed end to end these casks would stretch over 35 miles'.) Penfolds also invested heavily in new equipment, including refrigeration and high-grade stainless steel vats.

The backbone of Penfolds red wine portfolio, Bin 389, Bin 707, Bin 28 and Bin 128, was introduced during the 1960s. These wines, together with Grange, would define the Penfolds brand as the market for fortified wines slowed down. The acquisition of prime *terra rossa* vineyards in Coonawarra during this period resulted in more reliable grape quality for the growing table wine market.

During this decade the first Penfolds Coonawarra wines were crushed at Redman's Rouge Homme Winery and then tankered up to Magill and Kalimna for fermentation and maturation. Experimental work with fruit from Coonawarra and the Kalimna Vineyard resulted in the legendary 1962 Bin 60A, a blend of Coonawarra Cabernet and Kalimna Shiraz, and regarded as one of the greatest wines of the twentieth century. Max Schubert and his team also entered a bewildering number of experimental or one-off wines into Australian wine shows. Even today old bottles with previously forgotten bin numbers or vintages turn up at auction, illustrating what a hive of activity it was at Penfolds during this period.

'We were always taught to use our imagination since the early days of winemaking at Penfolds. As opposed to going straight by the book, sometimes you have to throw the book away to be realistic and accomplish your aim of making top wines.'
—John Bird, senior red winemaker, with fifty-four Penfolds vintages.

In 1962, Max Schubert, now Penfolds Technical Director, was entrusted with the establishment of vineyards and production of Penfolds in New Zealand. This relatively short-lived venture is now almost forgotten, but it represents the beginning of the end of an era. Visionary, but a bridge too far, its potential was never realised. The acquisition of the Wybong Vineyard in the Hunter Valley in 1960 was also considered a mistake.

Gordon Colquist, a fifty-vintage veteran who joined Penfolds in 1938, said, 'Penfolds was very close knit. After the war everyone came together and worked, sometimes on Saturdays without pay, just to get the wines corked and bottled for the wine shows. We really wanted to win.' Taking the extra care and time was steeped in the culture of Penfolds, even during difficult times.

Penfolds became a public company in 1963 and all employees over the age of sixty-five were forced to retire, including the legendary Alfred Scholz, the hugely respected manager of the Nuriootpa winery. Despite mixed fortunes during the 1960s, Penfolds enjoyed a prominent market share in Australia for its fortified wines, vermouths, cocktails and table wines. Advertising and marketing reached a new level of sophistication with the considerable flair of Rada Penfold Russell (1923–80), who was instrumental in bringing about a change from fortified wine to table wine after World War II. A Sydney socialite, described as 'a lavish and flamboyant hostess', she regularly entertained media opinion leaders, wrote books on food and wine and brought the Penfolds brand to an even wider general audience. She encouraged consumers to steer away from 'the elaborate hocus pocus of certain self-appointed knights of the wine table' and introduced innovative advertising, such as 'What a wine label tells you about the wine it labels'.

By the 1970s, the Australian wine market was rapidly changing, with new multi-national entrants. Penfolds market share in Australia slipped from 16 per cent to 5 per cent in less than a decade. The Penfold Hyland family, having had a majority stake in the Penfolds business, finally lost control in 1976, beginning a new era of corporate ownership, new investments and, before too long, renewed vigour. Behind the scenes, the politics of a dying family dynasty, divided loyalties and ill health led to a change of guard. Ray Beckwith, John Davoren and Max Schubert all stepped down between 1973 and 1976.

The 1970s was also a period when Penfolds Grange slipped out of its revolutionary clothing into the respectable garb of an established and venerated fine wine. The release of 1971 Grange created a long-lasting waypoint. It is still regarded by many as the greatest vintage during this period. In 1973 the Magill cellars temporarily closed its doors as a crushing facility. It was also the same year that Jeffrey Penfold Hyland received an OBE for his services to the wine industry.

Although Penfolds launched new commercial lines – such as the ill-fated Blue Rhapsody, a blue-coloured sparkling wine – in general, conservatism and tradition underpinned winemaking during the post–Vietnam War era. When Grange production was moved from Magill to Nuriootpa in 1974, Max Schubert asked Robin Moody, an assistant production manager, to keep an eye on Grange and 'ensure there weren't any changes'. The winemaking team intuitively understood that it would be 'crazy to fiddle with a wine of this quality and style'.

Initially there was a period of consolidation under the new corporate ownership. Penfolds had already scaled back its vineyards in New South Wales during the 1960s, with the sale of Dalwood Estate to veteran Penfolds winemaker Perc McGuigan. This was followed by the sale of the ill-fated Wybong Vineyard in 1977 and the HVD vineyard to Tyrrell's in 1982, ending a seventy-eight-year association with the Hunter Valley. Even so, these assets would enjoy a new lease of life

under their new ownership during the coming years.

The continuing expansion of Adelaide's and Sydney's suburbs engulfed other Penfolds vineyards, leading to the controversial disposal or government acquisition of Minchinbury (1978), Auldana (1980), Modbury (1983) and partially Magill (1984) vineyards. These historic properties, which formed the backbone of Australia's colonial wine industry, now lie under modern housing estates.

Don Ditter, production manager in New South Wales, was formally appointed Penfolds Chief Winemaker in 1975. First hired as a laboratory assistant at Magill in 1942, he had been transferred to Sydney in 1953. As Chief Winemaker he introduced a more consultative and ordered winemaking management style. During this period of consolidation the Penfolds brand 'returned' to South Australia. The launch of the now legendary 1976 Koonunga Hill Claret marked the beginning of a golden period for Penfolds.

The hugely successful 'all you need to know about red wine' marketing campaign in the mid-1980s coincided with a buoyant domestic market and strengthening export sales, especially to the UK. Don Ditter reintroduced Bin 707, after a six-year hiatus, with the launch of the 1976 vintage: it realised Max Schubert's commitment to making an ultra-fine Cabernet Sauvignon. Within just a few years Bin 707 would be recognised as one of Australia's leading wines – illustrating that taking a step backwards (by discontinuing the bin number after the 1969 vintage) in favour of improving vineyard resources was a move forwards. Penfolds also released the first homage wines to the 1962 Bin 60A and 1966 Bin 620, a new chapter in wine marketing: the 1980 Bin 80A and 1982 Bin 820 were instantly accepted by the public and enjoyed great prestige at auction.

The launch of 1983 Magill Estate marked the end of several years of anguish. The destruction of this historic vineyard had been a huge blow to Penfolds winemakers, staff and the local community. Only twelve acres of vineyard, mostly replanted to shiraz in the early 1950s, remained. Don Ditter and Max Schubert believed that this small patch of vines could only be preserved and protected by making one of the world's great 'monopole' wines from it. The idea of a single-vineyard shiraz – from where Dr Christopher and Mary Penfold first planted their vineyard in 1844 – would provide a link with the past. Centred on the original Grange Cottage, the remnants of this historic vineyard still enjoyed 'the choicest' aspect. The 'Château Magill' project was mooted to Jim Williams, the pugnacious but fiercely loyal Penfolds General Manager.

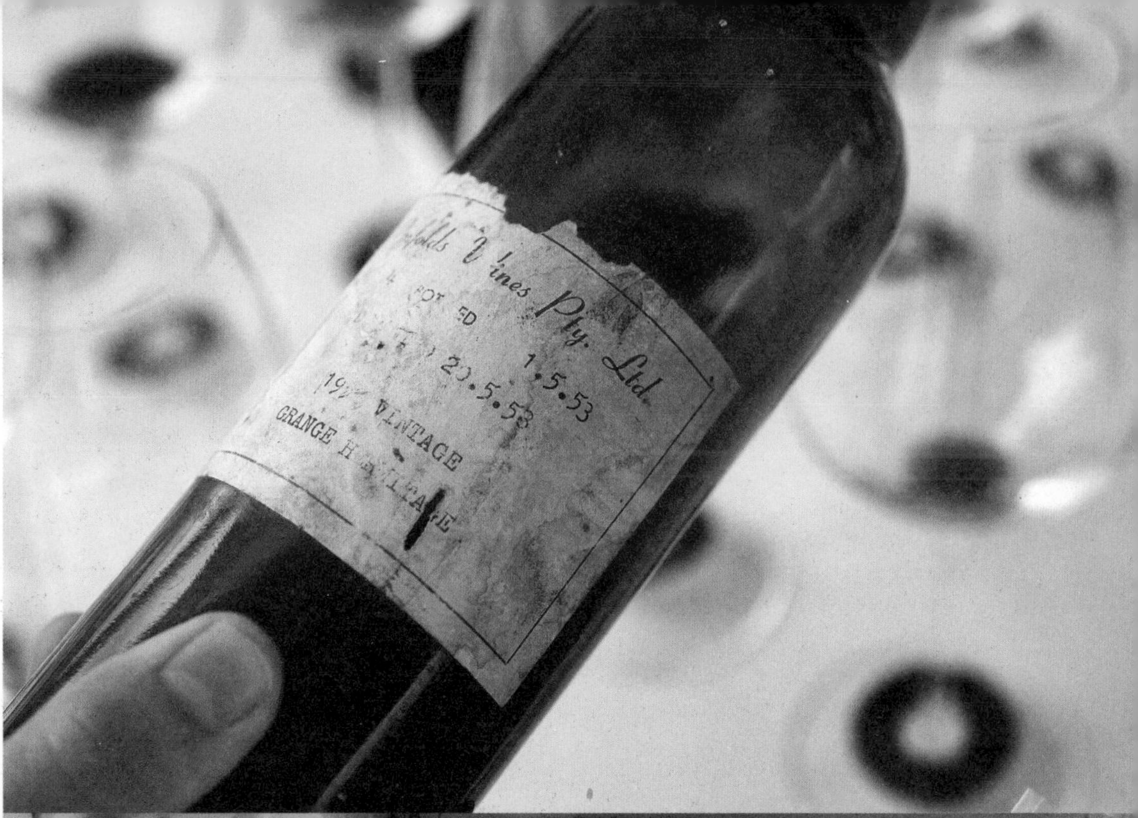
ines Pty. Ltd.
1.5.53
20.5.53
VINTAGE
GRANGE H

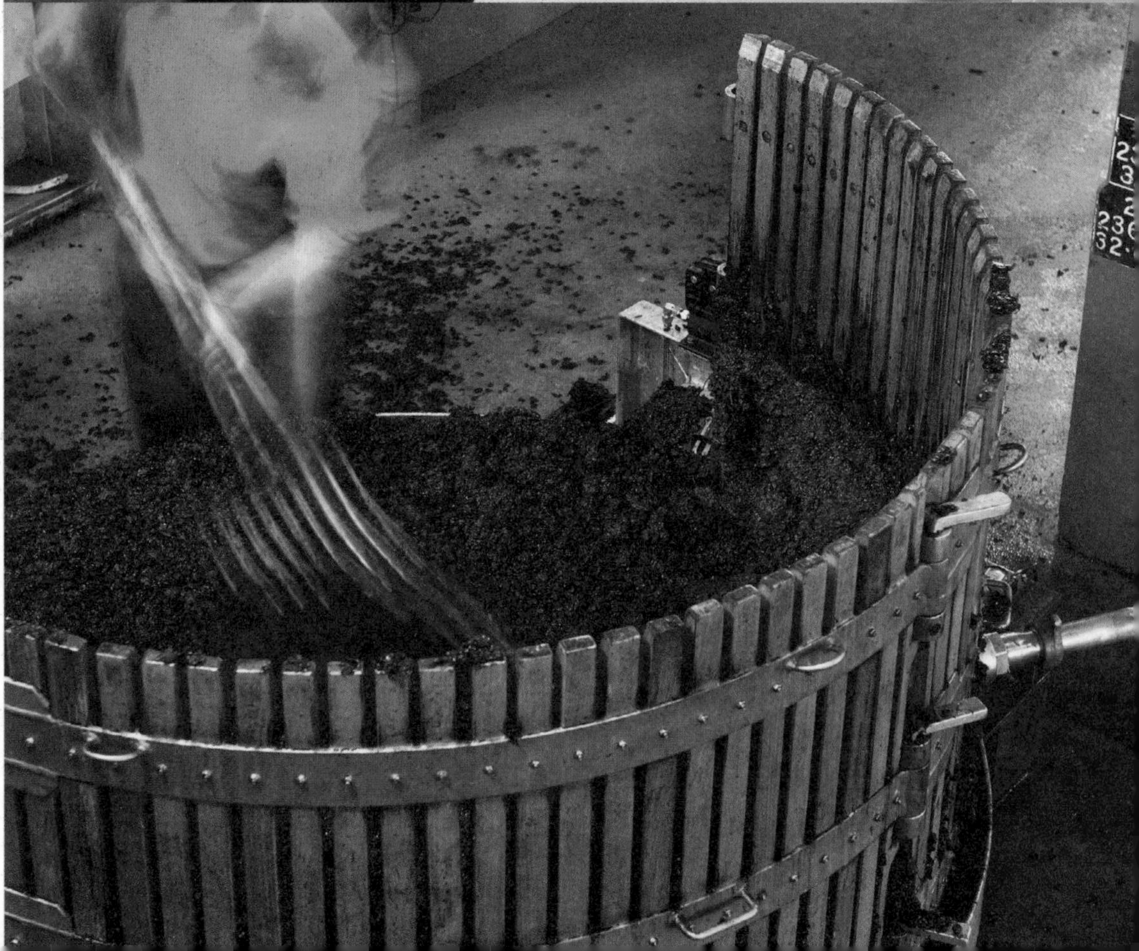

'I discussed it with Jim Williams, and I reckoned we could make something along the château line … he was enthusiastic about it, so we went into the next board meeting with this proposition that I would design a wine that would be different to a Grange and somewhat different to our other wines in the main and it would be more in keeping with what was then termed as the modern style.'
—Max Schubert, in an interview with Australian wine writers David Farmer and Chris Shanahan, 1992.

The board 'surprisingly went along with it'. Don Ditter and Max Schubert were delighted. They believed that the Tonkin Government of South Australia and the corporate accountants did not have the political will to save the historic Penfolds site. Over twenty years in the making, the dream of a single-vineyard wine, without compromise, is now being realised.

John Duval, an Adelaide University graduate in agriculture and understudy of Max Schubert and Don Ditter, was appointed Chief Winemaker in 1986. His family had for many years run a world-famous sheep stud and high-quality vineyards at Morphett Vale, south of Adelaide, coincidentally supplying shiraz grapes and vine cuttings to Penfolds. He oversaw one of the most dynamic periods of change in the Australian wine industry. It was also a period of unprecedented international accolades, reflecting Penfolds reputation as one of the world's great wine producers. Duval was awarded the International Wine and Spirit Competition's 'Winemaker of the Year' in London (1989) and the International Wine Challenge's 'Red Winemaker of the Year' in London (1991 and again in 2000). In 1995, the prestigious US magazine *Wine Spectator* announced 1990 Penfolds Grange as its Wine of the Year. Since 1990, Penfolds Grange has held pride of place at the head of Langton's Classification of Australian Wine, the most important benchmark of Australia's ultra-fine wine scene.

… a leading candidate for the richest, most concentrated dry table wine on planet Earth …
— Robert Parker, the world's most influential wine critic, writing about Grange in his *Wine Advocate*, 1995.

Max Schubert died peacefully on 6 March 1994. During his lifetime he was recognised for his achievements, being awarded the inaugural Maurice O'Shea Award and becoming a Member of the Order of Australia for his contribution to the Australian Wine Industry. He was also named by the UK's *Decanter* magazine as Man of the Year in 1988. His name, however, will be inseparably linked with Penfolds Grange for as long as it exists in the world of fine wine. In the year of his death, the South Australian Parliament created a new political electoral district called Schubert, in memory of Schubert and his extraordinary contribution to the prestige of South Australia and its wine industry. His legacy lives on: in 2001 *The Sydney Morning Herald* named Max Schubert in its top 100 most influential Australians of the twentieth century.

The 1990s, under the ownership of South Australian Brewing Holdings (later renamed Southcorp), saw the introduction of an unprecedented number of new styles within the Penfolds portfolio. This was also a period of intense winemaking trials and commercial opportunity. The search for new wine expressions and experiences was reminiscent of the 1950s and 1960s. The 'white Grange' project spawned the release of Penfolds Bin 144 Yattarna Chardonnay in 1995; the barrage of front-page media attention was extraordinary, illustrating Australia's national interest and pride in the Penfolds brand. Wine trials, including the 'experimental/homage' releases of 1990 Bin 90A, 1990 Bin 920 and 1996 Block 42, also led to the launch of RWT Shiraz, an ultra-fine Penfolds Barossa Shiraz, matured in French oak.

'Corporate ownership has always pushed us to redefine what we think is possible for the Penfolds brand. In the 1990s, to deliver a corporate result, we brought the Bin & Icons release forward to March and May, with the ambition to sell through an entire vintage in a few months … And it worked. It shocked the industry so much that brands now plan around the annual Penfolds release.'
—Andrew O'Brien, veteran Penfolds sales executive and General Manager Asia.

During the 1990s, Penfolds introduced the Penfolds Red Wine Re-corking Clinics, a unique and revolutionary after-sales service, where wine collectors can have their aged wines – particularly Grange – assessed, re-corked, topped up and re-capsuled. This ongoing project, now running for two decades, emphasises the ageing qualities and secondary market importance of the Penfolds brand. Over 120 000 bottles have been 'clinic-ed and certified' throughout Australia, New Zealand, the UK, Europe, the US and Asia, including China, Hong Kong and Singapore.

Penfolds also celebrated the fiftieth anniversary of Penfolds Grange, announcing plans to return some of the winemaking of Grange back to its original home at Magill. The following week a rare bottle of 1951

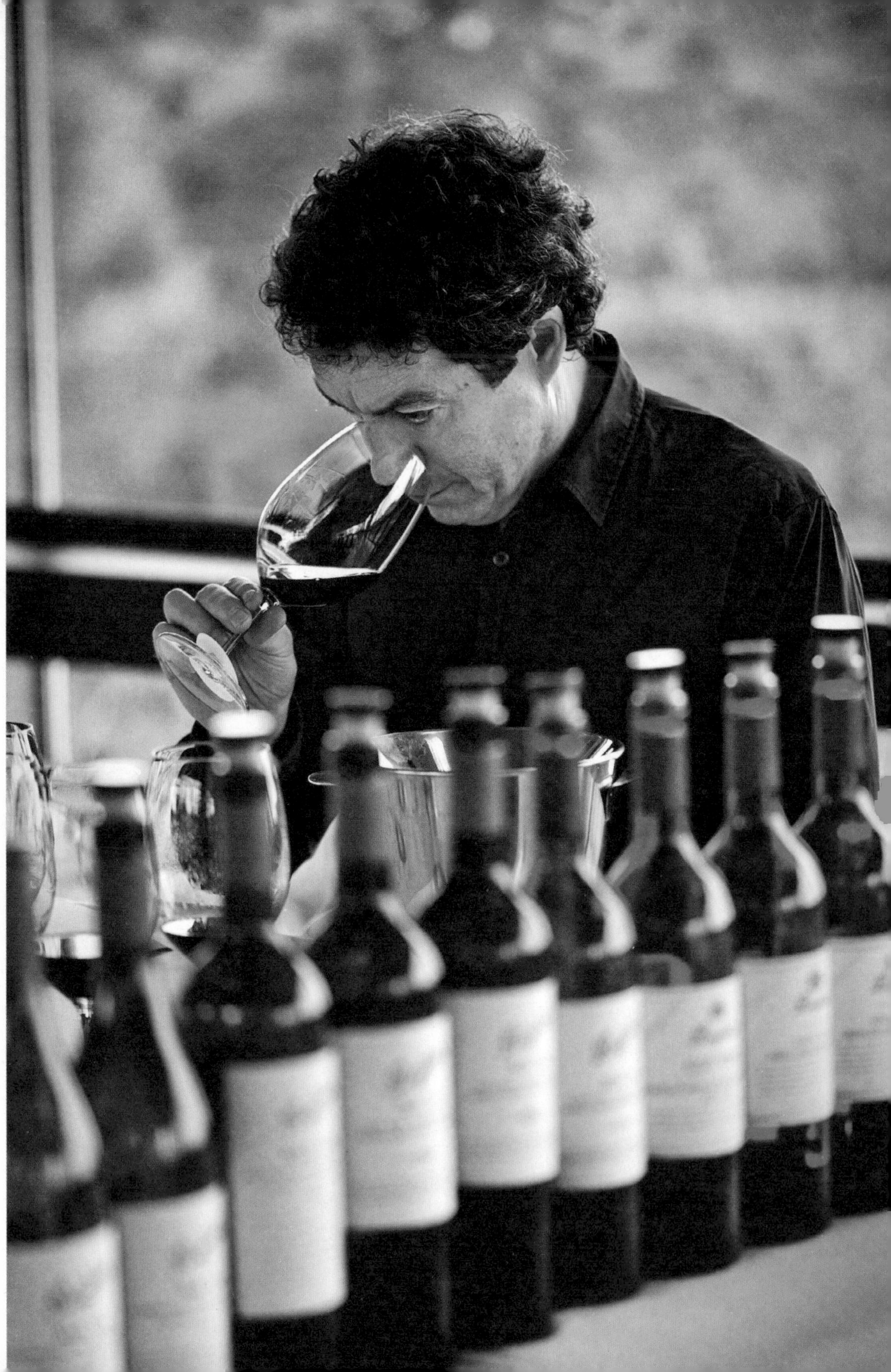

Grange Hermitage – in perfect condition – sold at auction for a record AU$52 211. In the same year the National Trust of South Australia, in a remarkable gesture, listed Penfolds Grange as a Heritage icon. In 2002 grapes destined for Grange were crushed and vinified at Magill Estate for the first time since 1973. In the same year, John Duval stepped down as Chief Winemaker and was succeeded by Penfolds oenologist Peter Gago, a Roseworthy Graduate in Oenology (Adelaide University) and a former teacher of mathematics and chemistry.

Down-to-earth, likeable and intelligent in equal measure, Gago is a natural and brilliant communicator who has effortlessly morphed into Penfolds ambassador-in-chief.
—John Stimfig, *Financial Times*, UK, 2012.

Peter Gago has reinvented the position of Chief Winemaker. Under his watch Penfolds has enjoyed unprecedented international attention, a renewed fascination in the people behind the scenes and a new golden age in winemaking, experimentation and engagement with the consumer. All of these things would not be possible without visibility, collaborative energy and imagination.

Although the ownership of Penfolds has changed several times since 2001 – first brewing company Fosters and then Treasury Wine Estates – the winemaking team has hardly changed in decades. There is an unbroken connection with the past. John Bird, who began as a laboratory technical assistant, has worked every Penfolds vintage since 1960. He was trained by winemakers who were taught by Alfred Vesey, Jim Warner, Alfred Scholz and cellar staff, some of whom had worked for Mary Penfold! Steve Lienert, Peter Gago's second-in-command and leading senior red winemaker, joined Penfolds in 1978 as a cellar hand. He now supervises all red wine production prior to classification and blending.

The 2000s have seen an expansion and consolidation of the Penfolds wine portfolio, to meet the needs and interests of a global premium wine brand. Peter Gago has overseen winemaking during the most productive and commercially successful period in Penfolds history. The range today reflects Max Schubert's ambitions of a dynasty of wine without compromise.

Experimentation and innovation are expressed in the release of Special Bin, Reserve Bin and Cellar Reserve wines. Highlights have included the releases of 2004 Bin 60A Cabernet Sauvignon Shiraz, 2004 Block 42 Cabernet Sauvignon, 2008 Bin 620 Cabernet Shiraz, 2010 Reserve Bin Block 25 Mataro

and 2010 Bin 170 Kalimna Shiraz. The new bin numbers reflect progress in white and red winemaking.

The extraordinary national and international show record of the Reserve Bin A series Chardonnay, and the acceptance of Yattarna as a credible foil to Grange, illustrate Penfolds cutting-edge winemaking. The release of Bin 150 Marananga Shiraz, Bin 169 Coonawarra Cabernet Sauvignon and Bin 23 Pinot Noir demonstrates market interest in Australia's classic regional styles. The historic backbone of Penfolds has been protected through rigorous selection, traditional craftsmanship and an uncompromising fine wine culture. The Rewards of Patience tastings arranged for this seventh edition amply show that the Penfolds wines, including Grange, St Henri, Bin 707, Bin 389, Bin 28 and Bin 128, are as good as ever. At the other end of the spectrum, Koonunga Hill Cabernet Shiraz is nowadays the entry wine to Penfolds range; the red winemaking team prides itself for offering something 'out of the ordinary' at its modest price point.

'At this point, I find myself in a difficult position. When I say this is one of the greatest red wines Penfolds has made in the last fifty years, I can hear the cries, "Well, he would have to say that, wouldn't he?" from polite readers, and a great deal worse from cynical tweeters, bloggers and so forth. My only response is that I have tasted the wine, and my critics haven't.'
—James Halliday, on the first official tasting of 2008 Bin 620 Cabernet Shiraz, in Shanghai, China, 2011.

Penfolds received a swag of awards and accolades during the 2000s. In 2005 the US *Wine Enthusiast* magazine named Peter Gago its 'Winemaker of the Year' and in the same year Langton's Classification of Australian Wine, led by Grange, recognised the history and popularity of the company's core range of fine wines, with:

- Exceptional – Bin 95 Grange, Bin 707 Cabernet Sauvignon
- Outstanding – St Henri, Bin 389 Cabernet Shiraz
- Excellent – Magill Estate Shiraz, RWT Barossa Valley Shiraz
- Distinguished – Kalimna Bin 28 Shiraz, Bin 128 Coonawarra Shiraz, Bin 407 Cabernet Sauvignon.

In more recent times 2008 Penfolds Grange received a perfect score of 100 points by two of the world's most influential wine publications, eRobertParker.com and *Wine Spectator*.

In 2011 Penfolds Reserve Bin 09A Chardonnay won the International Wine Challenge Champion White Wine award. At the same event, senior white winemaker Kym Schroeter was named among the world's top three white winemakers. He was also nominated in 2012 as *Gourmet Traveller Wine*'s winemaker of the year. In 2011 the prestigious *Wine and Spirits* magazine included Penfolds as a 'top winery of the year, for the twenty-second consecutive year'. In 2012, Peter Gago was the recipient of the Institute of Masters of Wine/*the drinks business* 'Winemakers' Winemaker Award'. This very special accolade, judged by leading winemakers and Masters of Wine, acknowledges a career devoted to fine wine making, Penfolds and Australian wine. Recently he was voted number 29 on *Decanter* magazine's International Fine Wine Power List.

The Institute of Masters of Wine and the drinks business are delighted to present this award to Peter Gago, a man who commits heart and soul to everything he does. It cannot be easy to manage an iconic brand and continue to take it to a new level, but he seems to constantly innovate and raise the bar of one of the world's most appreciated and valued wine labels.
—Lynne Sherriff, MW, Chairman of the Institute of Masters of Wine, 2012.

In 2012, Penfolds launched its highly ambitious 'Ampoule Project', a limited-release offering of twelve hand-blown glass ampoules, containing the celebrated 2004 Block 42 Cabernet Sauvignon. Each was housed in a crafted jarrah timber cabinet, and the wine was launched at the Philippe Starck–designed Cristal Room in the Baccarat showrooms, Moscow. Listed for sale at $168 000, the Ampoules attracted worldwide attention and controversy, but all eleven available found a buyer. The publicity for the Penfolds brand name was priceless. However, such projects, designed to build brand awareness, also celebrate craftsmanship and create something unique and different – the essence of the Penfolds story. Pushing the boundaries for the sake of making something better is a Penfolds tradition.

Penfolds now reaches into every major wine market of the world. The wines are widely celebrated for their diversity, history and quality across many price points. Although Penfolds continues to enjoy a high profile with its traditional customers, China has emerged as a remarkable new market, especially over the last five years. Don St Pierre Jnr, chief executive officer of ASC Wines, China, said in an interview with the UK's *Financial Times*, 'Apart from Penfolds auspicious red logo, China loves brands and it loves great red wine with a genuine narrative, which is primarily what Penfolds is all about. As a result it is firmly established as one of the top three global wine brands here.'

'The style of Penfolds is instantly recognisable. The wines are widely known for their longevity and class and are unmistakable, even in a blind tasting. To this day we are all driven to think outside the box, push boundaries, challenge norms … This philosophy started at the top and worked its way down. Penfolds wines will never die – they will only grow and those after us will carry on the tradition.'
—Andrew Baldwin, senior red winemaker of twenty-eight vintages.

The strength of Penfolds is that the wine always comes first. The work of previous generations, in the vineyards, the wineries and the markets, is recognised through a sense of history and place. Penfolds enjoys a unique Australian heritage; its wines evoke an ageless generosity of spirit, craftsmanship and the beauty of the Australian landscape.

Bread and Wine
by Inez K. Hyland (1863–92)
Daughter of Georgina and Thomas Hyland

A cup of opal
Through which there glows
The cream of the pearl,
The heart of the rose;
And the blue of the sea
Where Australia lies,
And the amber flush
Of her sunset skies,
And the emerald tint
Of the dragonfly
Shall stain my cup
With their brilliant dye.
And into this cup
I would pour the wine
Of youth and health
And the gifts divine
Of music and song,
And the sweet content
Which must ever belong
To a life well spent.
And what bread would I break
With my wine, think you?
The bread of a love
That is pure and true.

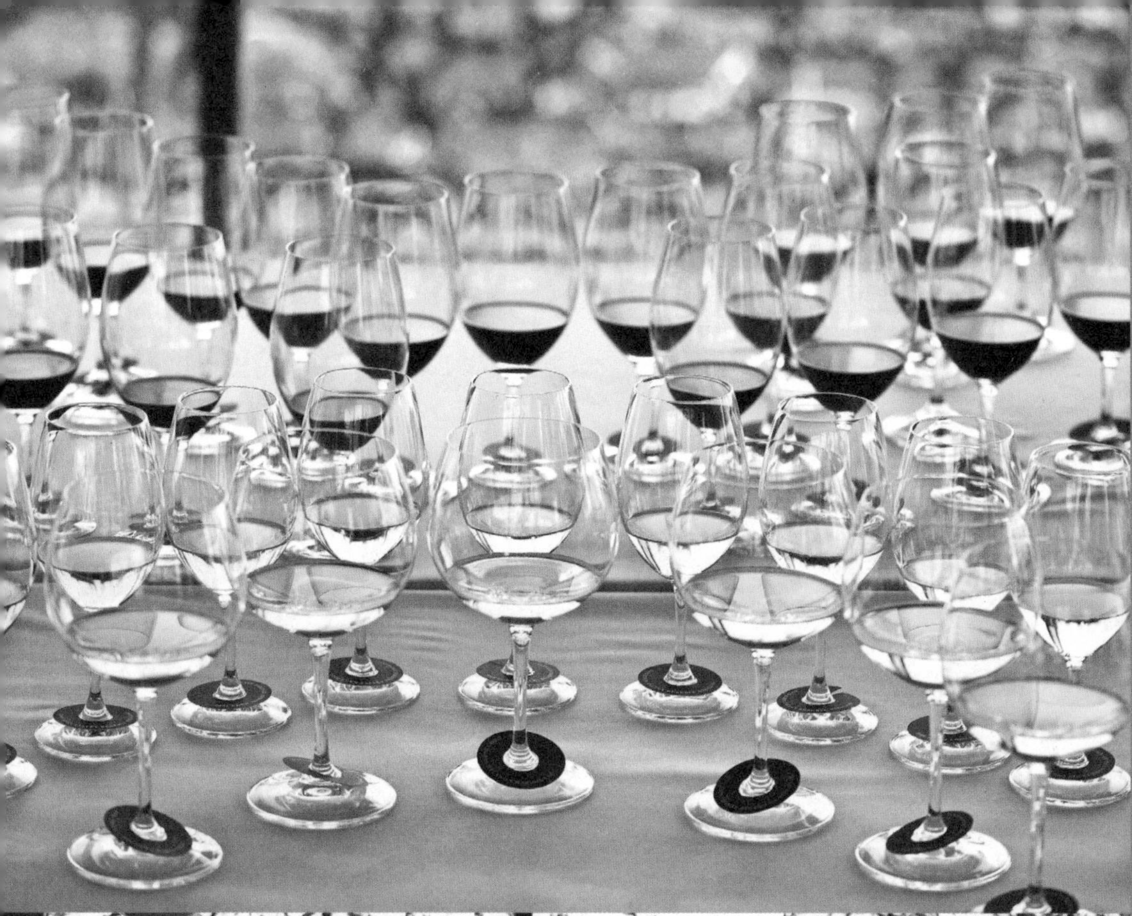

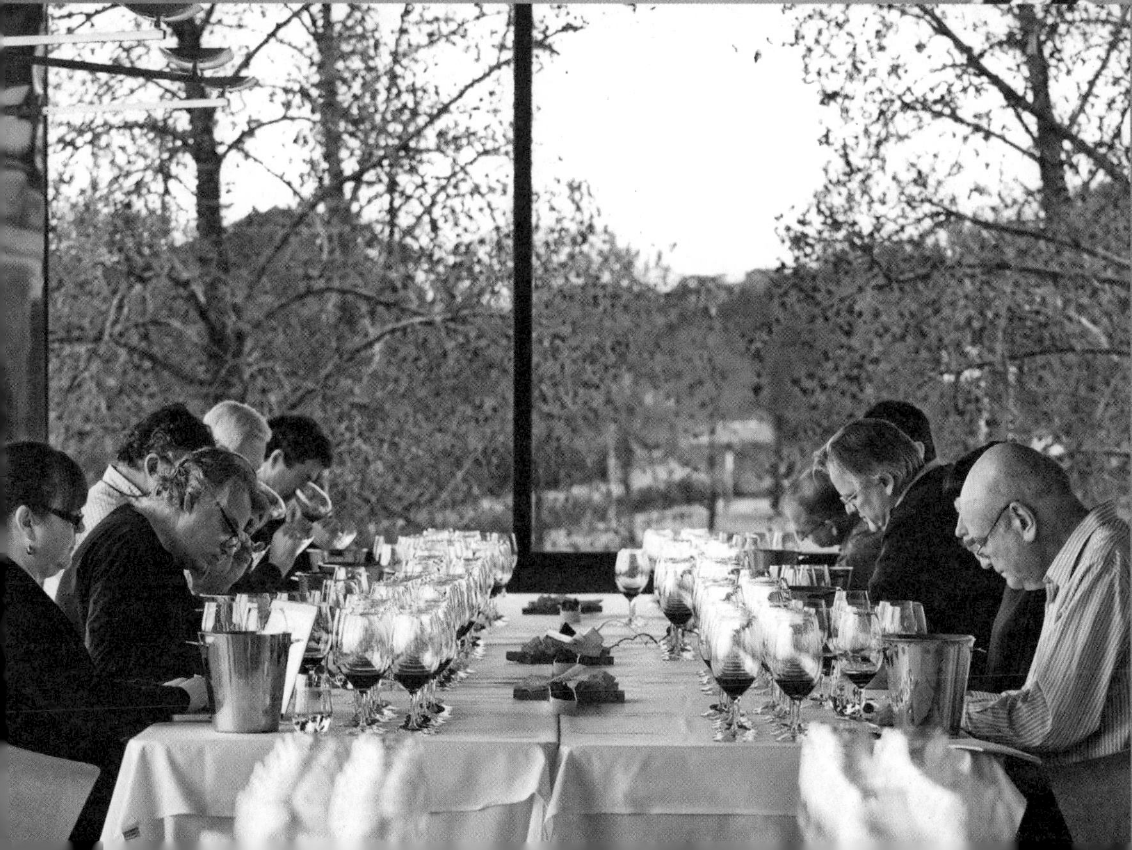

'We have some of the most experienced and dedicated
viticulturalists and grape growers in the world. We count
on them for the very best and we always receive it.'

STEVE LIENERT

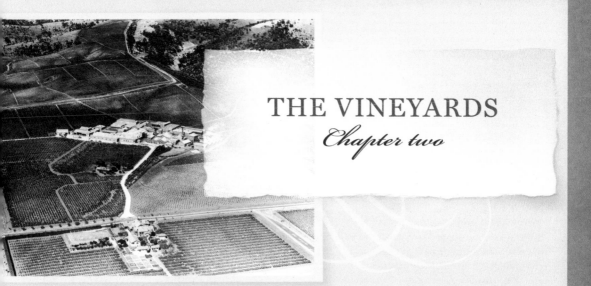

THE VINEYARDS
Chapter two

Dr Christopher and Mary Penfold planted their first vines at 'Mackgill' in 1844. Located approximately eight kilometres east of the centre of Adelaide, in the sheltered haunches of the Mount Lofty Ranges, 'Grange Vineyards' was regarded as one of the choicest sites in early colonial South Australia.

Penfolds later history of vineyard ownership follows the demographic and social development of Australia. Adelaide's and Sydney's post-war urbanisation saw the demise of great colonial vineyards, including Auldana, Grange and Minchinbury. The same period also saw the acquisition of new vineyards in the Barossa Valley, notably the nineteenth-century Kalimna Vineyard in 1945, Koonunga Hill in 1973 and prime *terra rossa* vineyards in Coonawarra during the 1960s.

Keeping one step ahead of progress has been the secret of Penfolds success since the 1880s when Messrs Penfolds & Co., the partnership of Mary Penfold, Thomas Francis Hyland and cellar master Joseph Gillard, was formed. Forecasting sales, fashions and the political landscape have shaped vineyard investments and the evolution of winemaking. It is a history of triumphs and mixed successes, steeped in ambition and grand visions. Today Penfolds vineyards are located on prime land throughout South Australia's viticultural regions. These holdings reflect the tradition of multi-vineyard and multi-regional sourcing, first introduced in the 1860s by Mary Penfold. Since those early days, Penfolds has purchased a large proportion of its grape intake from independent growers and has developed sustainable and innovative vineyard management practices to meet the demanding requirements of modern winemaking.

'Sustainability, consistency in quality and fruit sourcing are the key elements of Penfolds viticulture today.'
—Dean Willoughby, Penfolds viticulturalist.

During the 1940s, Penfolds still used horses in its vineyards for all manual work including ploughing and bringing in the harvest; long distances between vineyards and heat hampered the vintages' quality. However, vine research and technical advances in vineyard management led to significant progress in viticulture from the 1950s and 1960s. Mechanisation, clonal selection, vine architecture, canopy management, trellising systems, soil analysis, minimal tillage, precision irrigation, water deficit indexation, disease and yield control all contributed to improvements in fruit quality. Today, traditional dry-grown viticulture and a philosophy of organic and biodynamic farming, especially in small grower blocks, are increasingly common. Soil moisture monitoring, aerial infra-red photography and regularly walking the vineyards throughout the growing season give the modern grower a precise understanding of each block.

The magic of Penfolds begins in the vineyard. Winemakers cannot make good wine without the best quality fruit available. Dean Willoughby – a graduate in horticulture from New Zealand's Massey University and a former orchard grower – oversees the Penfolds crop from the Barossa and Clare Valley. He works closely with a 'field team' of grower liaison officers and vineyard managers, including Barossa vineyard manager George Taylor and Coonawarra viticulturalist Allen Jenkins, a leading light of Coonawarra's fine wine renaissance.

'We have some of the most experienced and dedicated viticulturalists and grape growers in the world. We count on them for the very best and we always receive it.'
—Steve Lienert, senior Penfolds winemaker for thirty years.

The best vineyard blocks are planted on soils with slow moisture and nutrient-releasing properties, including chalk, ironstone, *terra rossa* and medium-to-heavy red clays. The vines are regulated by pruning, crop thinning and using natural vegetation within the rows. Dry-land viticulture is widely practised by Penfolds throughout South Australia, with only minimal watering, sufficient to maintain vine health and optimum fruit quality. A naturally low-yielding vine that struggles always bears the most intensely flavoured fruit. Thus vineyards with low nitrogen-containing soils in areas of low growing-season rainfall produce the best quality grapes.

On a typical spring day the field team is 'making sure the vines are in balance with the character of the season'. Shoot and bunch thinning, done before fruit set, determines the potential quality and flavour con-centration of the fruit. In summer viticulturalists drive around the vineyards and monitor the development of the crop. It is also a time when grape yields can be more accurately estimated. This helps with harvest planning and addressing potential shortfalls or surpluses. The vineyards are regularly assessed and monitored leading up to harvest. A spell of wet weather, for instance, can cause disease pressure and affect the quality of the fruit.

Prior to vintage, a field classification takes place in each vineyard. Dean Willoughby and his team will typically look at the berry size, configuration of grape bunches and colour intensity of the skins. The fruit is then tasted for 'flavour intensity, skin sweetness, bal-ance of ripeness in fruit pulp and seed tannins'. The widely planted shiraz clone 1654 is particularly suited to warm, dry-land viticulture whereas cabernet sau-vignon usually performs well in the Barossa 'during a cool February'. Experience and technical knowledge gained over decades of vintages, together with long-standing relationships with growers and winemakers, bring an intangible but meaningful difference to the assessment of the grapes and decisions about harvest-ing. Dean Willoughby says, 'Clones and vine age are important but not as essential as *terroir* – a vine will not express itself if it's grown in the wrong place.'

'I love the fact that no two vintages are the same – every season is unpredictable and every season presents its own set of challenges.'
—Dean Willoughby, Penfolds viticulturalist.

At the start of harvest Dean Willoughby is 'excited, impatient and slightly anxious'. A usual day comprises an early-morning round with winemakers 'to check the ferments and flavour ripeness and get some feedback from winemaking'. Before heading out to the vineyards, the viticulturalists download the latest maturity results on their iPads and use this information to prioritise the vineyard visits of the day. Harvest day, potentially chaotic, is extremely well organised. 'Winemakers and viticulturalists can travel to as many as forty blocks during the day to grade or classify the fruit. After a random selection and tasting of berries, lots and lots of phone calls are made to booking officers, harvest contractors and growers to coordinate the harvesting.'

Penfolds uses an alpha-numerical classification system to grade fruit. Growers who provide A1 fruit for Grange or Yattarna receive a greater dividend than fruit destined for Koonunga Hill wines. Typically, A grade fruit has greater aromatics, colour intensity, flavour intensity, tannin ripeness and overall balance than C grade fruit.

The 'Grange Growers Club' is an exclusive group of growers who regularly contribute A1 fruit to Penfolds. Frank Gallach, John Gersch and Jamie Nieschke in the Barossa and the Oliver family in McLaren Vale have been growing Grange-grade fruit for many years. Paul Georgiadis, a previous head of Penfolds viticulture and nowadays a consultant, is also an important supplier. His Marananga property, in the heart of the Barossa, is one of the region's finest vineyards. A pioneering grower of sangiovese, he has contributed fruit to many of Penfolds luxury, icon and special bin wines.

'Soil type and depth are primary drivers of terroir *as they determine how vines grow and ripen.'*
—George Taylor, Penfolds viticulturalist.

Over winter, the newly completed vintage is assessed and plans for the next growing season are discussed. Work in the vineyard is never finished. Some vine-yard blocks may need replanting or retrellising. In response to climate change, new vineyards have been planted on south-facing slopes with an east–west row orientation, to deal with the increasing frequency of hot weather. Some vines may need to be replaced or treated for eutypa dieback – also known as 'dead arm disease'. By mid-winter the pruning season is well underway. It's also a time when soil moistures, friabil-ity and health are assessed.

Viticulture is a constant work in progress. The cycle of life in the vineyard never ends. The patterns of weather, the condition of the soil and the character of the season bring new challenges every year. Working with nature is the underlying modus operandi in Penfolds-owned and -managed vineyards. A balance between the practical and sustainable is achieved through understanding the micro-landscape and identity of each vineyard block.

Low-input dry-land viticulture is a characteristic of all Penfolds-managed vineyards. Methods of cultivation vary, however, reflecting the character and growing principles of each vineyard site. Organic and biodynamic viticulture are becoming important growing philosophies and will increasingly play a part in Australia's ultra-fine wine scene. The protection of Penfolds old vine heritage is also another concern. There are increasing incidents of eutypa dieback, while the old threat of phylloxera persists. Sustainability, security and conservation are underlying themes for the future.

Penfolds Vineyards

SOUTH AUSTRALIA
NEW SOUTH WALES
VICTORIA
TASMANIA

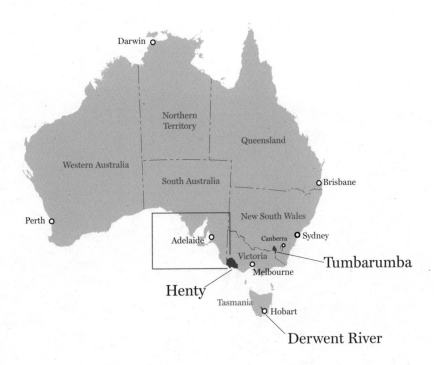

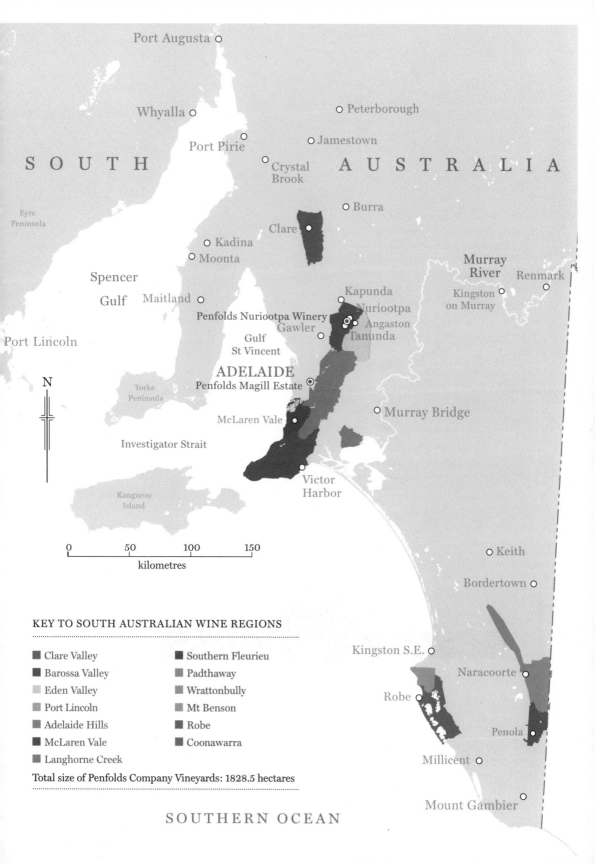

Port Augusta

Whyalla

Port Pirie

Crystal
Brook

SOUTH

Peterborough

Jamestown

AUSTRALIA

Eyre
Peninsula

Burra

Clare

Kadina

Moonta

Spencer

Gulf

Maitland

Kapunda

Penfolds Nuriootpa Winery

Gawler

Nuriootpa

Angaston

Tanunda

Murray
River

Kingston
on Murray

Renmark

Port Lincoln

Gulf
St Vincent

ADELAIDE

Penfolds Magill Estate

N

Yorke
Peninsula

McLaren Vale

Murray Bridge

Investigator Strait

Victor
Harbor

Kangaroo
Island

0   50   100   150

kilometres

Keith

Bordertown

KEY TO SOUTH AUSTRALIAN WINE REGIONS
....

■ Clare Valley

■ Barossa Valley

■ Eden Valley

■ Port Lincoln

■ Adelaide Hills

■ McLaren Vale

■ Langhorne Creek

■ Southern Fleurieu

■ Padthaway

■ Wrattonbully

■ Mt Benson

■ Robe

■ Coonawarra

Total size of Penfolds Company Vineyards: 1828.5 hectares
....

Kingston S.E.

Naracoorte

Robe

Penola

Millicent

Mount Gambier

SOUTHERN OCEAN

Barossa Valley

The Barossa is 70 km north of Adelaide and is comprised of two distinct sub-regions: the Barossa Valley (shown here) and Eden Valley. Penfolds largest winery is close to the township of Nuriootpa. The Barossa Valley is home to many of Penfolds most celebrated company and grower vineyards. For more information, see page 32.

Statistics

TOTAL AREA	Total vineyard area for entire Barossa Valley GI: 10 350 ha
ALTITUDE	230–270 m
RAINFALL	Growing season rainfall, October–April: 160 mm
TEMPERATURE	Mean January temperature: 21.4°C
VARIETALS SOURCED	Penfolds varietals sourced: shiraz, cabernet sauvignon, grenache, mataro and sangiovese

BAROSSA ALTITUDE

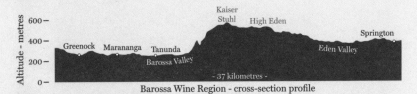

Barossa Wine Region - cross-section profile

BAROSSA VALLEY SOIL TYPES

The Barossa Valley soils are ancient, up to 200 million years old. The soils vary throughout the sub-regions ranging from red-brown earth over heavy clay at Koonunga Hill to duplex sandy loams over clay at Kalimna and red-brown soils over limestone in Gomersal.

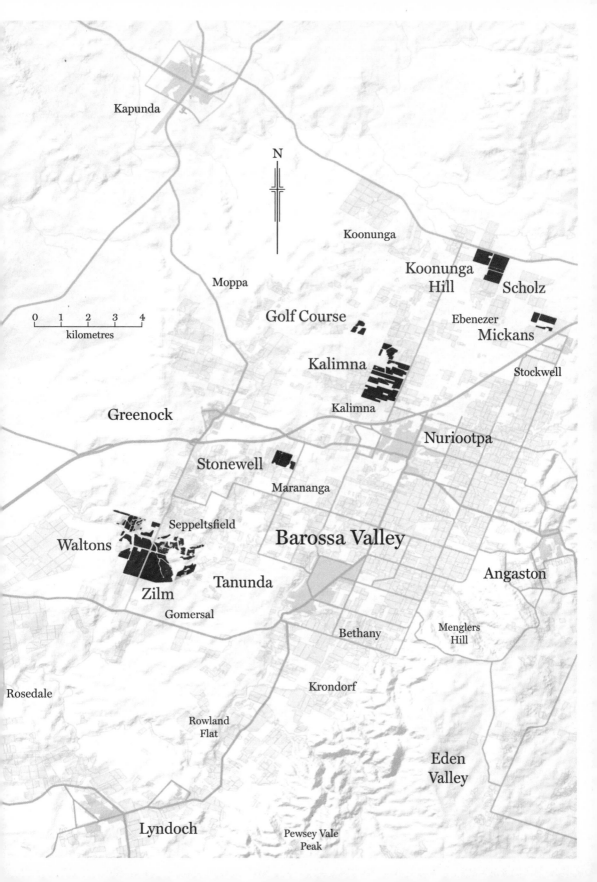

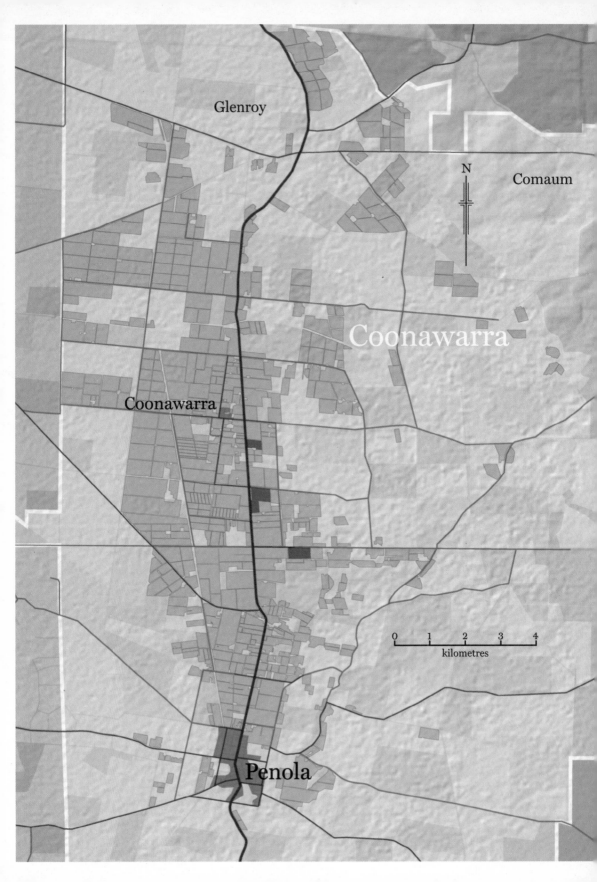

Coonawarra

Coonawarra is 390 km south-east of Adelaide and approximately 100 km inland from the coast of South Australia. Famed for its *terra rossa* soils, Coonawarra produces cabernet sauvignon and shiraz grapes that are used in some of Penfolds finest wines. For more information, see page 41.

Statistics

TOTAL AREA	Total vineyard area for entire Coonawarra GI: 5720 ha
ALTITUDE	60 m
RAINFALL	Growing season rainfall, October–April: 220 mm
TEMPERATURE	Mean January temperature: 19.6°C
VARIETALS SOURCED	Penfolds varietals sourced: cabernet sauvignon and shiraz

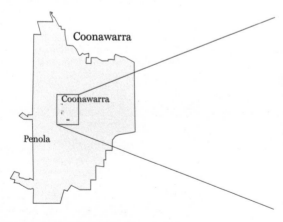

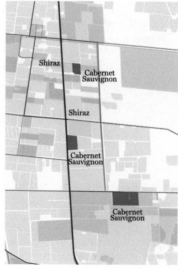

COONAWARRA SOIL TYPES

The famous *terra rossa* soil of Coonawarra is comprised of friable, vivid red clay loams over well-drained limestone subsoil.

Vineyard holdings

Penfolds has owned and leased vineyards throughout its history. It has also supported and received grapes from independent growers since the 1860s. Penfolds Magill Estate, Kalimna and Coonawarra vineyards represent a connection with South Australia's colonial past and Penfolds modern beginnings. Other regional vineyards are important contributors to Penfolds portfolio. Although South Australia provides almost all the vintage crop, Tumbarumba in New South Wales, Henty in southern Victoria and Tasmania are significant sources for Penfolds Yattarna grapes. Altogether, grapes are sourced from over 220 vineyards and grape growers, but below are details of some of Penfolds key vineyards.

Adelaide environs

The first vineyards were planted around Adelaide for commercial production around 1838. The Grange Vineyards were planted in 1844 by Dr Christopher Rawson and Mary Penfold; Magill Estate is a surviving remnant. It is one of the few urban vineyards in the world. It has a warm, dry climate with moderating sea breezes and gully winds.

Magill Estate Vineyard

LOCATION	Magill – Adelaide metropolitan – 8 km (4 miles) east of CBD.
VINEYARD	Originally established in 1844. A remnant of the original Grange Vineyards. Area is 5.24 ha (12 acres).
	Gentle, west-facing slopes ranging between 130–180 m (430–600 ft), situated at the base of the Adelaide Hills.
SOILS	Rich chocolatey red-brown soils over limestone.
RAINFALL	Average rainfall is 502 mm (20 in). About 220 mm (9 in) falls during the growing season.
IRRIGATION	All dry-grown.
VITICULTURE	Original shiraz vines were planted in 1951, with most recent plantings in 1985. Hand-pruned and hand-harvested. Yields on average less than 35 tonnes of grapes.
GRAPE VARIETIES	Shiraz only.
COMMENTS	Contributes fruit to Magill Estate and Grange.

Barossa

The Barossa was settled by Silesians (from east Germany) and English immigrants during the late 1830s. Colonel William Light, the South Australian colony's Surveyor General, named the region after a battle site, Barrosa, near Jerez in Spain, where British and French forces clashed in 1811, during the Peninsular War. The region comprises two distinct sub-regions: the cool-to-warm Eden Valley, with schistic soils at 450 metres, and the warm Barossa Valley with its complex array of rich brown-red soils and alluvial sands at 270 metres. The overall region is dry, with rainfall predominating in winter. Cool afternoon breezes from the Gulf of St Vincent moderate temperatures during summer.

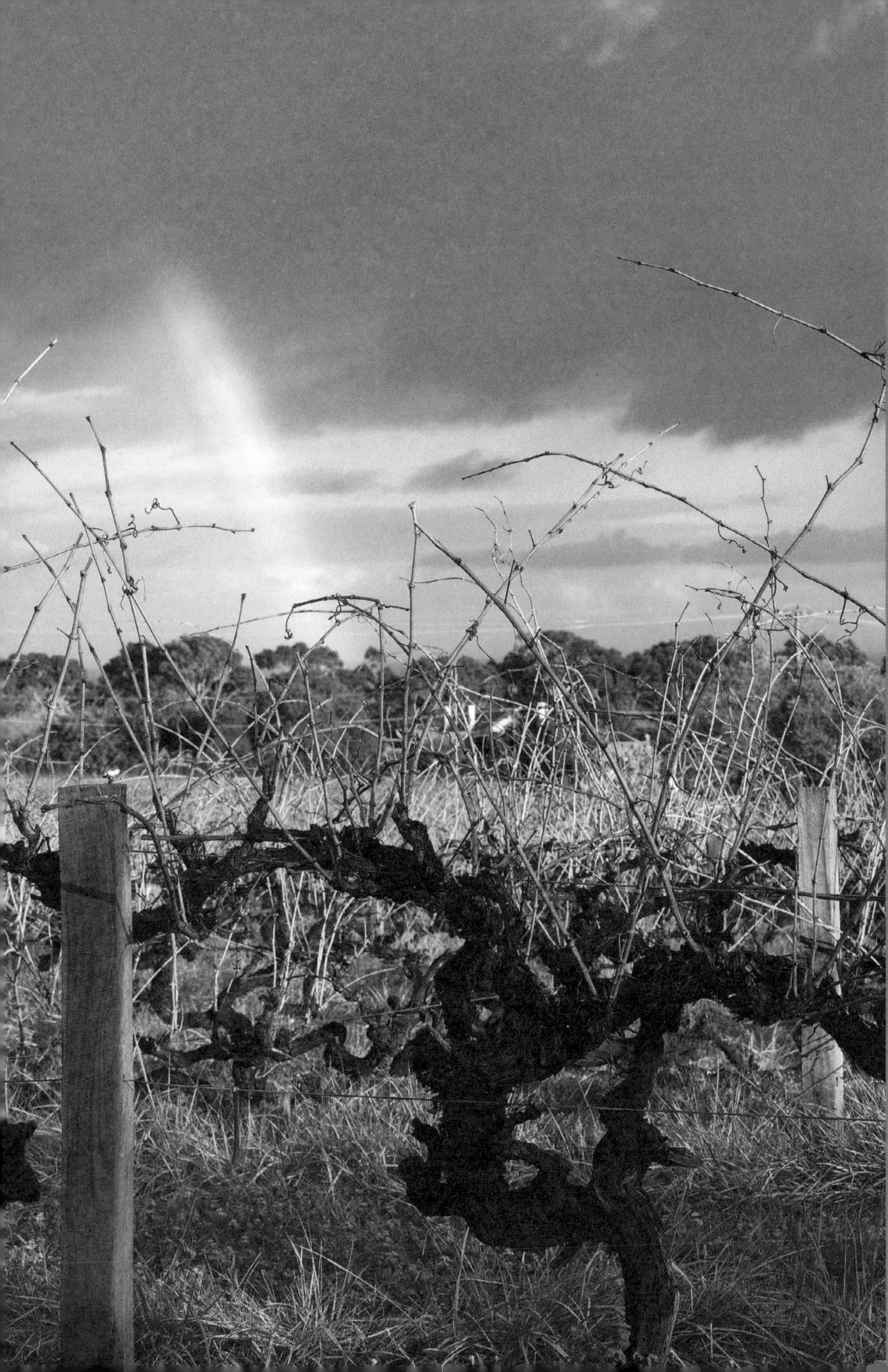

Kalimna Vineyard – Barossa Valley

CLIMATE	Warm and dry.
LOCATION	Northern Barossa – about 4 km (3 miles) north of Nuriootpa. Locality – Moppa.
VINEYARD	Established in the mid-1880s by George Swan Fowler. Acquired in 1945.
	Undulating slopes and flats with elevations to 340 m (1100 ft). Comprises the historic 1880s Block 42, thought to be the oldest pre-phylloxera cabernet sauvignon vineyard in the world. Also Block 3 shiraz (1948), Block 4 shiraz (1949), Block 14 shiraz (1948), Block 41 cabernet sauvignon (1955), Block 25 mataro (1964).
SOILS	Deep alluvial sands to sandy loams and heavy, red-brown clays.
RAINFALL	Average rainfall is 508 mm (20 in). About 200 mm (8 in) falls during the growing season.
IRRIGATION	Dry-land farmed but supplementary irrigation is available.
VITICULTURE	Original cabernet sauvignon vines were planted in 1888. Succession of plantings and retrellising has followed since acquisition in 1945. Some blocks are mechanically pruned and harvested. Many blocks – particularly Block 42 which comprises century-old bush vines – are hand-picked. Yields at 2.25–9 tonnes per ha (1–4 tons per acre).
GRAPE VARIETIES	Shiraz, cabernet sauvignon, mataro (mourvèdre) and 8 rows of sangiovese (planted in 1984).
COMMENTS	Contributes fruit to Grange, Bin 707, RWT, Bin 389, Bin 407, Bin 28 and Cellar Reserve wines. Also Special Bins, especially Block 42 and Bin 60A.

Koonunga Hill Vineyard – Barossa Valley

CLIMATE	Warm.
LOCATION	Northern Barossa – about 5 km (3 miles) north-east of Kalimna. Locality – Koonunga.
VINEYARD	64 ha (158 acres) of original flat grazing land at an elevation of 280 m (920 ft).
	32 ha (79 acres) planted in 1973. The 28 ha (69 acres) Scholz Block planted in 1999 and Stage 2 Koonunga Hill planted in 2007.
SOILS	Red-brown earth over heavy red clays.
RAINFALL	Average rainfall is around 508 mm (20 in). About 200 mm (8 in) during the growing season.
IRRIGATION	Supplementary drip irrigation is available.
VITICULTURE	Yields are about 3–4 tonnes per ha (1.2–1.62 tons per acre).
GRAPE VARIETIES	Shiraz and cabernet sauvignon.
COMMENTS	Contributes fruit to Grange, Bin 707, RWT, Bin 389 and Bin 28. Also Cellar Reserve and Koonunga Hill wines.

The Waltons Vineyard – Barossa Valley

CLIMATE	Warm.
LOCATION	Central Barossa – about 3 km (2 miles) south of Tanunda. Locality – Marananga.
VINEYARD	130 ha (321 acres).
SOILS	Sandy loams and red-brown soils.
RAINFALL	Average rainfall is around 508 mm (20 in). About 200 mm (8 in) during the growing season.
IRRIGATION	Supplementary drip irrigation is available.
VITICULTURE	Planted in 1999. Yields are about 3–4 tonnes per ha (1.2–1.62 tons per acre).
VINE DENSITY	1667 vines per ha (675 vines per acre).
GRAPE VARIETIES	Shiraz, cabernet sauvignon and mataro (mourvèdre).
COMMENTS	Contributes fruit to Bin 28, Bin 150 and Bin 2.

Zilm Vineyard – Barossa Valley

CLIMATE	Warm.
LOCATION	Central Barossa – about 3 km (2 miles) south-west of Tanunda. Locality – Marananga.
VINEYARD	168 ha (415 acres).
SOILS	Sandy clay loams and light to medium-heavy clays.
RAINFALL	Average rainfall is around 508 mm (20 in). About 200 mm (8 in) during the growing season.
IRRIGATION	Drip irrigation is available.
VITICULTURE	Planted in 2005–06. Yields are about 5–6 tonnes per ha (2–2.6 tons per acre).
VINE DENSITY	1851 vines per hectare (749 vines per acre).
GRAPE VARIETIES	Shiraz.
COMMENTS	Contributes fruit to Bin 150, Bin 28 and Bin 389.

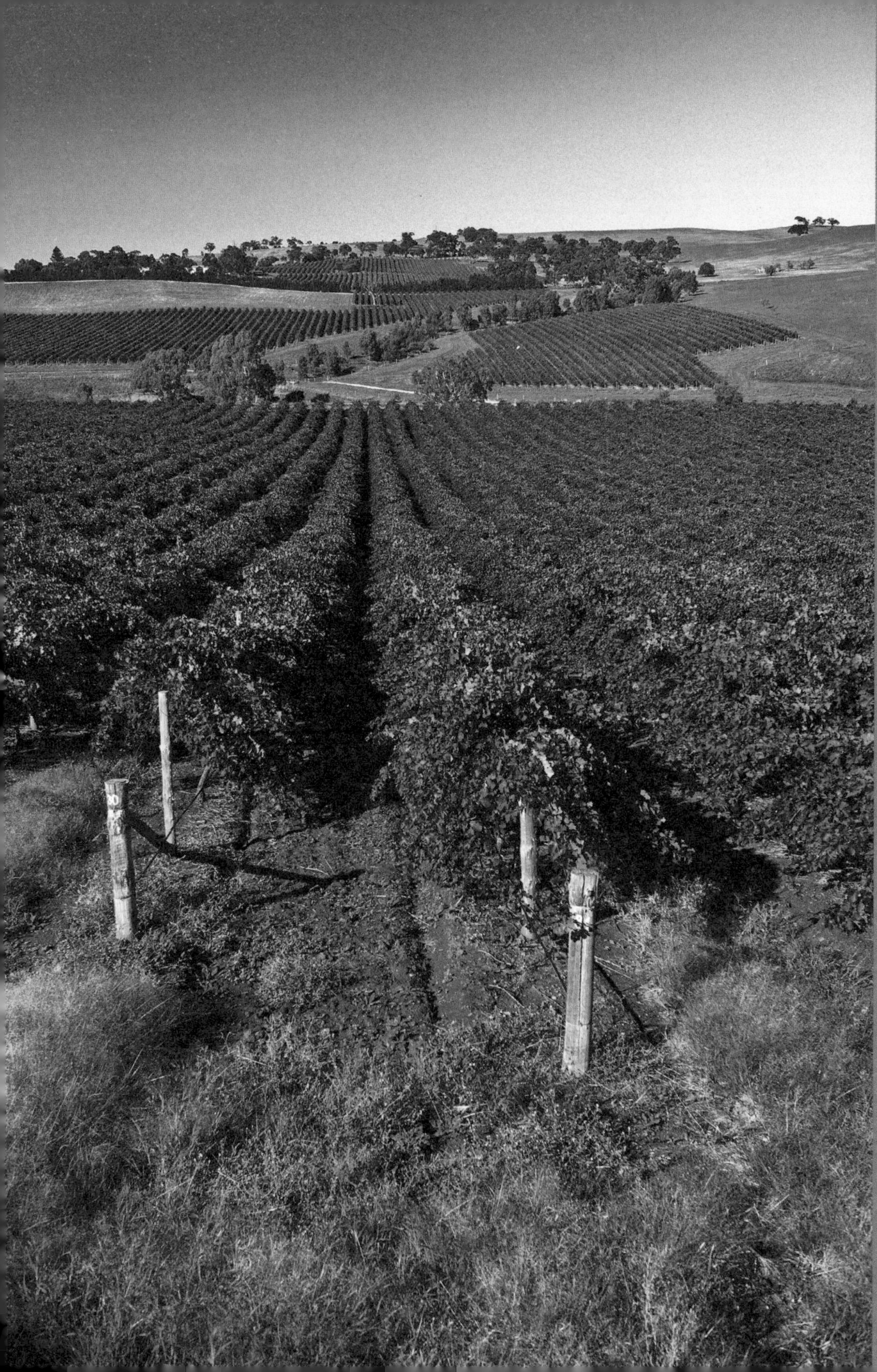

Stonewell Vineyard – Barossa Valley

CLIMATE	Warm.
LOCATION	Stonewell Road, central Barossa – 4 km (3 miles) west of Nuriootpa. Locality – Stonewell.
VINEYARD	33 ha (81 acres) on undulating slopes.
SOILS	Sandy loams and red-brown soils.
RAINFALL	Average rainfall is around 508 mm (20 in). About 200 mm (8 in) during the growing season.
IRRIGATION	Supplementary drip irrigation is available for use.
VITICULTURE	First planted in the early 1970s, but replanting programme put in place when Penfolds purchased the vineyard in the early 1990s. Yields are about 5.5–6 tonnes per ha (2.2–2.6 tons per acre).
GRAPE VARIETIES.	Shiraz, cabernet sauvignon.
COMMENTS	Contributes fruit to Bin 389, Bin 28, Bin 2.

Gersch Vineyard – Barossa Valley
a Penfolds grower vineyard

CLIMATE	Warm.
LOCATION	5 km (3 miles) from Nuriootpa. Locality – Moppa.
VINEYARD	20 ha (49 acres) of flat land at an elevation of 300 m (984 ft).
SOILS	Some areas of drift sand over heavy red clay and calcareous clays. Other areas consist of clay loam topsoil over deep red and yellow clays.
RAINFALL	Average rainfall is 500 mm (20 in).
IRRIGATION	All dry-grown; no irrigation.
VITICULTURE	The vines are grown on a single wire – some of the grenache is grown as bush vines with no trellis. Vine age from 10 years to over 100 years.
GRAPE VARIETIES	Shiraz and grenache.
COMMENTS	Contributes fruit to Grange, RWT, Bin 138 and special bins.

Clare Valley

John Horrocks was the first settler in the region and encouraged his servant James Green to plant the first vines in 1842 at Penwortham. The climate is warm to hot and dry with cool to cold winters. Most vineyards are located at higher elevations to take advantage of cooler nights and the moderating cool breezes that funnel up the Clare's corrugation of hills and gullies from the south.

Sevenhill Vineyard
a Penfolds grower vineyard

CLIMATE	Warm.
LOCATION	Sevenhill region, 5 km (3 miles) south of Clare.
VINEYARD	55 ha (135 acres) of undulating land at an elevation of 500 m (1640 ft).
SOILS	Siltstone, sandstone reefs and calcareous clays.
RAINFALL	Average rainfall is 600 mm (24 in).
IRRIGATION	Supplementary drip irrigation is available. Semi dry-grown.
VITICULTURE	Most of the shiraz and grenache is dry-grown on single trellis with catch wires. Natural yields of 4 tonnes per ha (1.7 tons per acre) are common. Vine age is 15 years up to 100 years.
GRAPE VARIETIES	Cabernet sauvignon, shiraz, merlot, chardonnay, grenache and cabernet franc.
COMMENTS	Contributes to Grange, St Henri, Bin 28, Bin 138 and Cellar Reserve wines.

Coonawarra and Limestone Coast

Coonawarra's first commercial vineyards were established by John Riddoch's Coonawarra Fruit Colony in 1890. Weathered limestone under *terra rossa* soils, relatively cool climate and overall water availability make it one of Australia's most important fine wine regions. Extremely flat and unprotected, it is exposed to the swinging influences of the cool Great Southern Ocean and hot, dry northerly winds. Spring frosts are a major problem and have been known to wipe out crops. Mechanical and machine harvesting is widely used in Coonawarra, although smaller producers prefer to tend their vines by hand. The region is best known for Cabernet Sauvignon and Shiraz. Robe is located very close to the coast, but it is protected by the low-lying Woakwine Ranges. The climate is warmer and drier at Bordertown.

Coonawarra Vineyards

CLIMATE	Cool to temperate, with significant cooling influence of Southern Ocean.
LOCATION	Coonawarra, Limestone Coast, far south of South Australia.
VINEYARD	First acquired Sharam's Block from Redman's in 1960. Now 100 ha (247 acres) of prime vineyards. 1962 Bin 60A, 1964 Bin 707, 1966 Bin 620 and 1967 Bin 7 began an important Penfolds tradition in Coonawarra. Old Penfolds blocks include the revered Block 20 cabernet and Block 14 shiraz.
SOILS	*Terra rossa* – friable, vivid red clay loams over well-drained limestone subsoil.
RAINFALL	Average rainfall is around 593 mm (23 in). About 210 mm (8 in) falls during growing season.
IRRIGATION	Supplementary drip irrigation is available.
VITICULTURE	Vines trained on high, single wire trellis with overhead sprinkler system for use in combating frost. Yields are approximately 4–5 tonnes per ha (1.8–2.2 tons per acre).
GRAPE VARIETIES	Cabernet sauvignon and shiraz.
COMMENTS	Contributes fruit to Grange, Bin 707, Bin 389, Bin 407, Bin 128 and special bins.

Blocks 5 and 6

VINEYARD	7.43 ha (18 acres).
VITICULTURE	Blocks 5 and 6 were planted in 1965 and trained on high, single wire trellis with overhead sprinkler system for use in combating frost. Yields are approximately 4–5 tonnes per ha (1.8–2.2 tons per acre); detailed two-bud hand-pruning.
VINE DENSITY	2151 vines per ha (870 vines per acre), 3 m (10 ft) row x 1.55 m (5 ft) vine spacing.
GRAPE VARIETIES	Block 5 – shiraz; Block 6 – cabernet sauvignon.

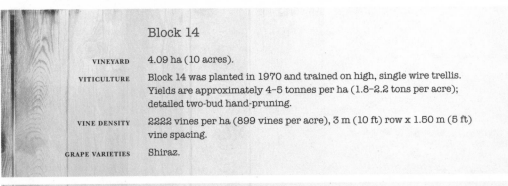

Block 14

VINEYARD	4.09 ha (10 acres).
VITICULTURE	Block 14 was planted in 1970 and trained on high, single wire trellis. Yields are approximately 4–5 tonnes per ha (1.8–2.2 tons per acre); detailed two-bud hand-pruning.
VINE DENSITY	2222 vines per ha (899 vines per acre), 3 m (10 ft) row x 1.50 m (5 ft) vine spacing.
GRAPE VARIETIES	Shiraz.

Block 10

VINEYARD	14.4 ha (35 acres).
VITICULTURE	Block 10 was planted in 1972 and trained on high, single wire trellis. Yields are approximately 4–5 tonnes per ha (1.8–2.2 tons per acre). Machine pre-pruning followed by hand-pruning to balance vine bud numbers and hand-harvesting are standard management practice.
VINE DENSITY	1477 vines per ha (598 vines per acre), 3.66 m (12 ft) row x 1.85 m (6 ft) vine spacing.
GRAPE VARIETIES	Cabernet sauvignon.

Block 20

VINEYARD	5.98 ha (14 acres).
VITICULTURE	Block 20 was planted in 1969 and trained on high single wire trellis. Yields are approximately 4–5 tonnes per ha (1.8–2.2 tons per acre). Machine pre-pruning followed by hand-pruning to balance vine bud numbers and hand-harvesting are standard management practice.
VINE DENSITY	2222 vines per ha (899 vines per acre), 3.03 m (10 ft) row x 1.5 m (5ft) vine spacing.
GRAPE VARIETIES	Cabernet sauvignon.

Robe Vineyard

CLIMATE	Cool maritime climate with close proximity to the Southern Ocean.
LOCATION	Limestone Coast. The Robe region is centred around the small townships of Robe and Beachport.
VINEYARD	First established in the mid-1990s. Formerly a grazing property for breeding horses for the British Army in India. 235 ha (580 acres) under vine.
SOILS	Shallow (5–45 cm/2–18 in) loamy sand over limestone with smaller areas of *terra rossa* over limestone and deep sandy loam.
RAINFALL	Average rainfall is around 667 mm (26 in). About 210 mm (8 in) falls during growing season.
IRRIGATION	Supplementary drip irrigation and electronic probes available.
VITICULTURE	Vines are trained on a single wire vertically positioned high trellis, with overhead sprinkler system and frost fans for use in combating frost. Yields are approximately 4–8.5 tonnes per ha (1.6–3.4 tons per acre), depending on grape variety.
GRAPE VARIETIES	Shiraz, merlot, petit verdot, cabernet sauvignon, pinot noir, semillon, sauvignon blanc and chardonnay.
COMMENTS	Contributes fruit to Bin 407, Bin 707 and St Henri.

Bordertown

CLIMATE	Warm and dry.
LOCATION	Limestone Coast. 265 km (165 miles) east of Adelaide and close to the state border of South Australia and Victoria.
VINEYARD	Established in the late 1990s. Planted on level ground elevated at 71–72 m (233–236 ft).
SOILS	Red-brown earth over limestone.
RAINFALL	Average rainfall is 520 mm (20 in). Majority falls during the growing season.
IRRIGATION	Supplementary irrigation available.
VITICULTURE	Single wire trellis. Machine-pruned with hand clean-up. Overhead sprinklers used for frost control. Permanent grass sward between rows.
GRAPE VARIETIES	Chardonnay, cabernet sauvignon, merlot, shiraz and tempranillo.
COMMENTS	Contributes fruit to Bin 389, Bin 407, Bin 28 and Cellar Reserve wines.

McLaren Vale

John Reynell began a tradition of viticulture and winemaking in 1838 by planting the first vineyard at Reynella. A classic Mediterranean climate with cool winters, warm summers, low rainfall and moderating sea breezes. There are three distinctive soil types: the sandy loams of Blewitt Springs; the darker soils of McLaren Flat; and the *terra rossa* over limestone soils of Seaview.

Blencowe Vineyard

CLIMATE	Mediterranean.
LOCATION	5 km (3 miles) north-east of the township of McLaren Vale.
VINEYARD	Planted in 1972. 25 ha (61 acres).
SOILS	Loam, sandy loam.
RAINFALL	Average rainfall is 550 mm (22 in).
IRRIGATION	Drip.
VITICULTURE	Gently undulating with north–south and east–west rows. Contour plantings in some of the steeper areas. Elevation is 196 m above sea level at the highest point. Plantings of shiraz in 1972, 2002 and 2008. Yields are usually less than 6 tonnes per ha (2.6 tons per acre).
GRAPE VARIETIES	Shiraz.
COMMENTS	Contributes fruit to St Henri, Bin 28 and Bin 389.

Bethany McLaren Vale Vineyard

CLIMATE	Mediterranean.
LOCATION	2 km (1 mile) south-east of McLaren Vale township, extending south through to Rifle Range Road.
VINEYARD	58 ha (143 acres). 76–82 m (249–269 ft) above sea level.
SOILS	Sandy clay, clay and Bay of Biscay soils.
RAINFALL	Average rainfall is 550 mm (22 in).
IRRIGATION	Drip.
VITICULTURE	Generally flat with north–south row plantings. Yields are less than 6 tonnes per ha (2.6 tons per acre).
GRAPE VARIETIES	Shiraz and cabernet sauvignon.
COMMENTS	Contributes fruit to St Henri, Bin 389 and Bin 407.

Booths Vineyard

CLIMATE	Mediterranean.
LOCATION	3 km (2 miles) north-west of the township of McLaren Vale.
VINEYARD	Established by Lindsay Booth, 1976–78. Acquired in 2000. 33 ha (81 acres).
SOILS	Sandy loam.
RAINFALL	Average rainfall is 550 mm (22 in).
IRRIGATION	Drip.
VITICULTURE	Vineyard is slightly undulating with both north-south and east-west running rows (mostly north-south). Elevation is approximately 120 m (394 ft) above sea level at the highest point. All shiraz with exception of 2 small cabernet sauvignon blocks. Yields are less than 6 tonnes per ha (2.6 tons per acre).
GRAPE VARIETIES	Shiraz and cabernet sauvignon.
COMMENTS	Contributes fruit to St Henri, Bin 389 and Bin 407.

'All winemakers should possess a good fertile imagination
if they are to be successful in their craft.'

MAX SCHUBERT

PEOPLE AND WINE

Chapter three

1875 was an average vintage in South Australia. 'Previous to the rains that fell in the last week of April the grapes were in very good order for wine, being uniformly ripe and more than usually free from bunches injured by excessive heat.' A hot October and November, followed by an unseasonably cool but dry summer, led to a moderately good harvest although the must weights (sugar density) were markedly down from the previous year. This was perceived as advantageous because, 'it tended to make the fermentations more complete.' Thomas Hardy, a pioneering South Australian vigneron, wrote in *The South Australian Register* (June 1875): 'a good many of the wines are already almost as bright as water – a circumstance generally considered as very favourable to their future preservation.'

Fortified wines dominated the colonial wine markets of Australia. Winemaking was rudimentary and wine science was basic. There was little understanding of why wines, during vinification or maturation, became spoiled. A huge percentage of wine was thrown away because of oxidation, volatility or disease. Hardy wrote, 'Some of the sweet red wines of the Barossa district, especially of vintages 1872 and 3 have suffered acid degeneration, the nature of which is certainly not yet understood by vignerons, and takes place often in wines between two and three years old and in spite of all the precautions taken to prevent it. As long as sweet wines are required this will continue to be the dread of winemakers unless some of our scientific men take up the question, and show us the cause.' Sweet wine disease, caused by the development of *Lactobacillus*, would damage severely

the reputation of Australian wine in export markets during the 1920s.

In 1936, Ray Beckwith, a young scientist working for Penfolds, cracked the problem by observing: 'pH might be a useful tool in the control of bacterial growth in wine'. In plain terms, Ray Beckwith discovered how to stabilise wine and protect it from spoilage. The implementation of pH meters and strict management standards revolutionised winemaking at Penfolds. This ground-breaking work and other technical innovations underpinned the creative development of Grange Hermitage by Max Schubert and the growing range of Penfolds table wines.

Veteran winemaker Ian Hickinbotham, an ex-Penfolds manager, writes in his autobiography *Australian Plonky*: 'Beckwith applied his unique knowledge to the making and husbandry of all wine types, with remarkable cost savings, when employed by Penfolds for the rest of his working life. In a nutshell, he saved the 25% wine component that previously had to be destroyed by distillation due to bacterial spoilage. From that time, around 1940, Australia became a world leader in the making of table wine.'

Vigilance, care and attention to detail were important to Beckwith. He developed systems that are now standard throughout the industry. This went beyond the laboratory, to practical winemaking solutions at every quality level. He designed and improvised new plant and equipment in stainless steel, and introduced new technology, including refrigeration and sterile filtration. He was probably the first person to introduce paper-chromatography as a test for completion of malolactic fermentation. He designed quality standards in a time before consistency and standardisation became an industry norm. His early work with the mass production of flor sherry and solving the issues of contamination were critical to the success of Penfolds during the 1940s and 1950s. Preventative winemaking, as pioneered by Ray Beckwith, is now a worldwide industry standard.

Alfred Scholz, Beckwith's first boss at Nuriootpa, was originally a mining engineer. He established Penfolds solera system and created Grandfather Port. He was an expert distiller and improved the quality of brandy spirit production. During the 1920s he excavated the Magill tunnels. Scholz worked as an assistant blender for Alfred Vesey, who had worked with Mary Penfold during the 1870s. Vesey, whose reputation as a taster and blender was unmatched, retired from Penfolds in 1952 at the age of eighty-nine. He died five weeks later.

Max Schubert (1915–94) joined Penfolds as a junior laboratory assistant in 1931 and worked with Alfred Scholz. In 1935 he was transferred to Magill where he also attended the Adelaide School of Mines to study chemistry. In 1938 he was appointed assistant winemaker. He rejoined Penfolds after the war and was promoted to senior winemaker at Magill in 1946 and then South Australia's production manager in 1948. Max Schubert was a self-taught winemaker with a natural inquisitiveness for winemaking theory. He had a whimsical and romantic view of winemaking, too:

I'd like to think that the wines with which I have been associated are descended from one ancestor vineyard established many years ago, marrying with another, and another, and even another if you like, thus creating and establishing a dynasty of wines. These may differ in character year by year, but all bear an unmistakable resemblance and relationship to each other … This whole approach and concept has been of great assistance to me, not only in the technical sphere, but as a means of stimulating my imaginative powers as far as winemaking is concerned … All winemakers should possess a good fertile imagination if they are to be successful in their craft.

After World War II John Davoren first worked as the manager of the Kalimna Vineyard, then in 1947 he was transferred to Auldana Cellars. His experimentation with sparkling wine and red wine fermentations is largely forgotten today. However, his discoveries revolutionised sparkling wine production, contributed to Max Schubert's winemaking theories and led to the development of St Henri Claret.

Penfolds House Style emerged from a fortified wine–producing culture and evolved as a winemaking philosophy. Many of the techniques developed by Penfolds have become standard winemaking around Australia. Max Schubert's ground-breaking work with Grange incorporated traditional open fermenters and multi-regional blending used for tawny port production; Ray Beckwith's science, new ideas and winemaking practices including barrel fermentation and new 'untreated' oak maturation.

Vineyards were an important consideration too. Max Schubert said, 'The development of a new commercial wine, particularly of the high-grade range, depends on the quality and availability of the raw material, the maintenance of standard and continuity of style.' He achieved this through identifying specific vineyard sites and developing relationships with growers. He once observed, while developing Grange, that using shiraz from two specific vineyards would 'result in an improved all-round wine'.

During the 1950s Schubert searched widely for suitable fruit, particularly in the foothills around Adelaide, Morphett Vale, McLaren Vale and the Barossa Valley. The Grange Vineyards, Auldana Vineyards and Quarry Block were early fruit sources before the urbanisation of Adelaide. During the 1960s Penfolds invested in Coonawarra to secure supply.

Grange and St Henri were modelled on Claret styles. The availability of Bordeaux grape varieties in South Australia was limited during the 1950s. Schubert soon began to favour shiraz instead because of its spectrum of ripe flavours, tannin structures and the relative ease of supply. He struggled initially with cabernet sauvignon because of its scarcity and capricious nature in the South Australian climate, despite the availability of fruit from the Kalimna Vineyard and Auldana Vineyards. John Davoren was also similarly constrained. His first experimental St Henri vintages were based on cabernet and mataro. Shiraz, however, would increasingly dominate this blend also. The release of 1960 Bin 389 Cabernet Shiraz reflects the winemaking attitude of the time: that cabernet sauvignon did not have the power or mid-palate intensity to be made as a single varietal wine.

Multi-regional and vineyard sourcing are the essential elements of Max Schubert's 'all-round wine'. Without the constraints of a single vineyard, winemakers could select the best possible fruit comprising 'the outstanding characteristics of each vineyard'. This idea gathered pace during the 1960s. The popularity

Right, Dr Ray Beckwith, 1911–2012

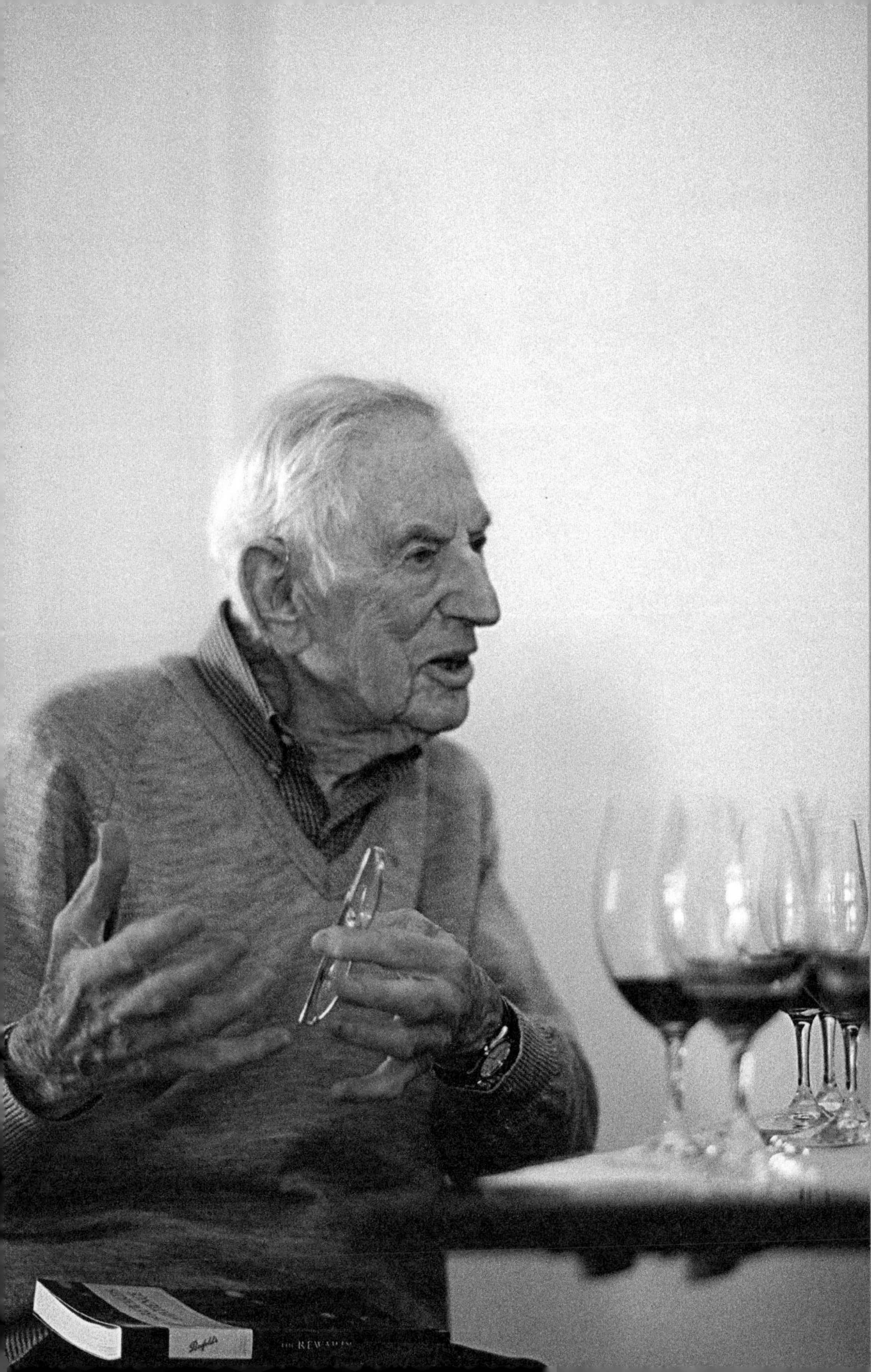

John Davoren (1915–91)

of Bin 28, Bin 389, St Henri and Grange and the increased volume of production led to new methods of classifying vineyards and identifying parcels of fruit.

Max Schubert's experiments with shiraz and American oak were profound. He discovered that if the wine completed fermentation in new American oak the two components would generate a tremendous 'volume of bouquet and flavour'. Schubert remarked, 'It was almost as if the new wood had acted as a catalyst to release previously unsuspected flavours from the Hermitage [sic] grape.'

Oak maturation, which follows vinification, was also an important feature of the emerging Penfolds House Style. Grange benefits greatly from being aged in American oak for around eighteen months. The aromas and flavours derived from the fruit and new oak evolve; the tannins polymerise and soften. At the other end of the spectrum is St Henri, which is matured in large fifty-year-old – or more – seasoned oak vats. A maturation effect takes place where the fruit develops complexity without the influence of new oak. Despite this, St Henri and Grange can look like stablemates after ten or fifteen years of age. The distinctive hallmarks of rich aromatic and plush fruit, beautifully integrated oak and chocolatey, round tannins are a common theme. The techniques developed by Max Schubert and his team were used extensively in the commercialisation of Penfolds table wine portfolio during the 1960s, including Bin 389, Bin 128 and Bin 28.

Many people assisted Max Schubert to achieve his aims. Murray Marchant, who started at Penfolds in 1946, was initially a laboratory assistant to the 'frighteningly fiery' John Farsch and helped prepare samples for Penfolds stalwart Alfred Vesey. He was also a protégé of Ray Beckwith and played a crucial role in protecting the early Granges from spoilage, as well as helping to hide the 1957, 1958 and 1959 from the view of management. A graduate of the Adelaide School of Mines and fascinated with organic chemistry, he initiated many of Ray Beckwith's scientific winemaking advances at Magill. Around 1960, he was appointed Magill's senior winemaker after Max Schubert became Penfolds national production manager.

Penfolds hired Gordon Colquist as an odd jobber in 1938 at age sixteen. His Swedish father, Herman Julius Karlvist, had worked at Penfolds as a boiler man until he was killed in a freak accident at Magill in 1936. In 1941, Gordon Colquist enlisted with the Royal Australian Air Force and served as an air observer in the Atlantic and Europe. He worked at Penfolds for fifty years, as did his brothers Laurie and Harold, and finished up as the cellar supervisor and winemaker at Magill. The Colquist brothers were all given 'a job for life' by the Penfold family because of their father's untimely death. Gordon assisted Max Schubert with the development of new red table wine styles, including Bin 389, Bin 28 and Bin 128. Together with Murray Marchant, he kept Schubert's dream alive by secretly looking after the hidden Granges.

Don Ditter, raised in South Australia's Barossa Valley, started work as a laboratory assistant at Magill in December 1942. From 1944 to 1945 he served in the RAAF and after the war studied winemaking at Roseworthy Agricultural College. After graduating in 1950 with first-class honours he was reassigned as an assistant winemaker at Penfolds Nuriootpa in the Barossa Valley. There was a shortage of suitable technical people in New South Wales, so in 1953 he was lured to Sydney to improve cellar operations and bottling at Penfolds Queen Victoria Cellars in the central business district and at Alexandria in Sydney's inner west.

The post-war influx of immigrants from Europe not only created market opportunities but also new employees familiar with winemaking: many families from Italy and Eastern Europe worked in the vineyards during vintage, and later worked directly for or supplied fruit to Penfolds. In 1960 a young Dutchman, Pieter Van Gent, newly arrived at Scheyville Migrant Camp on the western extremities of greater Sydney,

Opposite: Don Ditter, left, and John Duval, right.

took the train from Windsor to Tempe and applied for a job at Penfolds. After three months as a cellar hand 'hosing out casks in the table wine section', Don Ditter put a white coat over Pieter's shoulders and said, 'You'd better help me in the laboratory from now on.' At the time Van Gent had 'knowledge of six languages, but nothing of chemistry'. Over a ten-year period at Penfolds Pieter Van Gent learned the craft of winemaking, blending and tasting. He worked vintage at Dalwood with Perc McGuigan and was briefly seconded to Nuriootpa with Ray Beckwith. Throughout the 1960s he prepared samples for various national shows and participated in the annual Sydney Royal Wine Show. Coinciding with the retirement of Ivan Combet at Minchinbury Cellars in 1967, Pieter was appointed assistant winemaker manager, specialising in making champagne, sparkling burgundies and vermouth. During the 1970s he went on to work for one of Mudgee's pioneering wineries before establishing his own successful family winemaking enterprise.

Veteran senior winemaker John Bird started work as a laboratory technical assistant at Magill prior to the vintage in February 1960 on a wage of £5 a week. It did not start smoothly. He was told that he was 'a bit slow and casual' and warned that he was in danger of being relegated to the bottling department. Bird reacted by saying that he was 'a bit bored and needed more work'. He was promptly given a white lab coat in place of his grey general-purpose dust-coat, by Magill's chemist Karl Lambert and senior winemaker Murray Marchant. The 1960 vintage represented a significant step up in Penfolds table wine production. John Bird – 'awestruck by the enormity of it all' – found himself immersed in laboratory work, winemaking, general cellar work, bottling and warehousing.

During the 1960s John Bird was in the thick and thin of wine production at Magill, working closely with various winemaking trials and experimental vintages. He assisted Max Schubert with the final blending of the famous 1962 Bin 60A Coonawarra Cabernet Sauvignon, Kalimna Shiraz and many other experimental wines including 'the curatorship of 1950s Grange bottlings'. The first vintages of Bin 28, Bin 128, Bin 389, Bin 707 were also assembled at Magill. In 1970 John Bird was appointed Magill's senior winemaker and then became Penfolds senior winemaker from 1981 to 1996. Since 1997 he has worked as a consultant winemaker, bringing fifty-four vintages of valuable continuity, perspective and insight to the winemaking team.

In 1963 Don Ditter was promoted to New South Wales production manager and looked after winemaking operations in Sydney, Minchinbury, Griffith and the Hunter Valley. After his formal appointment as Chief Winemaker in 1975 (he took over from Max Schubert in 1973), he oversaw the consolidation of Penfolds winemaking to South Australia. This included a major overhaul of vineyard management, tracking of fruit, refinements in winemaking and bottling.

The 1970s were difficult times because of ownership changes and challenging market conditions. But by the time of Don Ditter's retirement in 1986, Penfolds was back on the ascendancy, its cultural values still unbroken. Ditter said, 'Every day we were thinking of new ways to make better wine. If you could do something better, why not? Back in my day it was all about making a lot of small, gradual improvements. Getting rid of things that were not working, as well as dreaming up things that have never been done before ...' Koonunga Hill, Magill Estate, Bin 707 and the first special bins since the early 1970s were released during this period.

During the early 1970s Robin Moody joined the production team as an assistant winemaker at Auldana and became very close to John Davoren. After a few years he was posted to Nuriootpa. He played a vital role in maintaining the values of Max Schubert's winemaking theories after Grange production closed down at Magill. John Bird would also travel up to Nuriootpa to provide winemaking instructions. One of the problems to begin with was instilling at Nuriootpa the radical new red winemaking techniques that had been developed at Magill during the previous decade. This included the way Grange was racked off into barrel towards the end of fermentation and also how to encourage and maintain the prescribed volatility levels in the wine.

Max Schubert's protégé Chris Hancock, a precociously talented winemaker, was appointed senior winemaker at Magill at twenty-three years of age. His winemaking career at Penfolds was cut short by a promotion and transfer to the executive team in Sydney. In later years, as Penfolds corporate custodian, he reinstated a component of Grange production at Magill. In 1973, Mike Press, 'a forceful and efficient manager', was appointed South Australia's production manager and charged with building a younger and more technical team of winemakers for the future. Kevin Schroeter, a senior winemaker specialising in fortified wine and brandy production, played an important role in modernising the winemaking facilities at Nuriootpa. Together with his brother

Right, Steve Lienert

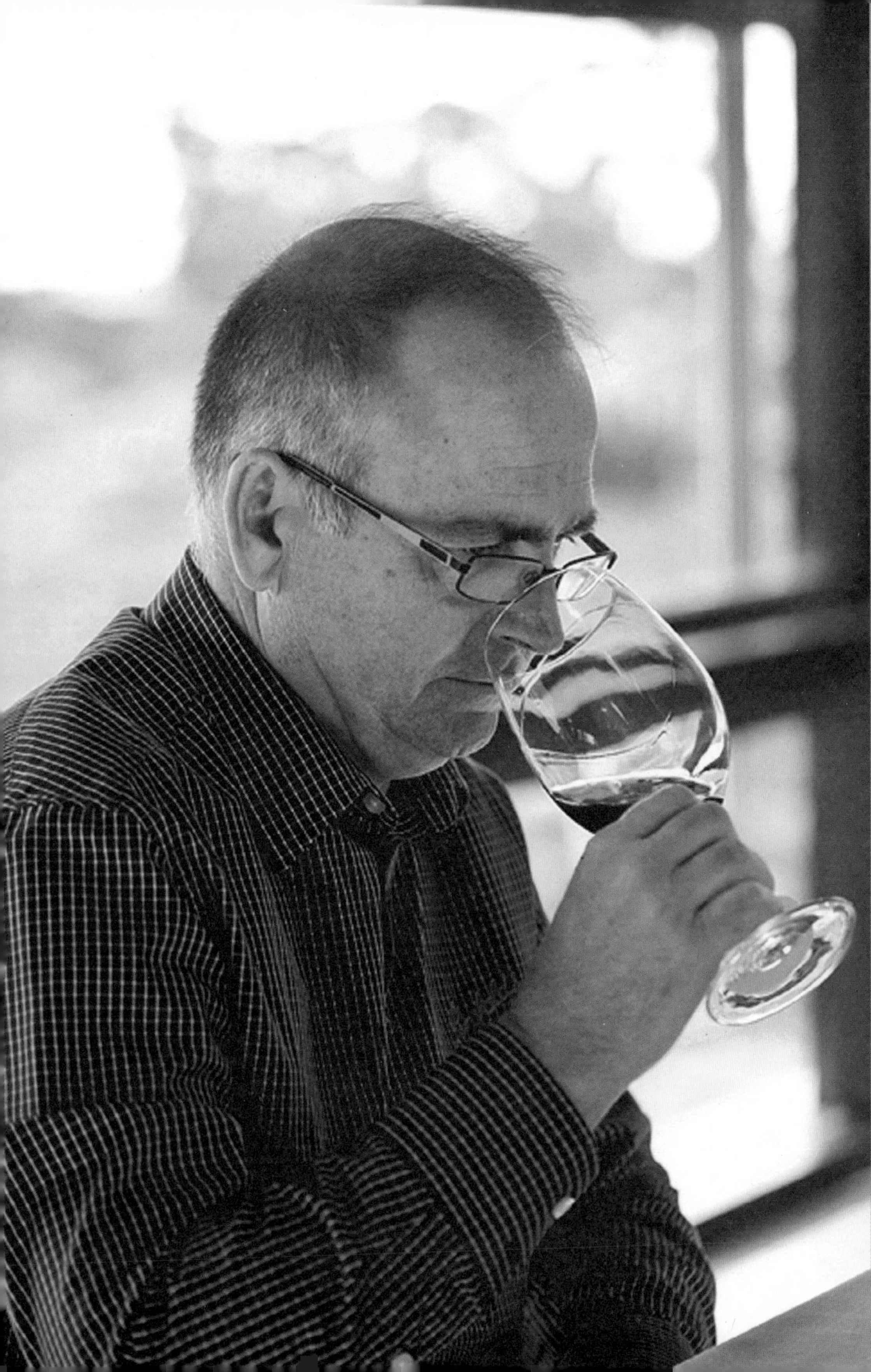

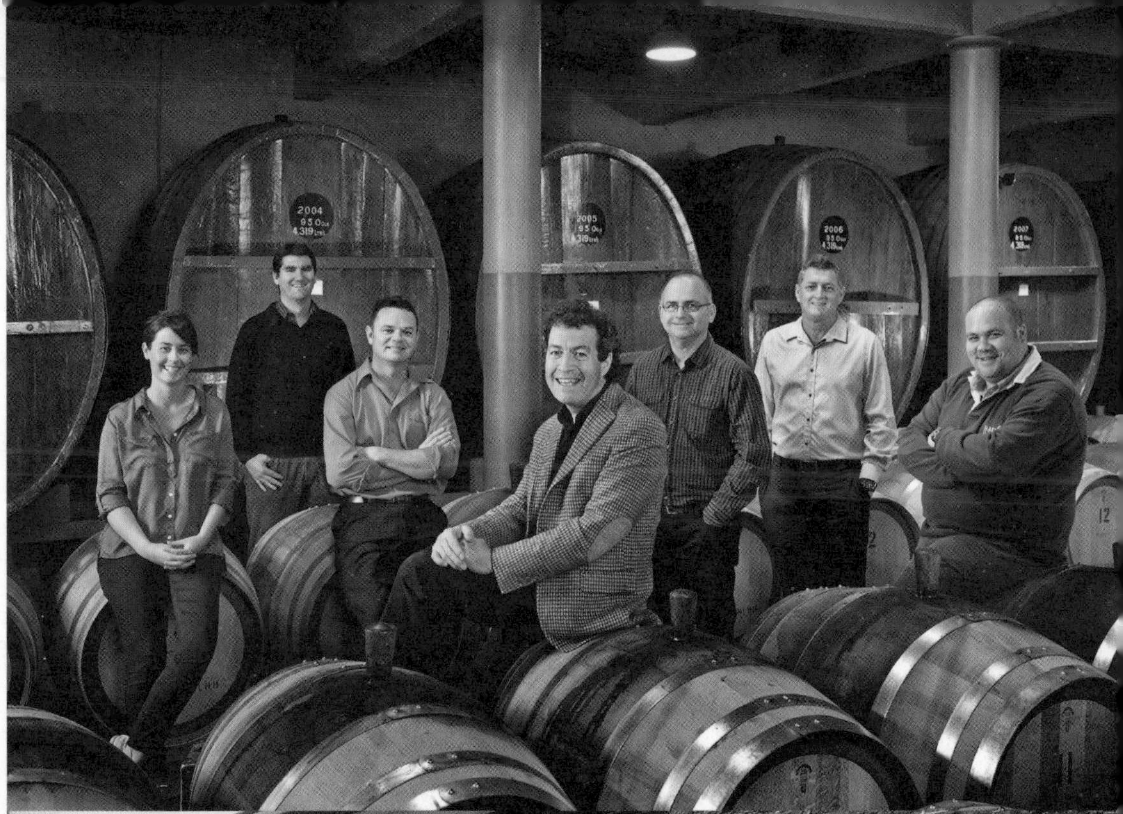
2004
2005
2006
2007

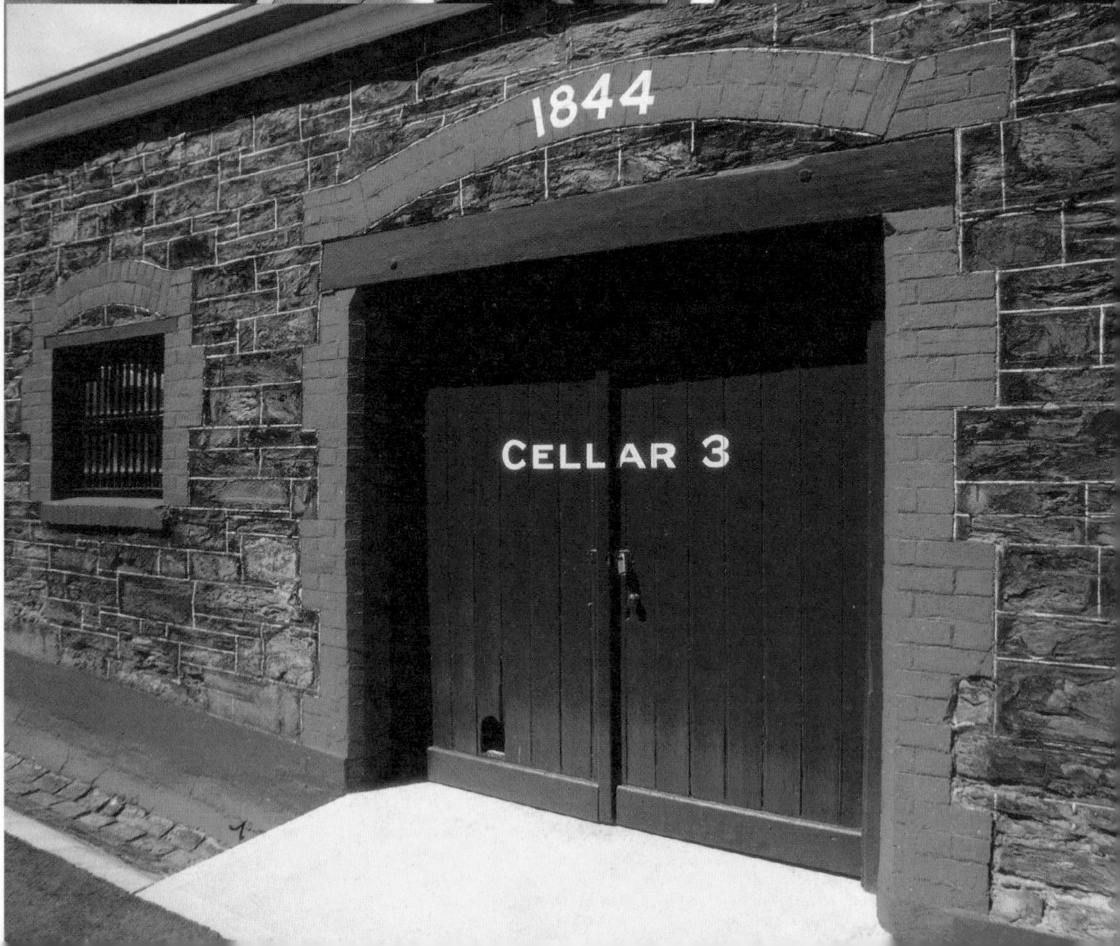
1844
CELLAR 3

Les (father of Kym Schroeter), they virtually ran the cellars during the 1960s and 1970s.

John Duval, a graduate in Agricultural Science at Adelaide University and a postgraduate in Oenology at Roseworthy College, joined Penfolds at Nuriootpa in 1974 where he worked with winemakers John Davoren and Kevin Schroeter. His natural ability, technical skills and leadership qualities, together with the support of an experienced team, led to a stellar career at Penfolds. 'With big shoes to fill' he was promoted to Chief Winemaker in 1986, just as Australia's wine export boom started. Duval presided over a golden period at Penfolds, which saw the development of new wines including Bin 138, Bin 407, RWT and Yattarna. Grange also reached a new level of refinement and fame. John Duval's outstanding technical ability and instinctive nature are decisively illustrated in the profoundly opulent and beautifully balanced wines of the late 1980s and 1990s. In 2002 he stood down as Chief Winemaker and established his own successful eponymous wine company, based in the Barossa Valley.

Steve Lienert joined Penfolds in 1978 as a cellar hand. His progress through the winemaking ranks exemplifies the Penfolds tradition and echoes the careers of many other notable Penfolds winemakers. A veteran of thirty-five vintages, Steve Lienert is responsible for the day-to-day management of winemaking and maturation. His intimate knowledge of individual wines, masterful eye for detail and remarkable vintage experience instils continuity of Penfolds winemaking culture.

Other winemakers have played an important role in developing new standards of winemaking. Peter Taylor, Nuriootpa's senior red winemaker during the mid-1990s, introduced new exacting assessments of the growing season to improve classification and earmarking of fruit. Roseworthy oenology graduate Kym Tolley, a member of the Penfold family, started at Penfolds Magill Cellars in 1973. He worked as a red and sparkling winemaker for around fifteen years, before establishing his own business in Coonawarra.

Although Penfolds has a history of keeping its most talented winemakers for significant periods, there are a few who have moved on to other horizons: Neil Paulette, Warren Ward, Frank Newman, Daryl Groom, David Slingsby Smith, Mick Schroeter, Garry Phipps, Rod Chapman, Dean Kraehenbuhl, Evan Savage, Tom Sexton, Mike Press and Mike Farmilo all worked at Penfolds as leading winemakers. Daryl Groom, a senior red winemaker of prodigious talent, was seconded to Penfolds joint venture at Geyser Peak in California in 1989 to 'resuscitate the winery'. Previously engaged with making many of Penfolds top wines, including Grange, he soon achieved substantial recognition for his winemaking skills in the US media, including the *San Francisco Chronicle*'s Winemaker of the Year. With a change of ownership, and many great wines in barrel, he stayed on in California.

Andrew Baldwin and Kym Schroeter have each notched up more than twenty-five vintages at Penfolds. These two senior winemakers join an elite group of Penfolds craftspeople who have made important differences and advances in wine quality. It is this type of loyalty, and experience gained by rising through the ranks, that gives Penfolds such a distinct personality and advantage. Being a Penfolds winemaker is a way of life, a vocation that can last a lifetime.

Recent additions to the team include Adelaide University oenology graduates Adam Clay and Stephanie Dutton and Charles Sturt University wine science graduate Matt Woo. The winemaking team also comprises former Magill Estate Restaurant sommelier Jamie Sach, a graduate in hotel management and media studies. As Penfolds Ambassador he contributes expertise in fine wine service and training. This contemporary role responds to the expectations and needs of the modern customer wanting advice, detailed wine knowledge and interaction with the Penfolds brand.

Peter Gago's appointment in 2002 as Penfolds Chief Winemaker heralded a new era for Penfolds. After completing a Bachelor of Science at Melbourne University, a few years later he duxed the winemaking course at Roseworthy Agricultural College and graduated with a Bachelor in Applied Science, Oenology. In 1989 he joined Penfolds as a sparkling winemaker, eventually becoming Penfolds red wine oenologist. Now, as Penfolds Chief Winemaker, he has brought a new spirit of excitement, energy and discovery.

Because of its great winemaking history, Penfolds is one of the most bankable, consistent and reliable wine labels in the world. It is an everyman's wine company, with a track record of excellence from the inexpensive early-drinking Koonunga Hill and Bin 28 Shiraz to the legendary Grange. Says Peter Gago, 'It is our responsibility to build upon the legacy of winemakers past. It can only be done by adding something *extra, different and unique*; the original vision of Max Schubert.'

Winemaking team, left to right, Stephanie Dutton – Red Winemaker, Adam Clay – Red Winemaker, Kym Schroeter – Senior White Winemaker, Peter Gago – Chief Winemaker, Steve Lienert – Senior Red Winemaker, Andrew Baldwin – Red Winemaker and Matt Woo – Fortified and Red Winemaker.

PART TWO

The Wines

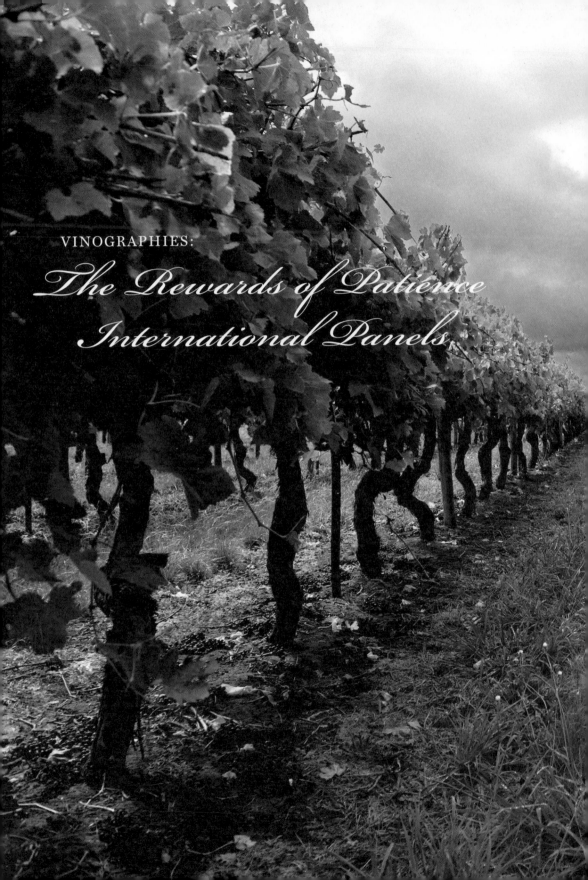

VINOGRAPHIES:

The Rewards of Patience

International Panels

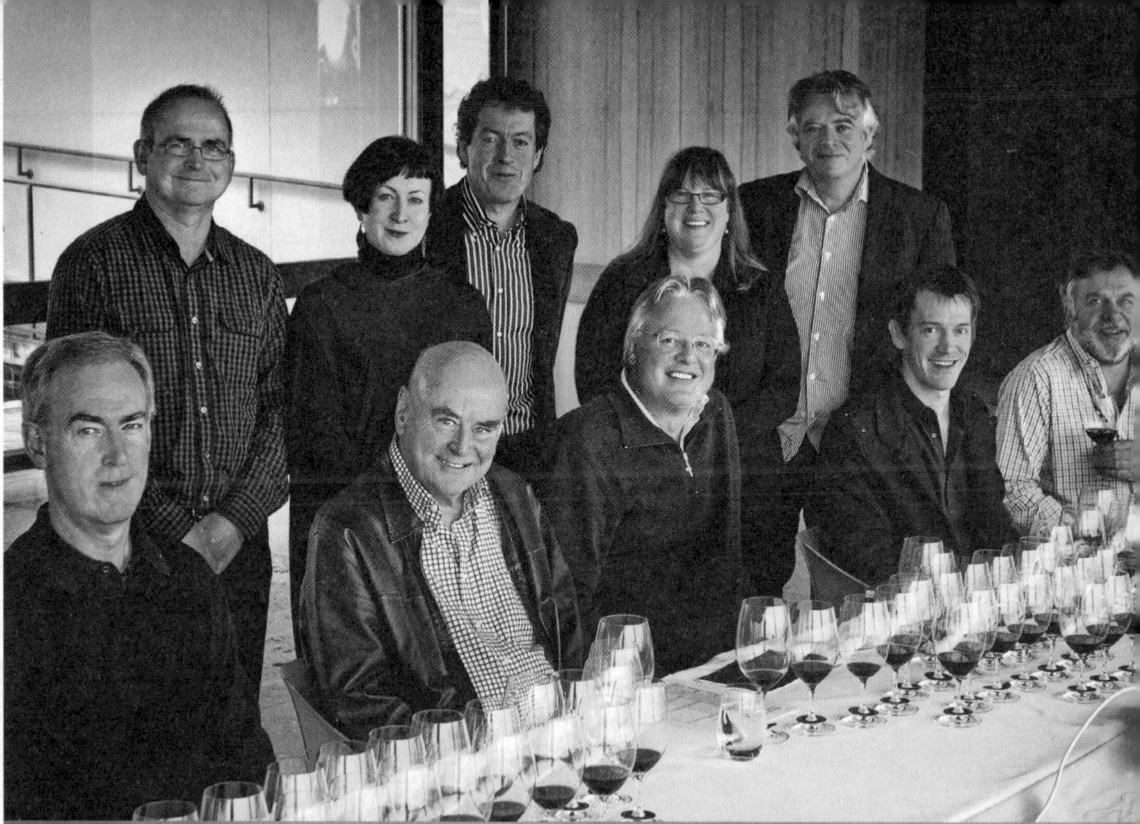

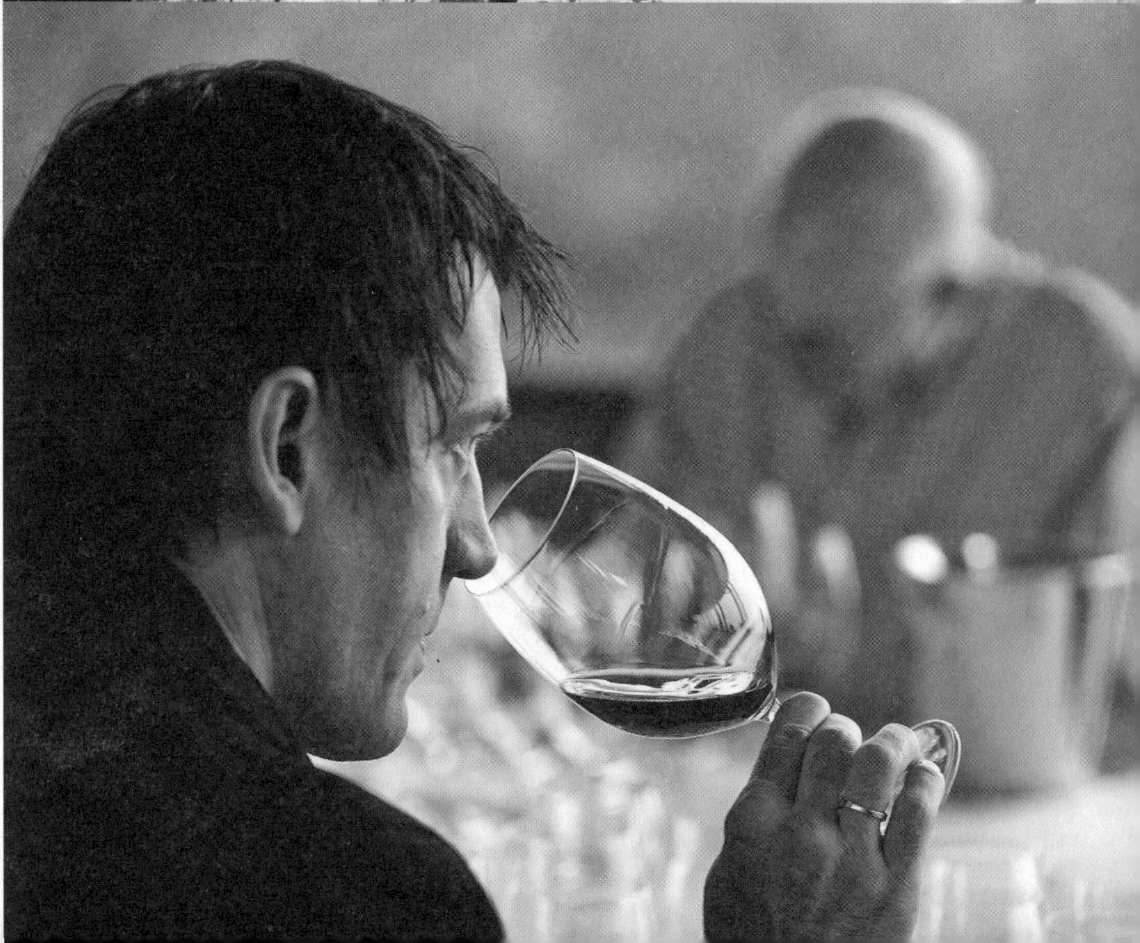

About the Panels

This edition comprises the opinions and reviews of four international wine panels. Drawing from a wealth of experience across Asia, the Americas, Europe and Australasia, this is the most ambitious, comprehensive and informed analysis of Penfolds catalogue of museum wines. Each panel member is a respected observer or authority on wine.

Tastings were held in Beijing, New York, Berlin and Adelaide. At every session, Peter Gago and various winemakers provided overviews of vintages and technical support. Tasting notes and reviews have been combined and compiled to reflect a balance in opinion and sentiment. This has been achieved through rigorous selection and editing of the tasting notes and post-tasting discussions. A panel review of the style, drinking potential and vintage highlights follows each 'vinography'. The individual tasting notes, which are found in Part Three of this book, are compiled by vintage timeline.

A brief biography of each panel member follows. Their initials are used in the text for attribution of tasting notes and opinions.

Peter Gago (PG)
Penfolds Chief Winemaker
Steve Lienert (SL)
Penfolds Senior Winemaker
Kym Schroeter (KS) (Adelaide only)
Penfolds Senior White Winemaker
Jamie Sach
Penfolds Ambassador
Andrew Caillard (AC)
Author and Tasting Editor

Asia Panel

Beijing, China – November 2011

Chantal Chi (CC)
Shanghai-based wine writer, wine educator, wine judge and author. Contributor to *Decanter China*, *Meininger's Wine Business International*. Co-author, with brother Noel, of *French Smile: The Soul of My Wine Cellar*.

Li Demei (LD)
Oenologist, academic, wine show judge, wine writer, visionary, winemaking consultant. Author of *Communications from a Chinese Winemaker*, contributor to many newspapers and magazines, including *Revue de Vin*, *Wine China* and *The Wine Review*.

Yu Sen Lin (YSL)
Freelance wine writer, author, wine educator, wine consultant, international wine judge. Contributing editor of *Decanter China*. Influential Taiwanese wine writer.

Dr Edward Ragg (ER)
Academic, wine writer, international wine judge, wine educator and poet. Co-founder of Dragon Phoenix Wine Consulting in Beijing. Associate Professor of English Literature and Poetry at Tsinghua University, Beijing.

Simon Tam (ST)
Hong Kong–based auction director, wine educator, wine judge, entrepreneur, wine marketer, linguist, food and wine expert. Head of Christie's Asia. Contributor to the *South China Morning Post*.

Ch'ng Poh Tiong (CPT)
Wine and food critic, pioneering bilingual publisher and entrepreneur, wine show judge, wine historian, educator and wine-video presenter. Publisher of Singapore's *The Wine Review*. Columnist and contributor to *Decanter* and *www.decanterchina.com*. Founder of ICCCW, the International Congress of Chinese Cuisine and Wine. www.vinovideos.com.

Fongyee Walker (FW)
Academic, wine expert, wine educator, entrepreneur, linguist, China observer, international wine judge and wine consultant. Co-founder of Dragon Phoenix Wine Consulting. Wine and Spirit Education Trust provider. Contributor to English- and Chinese-language magazines including *Decanter*, *Hong Kong Tatler* and *Caijing Ribao* (China Business News).

Americas Panel

New York, NY, USA – July 2012

Marcelo Copello (MC)
Wine journalist, wine educator, television presenter and publisher of Brazil's *BACO* magazine. Author of the *Wines of Brazil Yearbook*. Contributor to *GOSTO* magazine and *Revista de Vinhos* (Portugal).

Joe Czerwinski (JC)
Wine journalist, educator and international wine judge. Managing Editor of Digital and Print for *Wine Enthusiast Media*. Formerly production editor of Prentice Hall and Hearst Magazines.

Martin Gillam (MG)
Journalist, former ABC correspondent in New York, and documentary filmmaker. Feature writer for *Australian Gourmet Traveller Wine* magazine and *Food Arts* magazine (US). Documentaries include 'A Brief History of Wine' for the History Channel and 'Drink of Kings, A History of Champagne'.

Anthony Gismondi (TG)
Broadcaster, wine journalist and wine judge. Wine writer for the *Vancouver Sun*, Canada. Wine Consultant to Air Canada. Editor-in-chief of *Wine Access*.

Joshua Greene (JG)
Wine critic, wine judge, feature writer, editor, and publisher of New York–based *Wine and Spirits*. www.wineandspiritsmagazine.com.

Ray Isle (RI)
New York–based Executive Wine Editor of *Food and Wine* (US). Former deputy wine editor of American Express Publishing and managing editor of *Wine and Spirit Magazine*.

Dave McIntyre (DM)
Wine correspondent for *The Washington Post* and blogger at www.dmwineline.com. Public Affairs Officer at the US Nuclear Regulatory Commission.

Linda Murphy (LM)
Contributing wine writer to *The San Francisco Chronicle*, correspondent for *Decanter* magazine, wine writer for *jancisrobinson.com*. Wine judge and former sports writer of *The San Diego Union*.

Josh Raynolds (JR)
Wine critic, international wine judge and former retailer. Contributing editor to *Stephen Tanzer's International Wine Cellar* (IWC) and a contributor to *winophilia.com*.

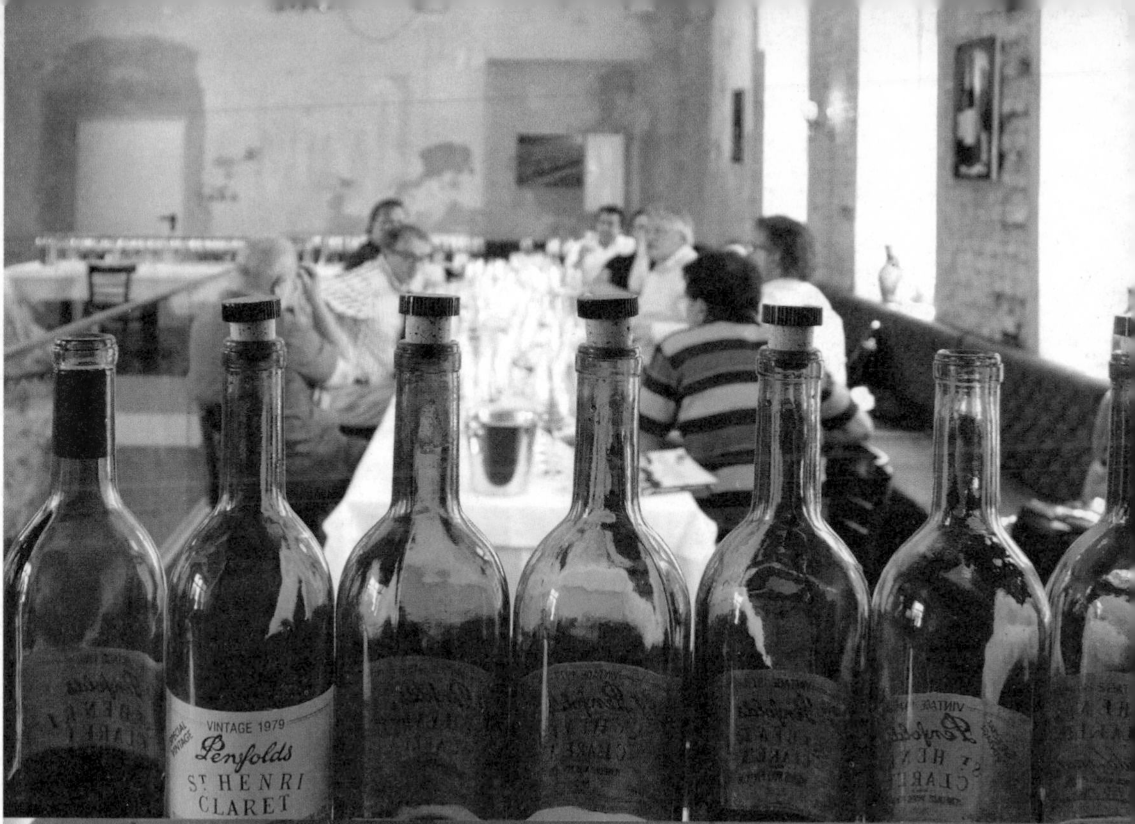
VINTAGE 1979
SPECIAL VINTAGE
Penfolds
ST HENRI
CLARET

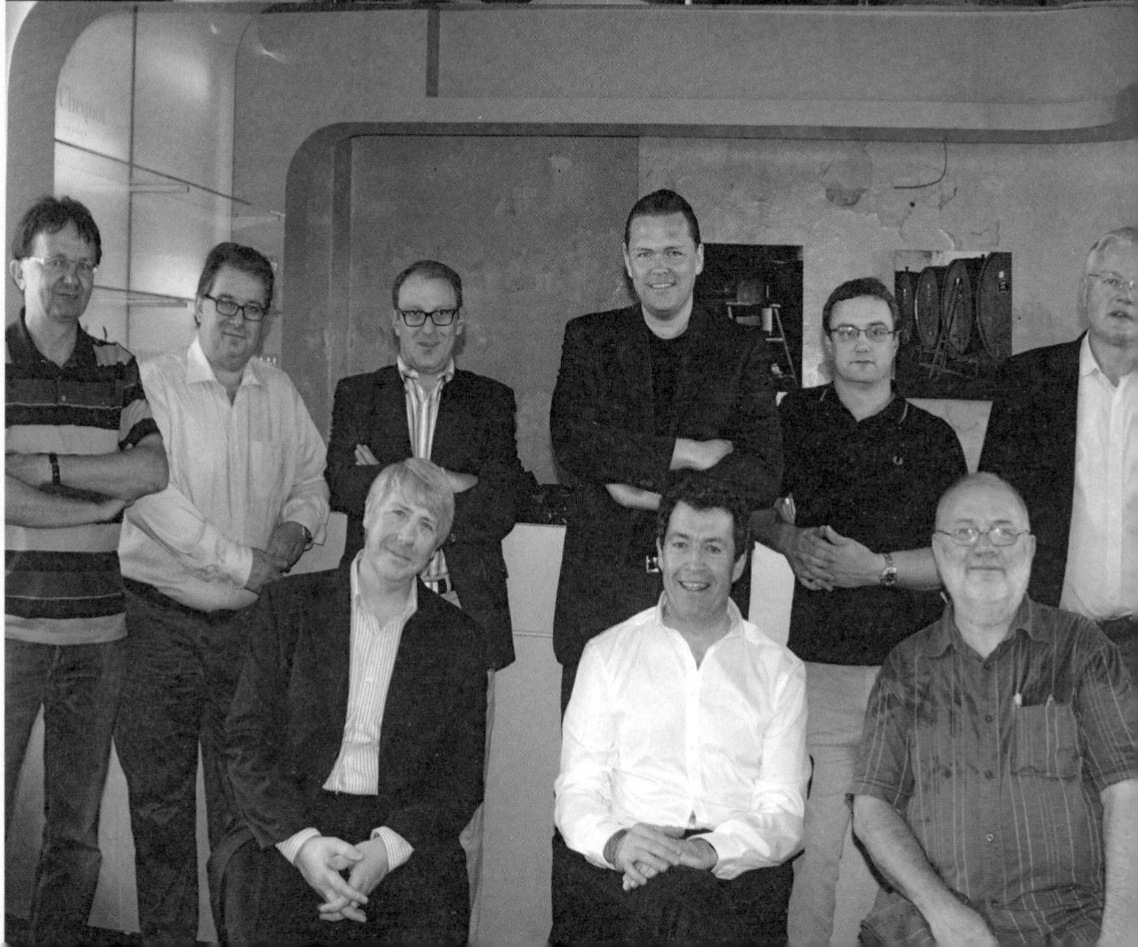

Europe Panel

Berlin, Germany – July 2012

Dr Neil Beckett (NB)
Academic in ancient languages, historian and wine journalist. Editor of the prestigious *The World of Fine Wine* and former editor of *Harpers*. Editor of *1001 Wines You Must Try Before You Die*. Member of the Grand Jury Européen tasting panel.

Frank Kämmer, MS (FK)
Freelance wine critic, wine educator, marketing consultant and Master Sommelier. Contributor to several magazines including *Gault Millau Wine Guide*, *Deutschland* and *Feinschmecker*.

Peter Keller (PK)
Wine journalist. Weekly columnist for Switzerland's leading national newspaper, *Neue Zürcher Zeitung*.

Andreas Larsson (AL)
Sommelier, wine writer, wine educator and consultant. 'Best Sommelier in the World 2007'. Wine director at the renowned restaurant PM & Vänner in Växjö, Sweden.

Neal Martin (NM)
Wine writer and author. Wine critic for *wine-journal.com* and reviewer for *eRobertParker.com*. The critically acclaimed *Pomerol* is his first book. A former English teacher and specialist Bordeaux and Burgundy importer.

Peter Moser (PM)
Wine journalist. Editor of *Falstaff Magazine*, one of Austria's most widely read fine wine magazines. Author of the *Ultimate Austrian Wine Guide*.

Mario Scheuermann (MS)
Wine and food critic, wine judge, author and blogger. Hamburg-based editor of web journal *The Drink Tank*. Author of numerous books and essays on fine wine and cuisine.

Australasia Panel

Adelaide, South Australia – September 2012

Jo Burzynska (JB)
Wine writer, international wine judge, educator and sound artist. Wine Editor of *The New Zealand Herald*'s *Viva Magazine*. Former associate editor of *Harpers Wine & Spirit Magazine* (UK).

Jane Faulkner (JF)
Wine, food and lifestyle journalist. Wine writer for Melbourne's *The Age*. Contributor to *Australian Gourmet Traveller Wine*, and *The Age* and *The Sydney Morning Herald* good food guides.

James Halliday, AM (JH)
Wine writer, international wine judge, doyen and author of *James Halliday's Annual Wine Companion*. Weekly columnist for *The Australian*. Contributor to *The Oxford Companion to Wine*. Author of more than 40 wine books.

Huon Hooke (HH)
Wine critic, author, international wine judge and observer. Columnist for *The Sydney Morning Herald* and editor of *huonhooke.com*. Contributor to *Australian Gourmet Traveller Wine* and international publications. Author of *Max Schubert, Winemaker*.

Ralph Kyte-Powell (RKP)
Wine writer, broadcaster, educator and wine judge. Wine writer for Melbourne's *The Age* and contributor to *The Sydney Morning Herald*, *Delicious* and *Cuisine*.

Tony Love (TL)
Wine journalist, wine judge and commentator. National wine editor of *Taste*, a syndicated weekly column for News Ltd's metropolitan newspapers in Brisbane, Sydney, Melbourne, Adelaide and associated websites. Contributor to *Gourmet Traveller* magazine.

Tyson Stelzer (TS)
Wine writer, author, publishing entrepreneur, screwcap advocate, wine judge and activist. Author of several books including *The Champagne Guide* and *Barossa Traveller*. Contributor to *Australian Gourmet Traveller Wine*, *Wine Spectator*, *Decanter* and other publications. www.winepress.com.au.

'It is hard to realise that in 1952, the first Grange vintage,
the use of small oak casks – hogsheads of 65 gallons –
to add to wine flavour was unheard of in this country.'

JAMES HALLIDAY

ICON WINES
Chapter four

Penfolds Grange is an Australian First Growth with a history, prestige and lasting quality that interconnects with the great wines of Bordeaux. For many wine collectors and observers it remains a defining modern wine.

The story of Penfolds Grange is an intriguing tale of one man's vision against almost insurmountable odds. Max Schubert was a winemaker who spent his war years fighting with Australian forces in North Africa, Greece and New Guinea. When he returned to Adelaide, South Australia's garden capital, he rejoined the family company Penfolds, a business managed like an army battalion. During the late 1940s, every man knew his position. Decisions were made from top management and every worker, including the senior winemakers, was expected to follow company rules. Key personnel were instructed to maintain secrecy

Bin 95 Grange Shiraz

FIRST VINTAGE	1951 experimental; 1952 commercial.
VARIETY	Primarily shiraz with a small percentage of cabernet sauvignon in most years (usually less than 8%).
ORIGIN	Usually a multi-district blend, South Australia. Significant shiraz contributions from the Barossa Valley, Clare Valley, McLaren Vale and Magill Estate; cabernet sauvignon from the Barossa Valley, Coonawarra, McLaren Vale, Padthaway and Robe.
FERMENTATION	10 tonne stainless steel tanks with header boards, at Nuriootpa. 3-7 tonne open fermenters with wax-lined/wooden header boards at Magill Estate. All components complete fermentation in barrel.
MATURATION	18-20 months in new American oak hogsheads (300 litres).
COMMENTS	Named after the Grange Cottage built in 1845 at Magill Estate and itself named for Mary Penfold's family home in England. Labelled Grange Hermitage until the 1989 vintage; Grange since the 1990 vintage. Made at Magill until 1973, then Nuriootpa, Barossa Valley, until 2001 and partially returned to Magill Estate in time for the 2002 vintage. 1979 was first vintage available in magnums. Packaged in laser-etched bottles with identification numbers since the 1994 vintage. Limited availability in all Penfolds markets.

about Penfolds pioneering winemaking practices and scientific discoveries.

After the war, there was a belief that the world would open up for fine wine as European settlers and returned soldiers brought back to Australia an acquired taste for dry red table wine. Penfolds had recently invested in vineyards and developed new technology in the full knowledge that the market was rapidly changing. The development of Penfolds Grange and its red table wines took place in an unprecedented period of prosperity and economic growth in Australia.

Penfolds Grange – created by Max Schubert in 1951 – is a beautifully seductive, richly concentrated wine which evokes the spirit of the Australian landscape, its natural affinity with shiraz and Penfolds remarkable winemaking philosophy.
— Citation: Exceptional classification, Langton's Classification of Australian Wine V.

'In the latter part of 1949' Max Schubert recalled that he was sent to France and Spain to investigate sherry-making practices and the production of port. Fortified wines still dominated the fine wine scene in Australia. On a side trip to Bordeaux the idea of producing a red wine 'capable of staying alive for a minimum of 20 years' first entered his mind. He was taken around the Médoc by Christian Cruse (1884–1975), 'one of the most respected and highly qualified wine men of the old school in France'. He visited many of the great vineyard estates, including first growths Château Lafite Rothschild, Château Latour and Château Margaux, where he enjoyed the 'rare opportunity of tasting and evaluating Bordeaux wines between 40 and 50 years old'. He would have witnessed the patchwork of close-spaced, low-trellised vineyards, the rows of fermentation vats and barrel cellars. After an abnormally hot 1949 vintage he would have observed basic cellar practices and inspected fermentations still bubbling away, most in headed-down wooden fermentation vats.

Penfolds Grange has a mythical status among wine collectors and is one of the few wines outside Europe that consistently influences market sentiment.

Max Schubert wrote, 'the basics of the whole [project] did not present any problems to my mind. However, there were details of climate, soil, grape variety, vinification, handling, maturation and bottling which appeared to me to be the most important factors in quality red wine production in the Médoc. I felt at the time, and I was ever the super-optimist,

that it would only be a matter of undertaking a complete survey of vineyards to find the correct varietal grape material, and with a modified approach due to differing conditions in Australia, it would be possible to produce a wine that would stand on its own feet throughout the world, and would improve with age year by year.'

Inspired and impressed by the French cellared-style wines, he dreamed of making 'something different and lasting' of his own. On his long five-day return flight to Adelaide, via the Middle East, India, Singapore and Indonesia, he made plans to make a great Australian red wine that would cellar for at least twenty years.

Back in Adelaide, in time for the 1950 vintage, Max Schubert set about looking for appropriate 'raw material'. He conducted a survey of all likely vineyards and varieties which would give maximum skin pigmentation and a sugar balance, resulting in a full-bodied, well-balanced wine 'containing the maximum extraction of all the components which make up the grape'.

The decision to use shiraz, also known as 'red hermitage' in Australia, was based on availability, quality, reliability and continuity of supply. Cabernet sauvignon was not widely planted in South Australia at the time. Although Penfolds had recently acquired the Barossa Valley's Kalimna Vineyard, including the 1880s' plantings of Block 42 cabernet sauvignon, its yields and quality were inconsistent.

Max Schubert sourced shiraz grapes from the Grange Vineyard at Magill and from the privately owned Honeypot Vineyard, planted in 1920 by Samuel Wynn, the founder of Wynns Coonawarra Estate, just south of Adelaide at Morphett Vale. He said: 'I had already observed that both vineyards produced wines of distinctive varietal flavour and character with a great depth of colour and body weight, and felt that by producing them together, the outstanding characteristics of both vineyards would result in an improved all-round wine eminently suitable for my purpose.' Combining traditional Australian techniques, new ideas from Bordeaux and precision winemaking practices developed at Penfolds, Schubert made his first experimental wine in 1951.

'The basic method adopted opened up a whole new concept of quality dry red production, in that fermentation was strictly controlled over a much more lengthy period than hitherto, maximum extraction was obtained by daily handling and maceration of juice and grape solids. When this had been achieved the partially fermented wine was separated from the

skins and the fermentation was then completed in the [five untreated] new hogsheads where the wine remained until the time of bottling some 18 months later. The objective was to produce a big, full-bodied wine containing maximum extraction of all the components in the grape material used.'

Although the 1951 vintage (which was never commercially released) was moderately successful, it did not fully reflect the ambitions of Max Schubert. It was the subsequent 1952 vintage, bolstered up with pressings from Nuriootpa, that gave the wine an extra richness and concentration. He called it Grange Hermitage after the house and vineyard established by Dr Christopher and Mary Penfold in 1844. Schubert used 'Hermitage' rather than 'Shiraz' to 'pander to the snobs in New South Wales' – an important market for Penfolds.

A bottle of 1951 Penfolds Bin 1 Grange recently fetched a record auction price of $53 936. This historic 1951 vintage is modern wine's equivalent to powered flight! — Langton's.

The commercial release of 1952 Penfolds Grange Hermitage was an historic moment for Australian wine. It marked the beginning of a 'dynasty of wines' that would capture the imagination of the Australian wine consumer. Within a decade, Penfolds Grange would become the reference point for Australia's emerging ultra-fine wine scene.

It is hard to realise that in 1952, the first Grange vintage, the use of small oak casks – hogsheads of 65 gallons – to add to wine flavour was unheard of in this country. — James Halliday, *National Times*, 1978.

Over the next five years, Max Schubert quietly and studiously developed the Grange style. Using ground-breaking technology developed by Penfolds scientist Dr Ray Beckwith, with which wine could be accurately controlled and stabilised, he established a unique wine based on warm-climate shiraz fruit, barrel fermented at the end of vinification and matured in American oak hogsheads.

In 1957, Max Schubert was asked to show his efforts in Sydney to top management, invited wine identities and personal friends of the board. To his horror and humiliation the Grange experiment was universally disliked. Even further tastings in Adelaide resulted in negative opinion. One critic observed, 'Schubert, I congratulate you. A very good, dry port, which no one in their right mind will buy – let alone drink.'

Gladys Penfold Hyland, the autocratic chairman of the board, finally ordered Schubert to stop making Grange. 'Stocky in build, but generally dressed in beautifully tailored suits', she ruled the family business with an iron rod. Defiance of such an order meant certain dismissal and severe loss of face.

Embarrassed, angry and dejected, Max Schubert's ambitions to make 'a great wine that Australians would be proud of' were completely destroyed. Experimental Grange vintages, already bottled and binned, would soon be sold off to clubs as house wine. The remaining stock would be blended away into oblivion. Grange was dead.

It was the happenstance of distance between senior management in Sydney and winemakers in Adelaide, 1400 kilometres apart, which saved Grange from imminent doom. With the help of Magill's assistant general manager Jeffrey Penfold Hyland, a visionary black sheep of the family (but who would later become Chairman of Penfolds during the 1970s) and Schubert's team of winemakers, all the experimental Grange was stashed away out of sight, in the underground cellars of Magill. The tradition of secrecy expected of every Penfolds employee was taken to a new level. From 1957 to 1959, the 'hidden Granges' were made without the knowledge of the Penfolds board. Max Schubert continued to source fruit and make his experiments in secret. Schubert's biographer, the Sydney-based wine writer Huon Hooke, says, 'The hidden Granges were made, matured and bottled in strict secrecy and word never leaked out to the powers-that-be.'

Without finance, he matured the wines in previously used American oak hogsheads, although the maturation time was halved to just nine months. However, his revolutionary winemaking techniques – comprising partial barrel fermentation – was continued. He then had to make do with abandoned bottles found lying in the cellars at Auldana. The 'hidden' Granges were stored away in the Magill drives built by Alfred Scholz.

The whole affair could not have been carried off without the support of Murray Marchant and Gordon Colquist, Schubert's senior winemakers, who helped care for the wines during these uncertain times. Max Schubert said, 'while undoubtedly Grange, they lacked that one factor [of new wood] which made the difference between a good wine and a great wine'.

Although management was kept away, friends and associates were occasionally brought in to taste the wines. Some bottles were even given away. Max Schubert and Jeffrey Penfold Hyland, both 'gluttons for punishment', would also show the early vintages to 'wine and food societies, Beefsteak and Burgundy Clubs, and wherever men congregated'. Although considered uncommercial in 1957, news was filtering out about Schubert's unique Grange Hermitage.

A second tasting with the same board members was organised in 1960 by Doug Lamb, a consultant wine merchant, Penfolds director and supporter of Schubert. This time the 1951 and 1955 vintages, both with bottle age development, were greeted with enthusiasm (the 1955 went on to have a very successful wine show career). The Penfolds board ordered production of Grange to restart, just in time for the 1960 vintage.

A sense of experimentation pervades across all of the 1950s. This is a fascinating study of a developing wine style.
— Dave McIntyre, *The Washington Post.*

Len Evans, in his last published book, *How to Taste Wine*, remembered: 'Great Granges were often quite volatile and the 1955 caused a show incident. I was on a panel of three, two of whom, including me gave the wine a gold [medal]. We recognized the acetic acid but also gloried in the flavour, depth and balance of the wine. The other judge gave it 13, a very low score, and wouldn't budge. The chairman, the late great George Fairbrother, a man of infinite patience and great charm, took one sniff of it and said to the dissenter, "Well if you won't budge, I'm afraid I'll have to overrule you and give it a Chairman's Gold." In the 25 to 30 shows I judged under his guidance he only did this with one other wine, the famous Stonyfell 1945 Vintage Port.' George Fairbrother – a doyen of Australian Wine Show judges – was clearly instrumental in garnering support for the style. So too was Tony Nelson, who created the Stonyfell label and pioneered Coonawarra.

In 1960 Schubert was promoted to National Production Manager – the equivalent position to Penfolds Chief Winemaker, being responsible for all the company's wineries and vineyards. This was a period of strategic expansion and consolidation in both New South Wales and South Australia. While the purchase of New Dalwood in the Hunter Valley would be short-lived, the acquisition of significant vineyard holdings on prime *terra rossa* soils in Coonawarra was an important step forward for Penfolds during the 1960s, while the urbanisation of Adelaide would shortly engulf several vineyards, including Modbury, Morphett Vale, Auldana and Magill.

The expansion of Kalimna, in the Barossa Valley, and the purchase of Coonawarra vineyards protected continuity of grape supply and provided the foundations of the modern Penfolds wine business. Several

of the old vineyard holdings – used for port or sherry production – were replanted to table wine varieties. This was an exciting period at Penfolds, where new ideas and practices were implemented at every level. John Bird said, 'Max Schubert adored his senior cellar staff. They had gone through thick and thin with him. I was completely awestruck by the scale of table winemaking and the great working atmosphere.'

One of the great strengths of the Penfolds Grange style, which established a strong following in the 1960s, is that it has never relied on the performance of a single vineyard. In the context of South Australia's vast landscape, this has allowed winemakers to select the best parcels of fruit from the vintage. Max Schubert recognised that Grange should be based on a riper spectrum of fruit. He knew – intuitively – that fruit power, concentration and ripe tannins were key components of optimum fruit quality. He was well ahead of his time.

Schubert's experimental work in the 1950s confirmed his view that cabernet sauvignon was unreliable. He observed: 'The imbalance of the fruit invariably manifested itself on the palate with a noticeable break in the middle and a thinnish, astringent finish.' Nonetheless, following the 1953 vintage he found a parcel of 'cabernet grown at Kalimna that showed all the characteristics and analytical balance required for Grange production'. That year, Schubert and his winemaking team made five hogsheads of 1953 Penfolds Grange Cabernet, based on Block 42's 1880s colonial vine stock. These rare bottles occasionally surface at auction, and when opened still show a compelling freshness, richness and balance.

1960s – a quietly beautiful decade. The best vintages reminiscent of classic First Growth claret.
— Linda Murphy, *Decanter* magazine, US.

Max Schubert instigated the annual vintage classification tasting, a convocation of Penfolds winemakers that continues to this day. At this forum all wines of the vintage could be tasted and classified according to style and Penfolds wine type. New technology was also introduced, including temperature control, improved winery hygiene, inert pipes, stainless steel, Willmes air bag presses and general quality control.

During the 1960s Penfolds Grange firmed up its position as Australia's most distinguished wine. Previously it had been a small-scale wine of 'only four or five barrels'. Vigneron and wine author Max Lake observed 'there has been quite an amount of experimentation with various wines and blends till today it is obvious the style is consolidated into a consistently

Decanter

MARCH 1988 · THE WORLD'S BEST WINE MAGAZINE · PRICE £1.80

MAX SCHUBERT
Decanter's Man of the Year 1988

outstanding wine'. The 1960 and 1962 Granges were highly successful wines: 'they have superb, magnificent bouquet and balance running right through the start to finish'.

The 1960 to 1977 Penfolds Grange vintages were regularly entered in Australian wine shows with considerable success. Both the 1965 and 1967 vintages won the prestigious Jimmy Watson Memorial Trophy at the Melbourne Wine Show. Both vintages earned a considerable swag of Gold, Silver and Bronze medals in Australian capital city wine shows. The 1962 won over fifty Gold medals in its show career! Len Evans – a hugely influential wine show judge and legendary Australian industry leader – wrote in his publication *The Wine Buyer*, in 1972, 'Most people who have been lucky enough to see the wines of the early 1950s and others like the '62 would agree that we have been treated to something quite extraordinary. And whatever comparisons made to the wines of Bordeaux, I believe that the real Granges represent a new, great wine style of the world.'

Over the last twenty years, the Rewards of Patience tastings have revealed a relatively consistent impression of the Grange style, even as the wine evolves and matures. It is a very perfumed, concentrated wine, which combines the intensely rich fruit and ripe tannins of shiraz with the fragrance and complementary nuances of new, fine-grained American oak. Partial barrel fermentation takes place at the tail end of primary fermentation. It weaves the two elements together, producing a 'meaty' complexity and roundness of flavours on the palate. A portion of cabernet is used in some years to enhance aromatics and palate structure.

The 1970s decade presented wines of contemplation. I really appreciated them for their weight, texture and complexity of flavour.
— Josh Raynolds, Editor, *International Wine Cellar*, US.

1971 defined the 1970s decade and is regarded by many as one of the greatest Australian wines of its era. (It created a sensation when it beat the best Rhône Valley wines at the Gault-Millau Wine Olympiad in Paris in 1979.)

The colour of the 1971 is huge, deep purple black; the nose as big and voluminous as ever, showing great oak and fruit, deep, rich, lasting; the flavour is extraordinary. In no other red wine in Australia is there so much depth and intensity. The richness of the wine means that it can be enjoyed now, but this is really vinocide. It will last for years and improve as it softens.
— Len Evans, *The Bulletin*, 1976.

During the 1990s Access Economics, an Australian economic think-tank, used the 1971 Grange as an alternative indicator of investment performance. It continues to be one of the most highly prized wines at auction. For a brief period in 1970 and 1971, Chris Hancock, who would become a leading executive in the Australian wine industry, supervised Grange production. By this time winemaking of Grange was a team effort, although the final blend was always Schubert's decision.

Max Schubert retired as Chief Winemaker in 1975 but remained involved as a consultant winemaker for another twenty vintages. His aura continued to be felt throughout the Penfolds production arena until his death in 1994.

Max Schubert's vision to make a great Australian wine parallels André Tchelistcheff's ambitions to make a great Californian wine. They were direct contemporaries and died within a month of each other.
— Ray Isle, Editor, *Wine and Food*, US.

Don Ditter, who had joined Penfolds as a laboratory assistant in 1942, was appointed Max Schubert's successor. His contribution to the Grange style is immeasurable. His technical eye for detail and gentle collaborative approach to management took Penfolds Grange into the modern era. This included a major overhaul of vineyard management and tracking of fruit. With the advice of his red winemaking team, particularly John Bird, he refined a number of techniques, including the controversial method of encouraging volatile acidity.

Don Ditter said, 'I insisted we keep the volatile acidity within legal limits to avoid being challenged by authorities. Further, it wouldn't have been good if our opposition had pointed it out!' A more flexible approach to maturation and bottling was also implemented. 'Maturation has always been an important aspect of Penfolds winemaking. Timing however is everything when it comes to getting the right balance of freshness, fruit and maturation characters. If the wine was a little worn – it can never be reclaimed.'

Particular attention was made to the seasoning of American oak. Under Ditter's leadership, the Grange style was improved with fresher aromas, more richness and ripeness of fruit and better oak selection. 1976, 1978, 1983 and 1986 are probably his most admired vintages. Around this time Hugh Johnson, the distinguished UK wine critic, called Grange, 'one of the only true First Growths of the Southern Hemisphere'.

The winemaking talents of John Duval were recognised early by Don Ditter and Max Schubert. After completing an Agricultural Science degree at Adelaide University and a postgraduate diploma in Oenology at Roseworthy College, he joined Penfolds in 1974 at Nuriootpa in the Barossa Valley. His calm analytical approach, family wine background, technical skills and understated charm were perfect 'raw material' for Penfolds. His family, sheep farmers and grape growers at Morphett Vale, were early suppliers of high-quality shiraz to Penfolds. When Magill Estate was replanted, the cuttings came from the Duvals' vineyard.

I was impressed by the consistency and freshness of the 1980s. They are in some sort of transition; not yet fully mature nor do they show the opulence of youth.
— Joe Czerwinski, *Wine Enthusiast Magazine*, US, after New York Rewards of Patience panel, 2012.

In the early 1980s, Penfolds stopped entering Grange into wine shows, largely because it is such a distinctive style that most wine judges could spot it easily in a blind tasting – there was nothing further to be gained. This was perhaps illustrated by the poor wine show results of the 1976 Grange, a wine that Max Schubert regarded as a classic Grange vintage and which eventually, as a mature wine, became one of Robert Parker's 100 point wines!

The 1986 vintage – Ditter's last – is generally regarded as one of the greatest Grange vintages of all time. John Bird said, 'Don Ditter will be best remembered for adding extra polish and finesse to Grange. While not always deeply involved in the nitty-gritty of the vintage cellar he had a wonderful palate and really understood how to make the best of each vintage.' Furthermore, Ditter steered Penfolds winemaking through the unchartered waters of consolidation and company takeover. While the Penfold Hyland family lost control of Penfolds in 1976, the quality of the wines – especially Grange – remained on track.

In 1988, a group of Australian businessmen tried to corner the market for 1983 Grange, the notorious bush fire, flood and locust year. By this time, Grange was a solid auction performer, but the expected price evolution of the 1983 during the 1990s never really eventuated because the secondary market only recognised it as a very good rather than exceptional vintage. So the coup failed and has not been tried again.

John Duval worked for John Davoren and for several years played an understudy role to Kevin Schroeter, Don Ditter and Max Schubert. He was appointed Penfolds Chief Winemaker at a remarkably young age, yet his contribution to the evolution of Grange has been critical. His stewardship saw some of the greatest developments and innovations in viticulture and winemaking, including Penfolds 'White Grange' project and the ground-breaking launch of Yattarna and Reserve Bin chardonnays.

The 1990, 1991, 1996, 1998 and 1999 Granges are regarded as the finest vintages of Duval's custodianship. These wines define an Australian decade of winemaking, through their superb fruit complexity, power, finesse and remarkable cellaring evolution. The American consumer advocate magazine *Wine Spectator* also conferred two important honours during Duval's time: the 1955 Grange was named one of the top twelve wines of the twentieth century in 2000, while the 1990 vintage was named 'Wine of the Year' in 1995.

1990s – Grange makes its own tradition. 1996 and 1998 Grange are like two New York skyscrapers. The 1998 is just one floor higher in stature.
— Marcello Copello, wine journalist and publisher, BACO Multimedia, Brazil.

In 2002 some grapes destined for Grange were once again crushed and fermented at Magill and since that year a significant proportion of the blend has been matured in hogsheads in the Magill underground drives. John Duval stepped down in the same year, after sixteen years as Penfolds Chief Winemaker. He now has a successful family wine business, based in the Barossa Valley, and works as an international winemaking consultant.

Peter Gago, his successor, is the fourth custodian of Grange in sixty years. His appointment as Penfolds Chief Winemaker coincided with the internet age and the expectations of a rapidly changing global market. Described by James Halliday as a 'perpetual-motion brilliant speaker, wine educator and winemaker', and 'Ambassador-in-Chief', Gago brings the Grange

narrative into the twenty-first century by 'raising the bar of expectations' and making it accessible, magical and engaging.

Although the patterns of weather and growing seasons perplexed scientists, Penfolds enjoyed several great Grange vintages during the noughties decade. 'With their mountains of power and purity' Grange continues to take Max Schubert's vision to a new and unprecedented level. 'Generous', 'opulent', 'beguiling', 'seductive', 'expressive', 'voluminous' and 'extraordinary' are just some of the words used to describe Grange from the 2000s. This decade is remarkably consistent in quality. Inevitably vintages are singled out and then followed religiously by the market. In this edition of *Penfolds The Rewards of Patience*, the panel identifies 2004, 2005, 2006, 2008 and 2010 as the best vintages of recent times; 2010 is arguably the greatest vintage since 1990.

2000s – The Grange story continues to unfold with consistent refinements across the decade. These are rich, powerful, intense and expressive wines with a promising future.
— Tony Gismondi, *Vancouver Sun*, Canada.

The Grange winemaking philosophy hasn't really changed over the last sixty years. Any refinements of style over the decades reflect new reference points, evolution of vineyard management and winemaking practices. Max Schubert originally aimed to make a wine of between 11.5 and 12.5 per cent alcohol; nowadays Grange sits around 13.0–13.5 per cent alcohol, not necessarily a result of climate change but because of incredible innovations and changes at every level of wine production. Tannin ripeness has become as crucial as fruit ripeness, the level of volatile acidity has been dropped and the quality of oak has improved. Nonetheless, many of Max Schubert's original winemaking practices are still central to the Grange style. The Penfolds winemaking team continues to identify the best and most exquisite fruit available, apply submerged cap/headed-down vinification and complete fermentation in new American hogsheads. Many of these techniques have become standard practice across Australia.

Penfolds Ambassador Jamie Sach says, 'Max Schubert did not copy the French, he created a new paradigm for what fine wines could be. Max chose to use American oak over French oak. He completed fermentation in the barrel – which was not done in Bordeaux. He selected fruit on merit, and blended across regions, as opposed to the single-vineyard offerings of Bordeaux ... This philosophy lives on at

Penfolds – we are always looking to take wines to the next level.'

Josh Greene, publisher of *Wine and Spirits* magazine, who attended the New York tasting as an observer, has written, 'What was once an individual winemaking style has been adopted by any number of others, including some whose caricatures of Grange may tend to blur the distinctions that set it apart. Gago, charged with maintaining those distinctions, may be the most idiosyncratic Chief Winemaker at Penfolds since Schubert. He's sensitive to Schubert's legacy, and is now charged with sustaining it as the oldest wines begin to disappear.'

Grange is one of the singular great wines of the world.
— Josh Raynolds, Editor, International Wine
Cellar, US.

Grange is uniquely heritage-listed by the South Australian National Trust. It also heads up the highly influential and internationally recognised Langton's Classification of Australian Wine – in recognition of Grange's cornerstone presence on the secondary wine market. It continues to generate considerable collector interest and millions of dollars of auction revenue per year. In recent times Grange has been ranked as one of the most tradeable wines in the world by Liv-Ex – the London Wine Exchange. *The Wall Street Journal* has even published a Dow Jones Grange Index; the accompanying text was, 'Wine lovers remember their first Grange the way they remember their first kiss!'

The Penfolds Red Wine Re-corking Clinic, first initiated in 1991, has further enhanced the reputation of Grange. This free re-corking service, offered to collectors in Australia and around the world, has effectively weeded out poorly stored bottles in the secondary wine market. It has also allowed collectors and media observers to appreciate the longevity of Grange, notwithstanding the diversity of cellaring environments across the globe.

*   *   *

Penfolds Grange is released five years after vinification to allow the wine to further develop in barrel and bottle. The time-lag is also an historical one, a legacy of its rejection in 1957 by Penfolds management in Sydney, when Max Schubert was accused of 'accumulating stocks of wine which to all intent and purposes were unsaleable!'

The early Granges were labelled under different but non-sequential Bin numbers. While the line started as 1951 Bin 1, the Bin numbers are seemingly ad hoc, as shown in the tasting notes, until 1970, when Grange was given a standardised bin number, Bin 95.

The current production of Grange is 7000–9000 twelve-bottle cases a year. In some vintages – such as 2000 – it can be less than a quarter of the average. Quality always comes first. Peter Gago says, 'Grange is Max Schubert's creation. It is not something to tamper with. Ultimately we are custodians of his vision.'

It's a beautiful thing – forget the arguments, the alloca-
tions, the speculation, the price – whenever I find a quiet
moment and think about this thing called Grange, I can't
help but think that without Grange the entire Australian
wine industry would be entirely diminished.
— Campbell Mattison, Wine Front, Australia.

In the late 1940s post-war Australia was embarking on a journey to modern nationhood. Max Schubert, a returned soldier, dreamed of making something different and unique in the world of wine. The development of Penfolds Grange reflects a national mood: a sense of purpose and an enthusiasm for progress. Australia is a young country and does not have the highly evolved traditions of the Old World. The future is its only reference point.

The stature of Grange has been achieved not through the hindsight of centuries of heritage and accumulated wealth, but through trial, error and persistence. Max Schubert described Grange as 'buoyant – almost ethereal', evocative of friendship, happiness and wonder – the essence of the Grange experience. It has proven to be a wine that can evolve and last more than fifty years, further enhancing this timeless story of personal triumph and extraordinary vision.

I hope that the production and the acceptance of Grange
Hermitage as a great Australian wine has proved that
we in Australia are capable of producing wines equal
to the best in the world, as long as we are not afraid to
put into effect the strength of our own convictions, that
we continue to use our imagination in winemaking
generally, and are not afraid to experiment in order to
gain something extra, different and unique in the world
of wine.
— Max Schubert, Penfolds Tempe Cellars, November
1975 – to mark the occasion of Grange achieving its
one-hundredth Gold Medal Award at Australia's state
wine shows.

New York City, NY, USA – Penfolds Bin 95 Grange Shiraz
Marcelo Copello (MC), Joe Czerwinski (JC), Anthony Gismondi (TG), Ray Isle (RI), Dave McIntyre
(DM), Linda Murphy (LM), Josh Raynolds (JR); Bob de Bellevue (collector), Martin Gillam (MG) and
Joshua Greene (JG) in attendance; Peter Gago, Steve Lienert, Andrew Caillard.

> *'People say that places matter in fine wine, yet Grange goes completely
> against the grain.'* — Tony Gismondi

1952–1959

An historic flight of wines including the 'hidden Granges'. 1952, 1953, 1955 and 1959 still holding up
very well. Nonetheless, a 'fascinating study on a developing wine style'. Many of these wines possess
relatively lower ripeness and alcohol levels with the oak sustaining flavours and lengthening out the
palate. Peter Gago apologises for the 1953 looking so young.

1960–1969

A remarkably consistent decade with more potent fruit and tannin presence. Many vintages possessed
understated power and complexity. Early vintages – 1961, 1962, 1963 and 1965 – were top performers.
A lull in the latter part of the decade is perhaps a reflection of only moderate growing seasons.

1970–1979

'Wines of contemplation.' A very textural decade with impressive weight, and fruit sweetness. The
1970s saw a gradual evolution from 'claret' to modern Grange. An upswing in power, richness and
freshness suggests a style shift around 1975; it might also reflect the move of winemaking from Magill
to the Barossa, the use of closed headed-down fermenters and evolution of winemaking techniques
generally. Star performers were 1971 and 1976.

1980–1989

Plush, concentrated and powerful but not fully mature. Still room for further development. 'Moved
from finesse to power; a factor of youth.' 1981, 1982 and 1983: 'a great trio from any producer in the
world'. 1986 perhaps the most beautiful wine in the decade. 1989 – 'overt and flirty' – was the most
surprising for its overall deliciousness.

1990–1999

More primary in nature, with unevolved red and black fruits. The quality of tannins and oak improve
greatly in this decade, making the wines quite drinkable at an earlier age, yet dense and balanced for
the future. In this decade 'Grange makes its own tradition'. An emphasis on vineyard management,
fruit selection and introduction of gentle membrane/bag presses combined with several outstanding
growing seasons brings Grange to a whole new level. 'The decade of the '90s has more hits than the
Beatles, with almost every vintage wining raves.' (MG) 'A lavish decade with six unforgettable vintages:
1990, 1991, 1994, 1996, 1998 and 1999.' (TG)

2000–2010

Grange is still in its elemental stage. There's power and purity, mountains of tannins, extraordinary
richness and supportive rather than dominant oak; a consistent refinement across all of the decade.
'My notes get shorter and shorter because what is there to say other than they're terrific!' (RI) 'How
beautifully the wines age over 50 years.' (LM) 2004, 2005, 2006, 2008 and 2010 were the strongest
performers.

Bin 144 Yattarna Chardonnay

FIRST VINTAGE	1995
VARIETY	Chardonnay
ORIGIN	Multi-district blend drawing fruit from Adelaide Hills (SA), Tumbarumba (NSW), Henty (Victoria) and Derwent River Valley (Tasmania). Source is dependent on vintage conditions.
FERMENTATION	100% barrel fermented in varying percentages of new and older French oak barriques, depending on vintage.
MATURATION	8–9 months on lees; sensitive *battonage* to build complexity and texture.
COMMENTS	Packaged in laser-etched bottles. Released entirely under screwcap since the 2004 vintage.

The release of Bin 144 Yattarna Chardonnay, after 144 winemaking trials, marked a new chapter in the Penfolds story. The inaugural 1995 vintage was the most talked about and eagerly anticipated white wine in Australian history. At the presentation of the Tucker–Seabrook Perpetual Trophy at the 1997 Royal Sydney Wine Show, the Chairman of Judges, the late Len Evans, described the wine as 'a revelation' and 'a step forward for Australian Chardonnay'.

The development of Yattarna inspired a revolutionary movement in Penfolds winemaking and the creation of superbly refined styles based on cool-climate fruit. Since inception it has progressively set the standard for ultra-fine Australian Chardonnay. Yattarna is an elegant, intense and linear style with pure fruit expression and crystalline freshness. It can develop in bottle for ten years or more.

'Yattarna' derives from an Indigenous word meaning 'little by little' or 'gradually'. This utterly unmistakable Australian name – which evokes organic momentum and vintage-by-vintage effort – captures the essence and culture of Penfolds winemaking philosophy. The aim was 'to create a style which shows restraint and fineness of structure when released at three years of age, and will continue to develop richness and greater complexity as it ages in bottle'.

With an incredible portfolio of vineyard resources around Australia, Penfolds gradually, through trial and error, identified suitable places where the best fruit could be grown. The Adelaide Hills district was an obvious starting point. This historic winegrowing region of the mid-to-late 1800s had slumped out of fashion by the early 1920s, but by the 1990s the Adelaide Hills had re-established its name as one of Australia's premier chardonnay districts.

The desire to make something singularly exquisite and lasting also took winemakers along the underbelly of the south-eastern Australian mainland and across the Bass Strait into the island state of Tasmania. Tumbarumba, Henty, Adelaide Hills, Fleurieu and the Derwent River Valley have all played a role in Yattarna's evolution of style. In recent times the coolest climate fruit has dominated the Yattarna blend, reflecting the precision and freshness of the style.

Apple/white peach flavour profiles and natural mineral acidities are sought by the winemaking team. Whole-bunch pressing, barrel fermentation including the use of wild yeasts, malolactic fermentation and yeast stirring (*battonage*) are important elements in its production. New oak is a significant component of the Yattarna style, but the proportions are matched against the weight and character of the fruit each year. Nowadays it is rare for the wine to be barrel fermented and matured in more than 50 per cent new oak, reflecting the characteristics of minerality, texture, layering and longevity in the wine. The Yattarna vintages are beautifully restrained and convincing examples of Australia's Chardonnay revolution.

 *Panel Review*

Magill Estate, Adelaide – Penfolds Bin 144 Yattarna Chardonnay

Jo Burzynska (JB), Jane Faulkner (JF), James Halliday (JH), Huon Hooke (HH), Ralph Kyte-Powell (RKP), Tony Love (TL), Tyson Stelzer (TS); Peter Gago, Steve Lienert, John Bird, Kym Schroeter, Andrew Caillard.

> *Yattarna sits neatly on the middle path between leanness and generosity.*
> — Huon Hooke, *The Sydney Morning Herald.*

1995–1999

All of these vintages have reached their tipping point. Some well-cellared bottles may still be all right to drink, but optimum enjoyment is well past.

2000–2010

The style 'comes of age' around 2000. An elegant, intense, refined and linear Chardonnay with beautiful fruit complexity, restraint and tightness of structure reflecting cool-climate vineyard sourcing. An interesting evolution of style, with the most recent decade showing consistency as an Australian 'Grand Cru' wine; a tribute to fruit selection and winemaking technique. The colours of younger vintages are incredibly pale and surprising for barrel-fermented chardonnay. The most recent vintages possess 'no break in the line' (JH) with 'an assured completeness, a multiplicity of aromas, compelling oscillation of flavours and dissolved mineral notes' (RKP). Highlights included 2004, 2006, 2007, 2008, 2009 and 2010, reflecting a strong consistency across the decade and the preservation of freshness by screwcap.

'The original Bin 707 was a marvellous wine; it comprised
mostly Block 42 cabernet. The first releases had the
richness and ripeness expected of warm- to hot-climate
fruit. A gradual move to Coonawarra during the 1980s
changed it to a more elegant cool-climate wine. During
the mid-1990s it seems to have reverted back to its
original style; a distinctive Penfolds wine divorced
from other Australian Cabernets.'

DON DITTER

LUXURY WINES
Chapter five

'Its name sounds like a plane, but Bin 707 Cabernet Sauvignon is Penfolds masterpiece,' says Chantal Chi, Shanghai's leading wine writer. Over the last decade wine collectors in China have enjoyed a prolonged love affair with cabernet sauvignon. The grape variety has captured the imagination of this nascent fine wine–drinking nation. Some of the greatest cabernet sauvignon dominant wines of the world, including Château Lafite Rothschild, Château Margaux and Penfolds Bin 707 Cabernet Sauvignon, have been offered as gifts or poured at elaborate banquets at China's most prestigious gatherings. Max Schubert, the wine's creator, would have been perplexed at the almost mythological status of Bin 707 in this important, growing market.

Bin 707 Cabernet Sauvignon

FIRST VINTAGE	1964.
VARIETY	Cabernet sauvignon.
ORIGIN	A multi-regional blend, all South Australia: Barossa Valley, Coonawarra, Padthaway, Robe and Wrattonbully.
FERMENTATION	Stainless steel tanks with wax-lined/wooden header boards. All wines destined for Bin 707 complete fermentation in new oak barrels.
MATURATION	18 months in 100% new American oak hogsheads (300 litres).
COMMENTS	Bin 707 was not made 1970–75, nor in 1981, 1995, 2000, 2003 or 2011. Available in all Penfolds markets. Packaged in laser-etched bottles from the 1997 vintage onwards. Released under screwcap in some markets since the 2005 vintage.

Bin 707 Cabernet Sauvignon is completely unadulterated. This blend offers one of the most distinctive expressions of this variety in the world.
— Peter Gago, Penfolds Chief Winemaker.

On a cold November day in China's capital, Beijing, under unusually crystal-clear blue skies, a group of distinguished wine critics and opinion leaders from the Asia region gathered together for an historic retrospective of Bin 707 Cabernet Sauvignon. The tasting took place in a private restaurant among the blue-grey brickwork and rooftops of an old *hutong* district near the Forbidden City. It was the first 'off-shore' Rewards of Patience tasting since the project

The Kalimna Vineyard's Block 42 is a venerable patch of pre-phylloxera cabernet sauvignon vines. It was planted in the mid-1880s; old newspaper reports suggest around 1884. These low-yielding vines produce small-berried black fruit of superb colour, intensity and flavour.
— Dean Willoughby, Penfolds viticulturalist.

Block 42 was the source of Max Schubert's 1948 Penfolds Kalimna Cabernet Sauvignon, a wine that was never commercially released although a small parcel was presumably labelled as a favour for Max Lake, a surgeon, pioneering vigneron and wine writer. In *Classic Wines of Australia* (1966), Lake observed, 'it is becoming magnificent and can only be compared to the big Cabernet wines of Europe'. The last two bottles, unearthed from his cellar, were sold at an auction in Sydney in 1987. Both were consumed, one at a Royal Sydney Wine Show dinner in 1994. Despite the evolution of winemaking practices and time, the wine, boasting a duck-egg coloured capsule, was unmistakably Penfolds: it had classic mature, sweet fruit characters, chocolatey tannins and superb flavour length. Huon Hooke described it as 'powerful, rich, concentrated and fleshy. It brimmed with the trademark aged Penfolds bouquet of crushed ants, leather, chocolate and mint.'

Max Schubert experimented frequently with cabernet sauvignon, especially on his return from the fateful side trip to Bordeaux in 1949. The experimental 1952 and 1953 Grange Cabernets showed that the variety was never far from his mind. He realised, through trial and error, that the Kalimna Vineyard's cabernet sauvignon was inconsistent as source material for a single varietal wine. During the 1950s and early 1960s the variety was therefore used mostly for blending, including early vintages of Grange and Bin 389. The 1960 Bin 630 Kalimna Cabernet–Adelaide Hills Mataro, 1961 Bin 58 Cabernet, 1963 Bin 64 Cabernet and 1963 Bin 511 Kalimna Cabernet Ouillade were also 'one-off' bottlings or experimental wines of the period.

During the post-war years Australian society rapidly changed with the influx of immigrants, industrialisation and a new level of sophistication. For the relatively few wine connoisseurs of the day, Bordeaux was the reference for fine wine. Through the 1950s and 1960s Australian 'claret', a style modelled on the wines of Bordeaux but often made with shiraz, spearheaded Australia's red wine revolution. Initially there was not much cabernet sauvignon grown in South Australia.

first began thirty years ago; held in China to reflect both Penfolds unique position as a globally recognised Australian wine brand and Bin 707's status in this important market.

Bin 707 Cabernet Sauvignon represents the Penfolds house red wine style at its most rich and powerful. The fruit is sourced mainly from South Australian vineyards in Coonawarra, Padthaway, Wrattonbully and the Barossa Valley. Everything about the Bin 707 is large scale. Typically it is immensely concentrated with dark berry/dark chocolate fruit, balanced and enhanced by well-seasoned new oak, plenty of fruit sweetness and strong, but not overwhelming, tannins.

The 1964 Bin 707 Cabernet Sauvignon was Penfolds first commercial release of a single cabernet-based wine. After a stop–start beginning, it is regarded today as one of Australia's most important cabernets; made in a distinct Penfolds House Style and a foil to the great regional cabernets of Coonawarra and Margaret River. But its genesis began in 1948, the year before the foundation of the People's Republic of China. Its history is interlinked with the story of Grange – Max Schubert's dream of making a great Australian red wine that would last at least twenty years – with the Kalimna Vineyard's Block 42 plantings of cabernet sauvignon and Penfolds investment in Coonawarra during the 1960s. If Penfolds had had access to a consistent supply of cabernet sauvignon during the 1950s, the story of Grange might have turned out differently.

In 1960, with a good gathering, you could drink Australia's entire annual production of cabernet sauvignon, in a weekend.
— Don Ditter, veteran Penfolds winemaker who succeeded Max Schubert as Chief Winemaker in 1973.

The inaugural 1964 Bin 707 Cabernet Sauvignon was well received by the Australian wine trade and media. Len Evans, the distinguished Australian wine show judge, said in note number 18 of *The Wine Buyer* (1968): '1964 Bin 707 is one of the best red wines I have tasted for some time, being light and balanced, yet it will undoubtedly improve for some years and should develop into a wine that will long be remembered.' Doug Crittenden, his family wine business in Melbourne celebrating fifty years, purchased parcels of the same wine and bottled it in Melbourne as Crittenden's Celebration Reserve 1964 Kalimna Cabernet Sauvignon.

Breaking with a tradition of using storage bin numbers to name wine releases, Bin 707 was named after the Boeing 707, the aircraft that brought Australia closer to the rest of the world during the 1960s! The wine was christened by Penfolds marketing director Rowan Waddy, an Australian war hero and former Qantas Empire Airways executive. A Bin 747, labelled as 'Dry Red', was also released around the same time but was discontinued in 1975.

The early Bin 707s, from 1964 to 1969, were typically open fermented under wax-lined header boards and matured in seasoned old oak rather than new oak. The wines did not have the power or richness associated with modern vintages and after a disappointing 1969 vintage Penfolds abandoned the line. Nonetheless, many of the first vintages have stood the test of time, a reflection of the Block 42 vineyard and natural balance in the wines.

The early Bin 707s have aged very gracefully and are a good example of how cabernet sauvignon develops.
— Li Demei, consultant winemaker and Lecturer in Oenology at Beijing University.

By a quirk of fate Bin 707 is revered by Chinese collectors as an Australian First Growth. Its popularity stems from its days as a cult wine among the Chinese community in Sydney. It is regarded so highly because of its compelling history, its reputation and because it is made from cabernet sauvignon, a grape variety that carries great respect in China. Many of the country's wine regions, including Shandong Province, Ningxia and Xinjiang Uyghur Autonomous Region, are planted with vast tracts of cabernet sauvignon. Some Chinese people believe that it has magical health-giving properties. First brought out to China in the late nineteenth century and planted around Penglai and Yantai in Shandong Province by Jesuit missionaries, this variety forms the backbone of China's fine wine scene and emergent wine industry.

In 1960s Australia, Max Schubert enjoyed significant wine show success for his special bins based on Coonawarra cabernet sauvignon. 1962 Bin 60A Coonawarra Cabernet Sauvignon Kalimna Shiraz, 1966 Bin 620 Coonawarra Cabernet Sauvignon and 1967 Bin 7 Coonawarra Cabernet Sauvignon Kalimna Shiraz, the most famous experimental wines of the period, showed that the spectrum of cool-climate cabernet fruit aromas, flavours and structure were complementary and suitable for Penfolds growing dynasty of wines. Nonetheless, the problem with inconsistent supply forced Max Schubert to stop making Bin 707 after the 1969 vintage. These difficulties were eventually surmounted with the development of new vineyards, better vineyard management and access to independently grown fruit.

Between 1970 and 1975, most of Penfolds intake of cabernet sauvignon, with the exception of experimental bottlings, was included in the Bin 389 Cabernet Shiraz blend. In 1976 Bin 707 was reinstated after a superb Barossa and Coonawarra growing season. Matured in 100 per cent new American oak hogsheads, it enjoyed instant recognition from wine critics and collectors. With a strong track record at auction for both the early and most recent vintages, it was listed in Langton's inaugural Classification of Distinguished Australian Wine in 1990.

The original Bin 707 was a marvellous wine; it comprised mostly Block 42 cabernet. The first releases had the richness and ripeness expected of warm- to hot-climate fruit. A gradual move to Coonawarra during the 1980s changed it to a more elegant cool-climate wine. During the mid-1990s it seems to have reverted back to its original style; a distinctive Penfolds wine divorced from other Australian Cabernets.
— Don Ditter, Penfolds Chief Winemaker 1973–86.

Since 1976, the winemaking blueprint for Bin 707 has been almost identical to Grange. The difference mainly lies in the source of fruit and varietal character. The wines reflect the discoveries and developmental work during the 1950s and 1960s. Vineyard management, the increasing maturity of cabernet sauvignon vines and evolution of winemaking technology have further improved fruit quality and vinification practices over the last forty years. Typically the best parcels of

fully ripe cabernet sauvignon are vinified in open stain-less steel fermenters with wax-lined wooden header boards to achieve the best possible extraction of colour, flavour and balance. All components are partially barrel-fermented in new American oak hogsheads for eighteen months before assemblage and bottling.

The style relies heavily on a combination of cool- and warm-climate cabernet sauvignon. While Coonawarra remains an important component of the style, the Barossa Valley, especially the Kalimna and Koonunga Hill vineyards, play an equally vital role. This explanation of sourcing explains why Penfolds will sometimes not make Bin 707 in difficult or more elegant years. If the elements cannot be balanced in a trial assemblage or the fruit profile is underpowered, sinewy or out of character, the wine is declassified and blended into other wines.

Bin 707 enjoys a distinctive presence in Australia's ultra-fine wine market. It is a style that has taken over thirty years of trial and error to perfect. Australian wine collectors have enthusiastically followed its jour-ney since it was re-introduced in 1976 and since the 1980s it has become one of Australia's top collectibles. Currently Bin 707 is rated 'Exceptional' by Langton's Classification of Australian Wine, the same rating given to Grange.

It enjoys strong currency in other markets too: Peter Gago says, 'Bin 707 is like a runaway train; not only in China but also Europe'. In China's post–Cultural Revolution society, the narrative of struggle and then triumph is greatly appreciated. Bin 707 offers these collectors, enamoured with cabernet sau-vignon, a 'new exoticism'. Enjoyed by some of China's leading figures, Bin 707 has become one of Australia's most highly prized wines in this market. It is a medium- to long-term cellaring wine. With time and a little patience, Bin 707 builds up into a wonderfully complex and interesting wine.

Panel Review

Beijing, China – Penfolds Bin 707 Cabernet Sauvignon
Chantal Chi (CC), Li Demei (LD), Yu Sen Lin (YSL), Dr Edward Ragg (ER), Simon Tam (ST), Ch'ng Poh Tiong (CPT), Fongyee Walker (FW); Peter Gago, Andrew Caillard.

> *Who said Australian wines can't age? The fifty-year history of Bin 707 overthrows this statement.* — Chantal Chi.

1964–1979

'Loved the tenacity and stubbornness of freshness.' (CPT) Some of the older vintages possessed Chinese tree bark/medicine shop aromas. The tannins are sometimes a touch extracted but never overtly. The best vintages, including 1964, 1966, 1976 and 1979, are lovely examples of aged cabernet. A House Style has yet to fully emerge.

1980–1989

'More expressive and powerful with layers of fruit and beautiful tannins.' (FW) Savoury characters are more dominant than primary cassis notes. Overall noticeable evolution of style with a strong cabernet DNA. 1982, 1986 and the 'stylish elegant' 1987 were stand-outs.

1990–1999

'A band of brothers rather than cousins; each vintage is closely related with more consistent aromatics, richness of fruit and supple tannins. The American oak is more noticeable but overall the Bin 707 style is "polished and elegant".' (ER) 'Since 1994 the wine is more powerful with greater fruit density and propensity for age.' (CPT)

2000–2010

'More precision, more fruit, more "Lo Ching" complexity, more concentration and more tannin integration; a seamlessness and winsome. Abundantly Australian with a brightness of fruit unique to this country.' (FW) The House Style is finally reconciled, with vintage character differentiating each year. 2002, 2004, 2005, 2006, 2008 and 2010 all stand-out vintages.

Vintages not made: 1969–1975 inclusive, 1981, 1995, 2000, 2003 and 2011.

RWT Barossa Valley Shiraz

FIRST VINTAGE	1997.
VARIETY	Shiraz.
ORIGIN	Barossa Valley, South Australia.
FERMENTATION	Stainless steel tanks with wax-lined/wooden header boards. Fermentation completed in barrel. The odd components vinified at Magill Estate in some years, e.g. Kalimna Block 3C and Gallasch material.
MATURATION	12–15 months in (50–70% new) French oak hogsheads (300 litres).
COMMENTS	Packaged in laser-etched bottles.

Penfolds RWT Barossa Valley Shiraz, first made in 1997, was released after several years of **R**ed **W**inemaking **T**rials. This distinctly modern wine articulates the Barossa *terroir* with Penfolds signature method of winemaking. Matured in French oak, it is a foil to the maturation-style St Henri and the opulent and powerful Grange. RWT Barossa Valley Shiraz is typically inky-deep in colour, with sumptuous fruit sweetness, mouth-filling flavours, underlying spice/savoury nuances and chocolatey tannins. Although beautiful to drink when relatively young, these wines are built for the long haul. They have the precision, concentration, persistency and balance to age for many years. The best vintages should last at least thirty years.

RWT Shiraz has established a very good reputation as a cellaring-style wine. In 2010 it was classified by Langton's as 'Excellent', in recognition of its secondary-market presence and popularity with wine collectors. 'The perfect balance of Barossa shiraz and old-style European structure' gives the wine wide appeal. RWT is a modern reference for versatility and regional expression. It has won several trophies at Australian wine shows, emphasising its credentials as one of the Barossa's best shirazes.

Panel Review

Berlin, Germany – Penfolds RWT Barossa Valley Shiraz

Neil Beckett (NB), Frank Kämmer (FK), Peter Keller (PK), Andreas Larsson (AL), Neal Martin (NM), Peter Moser (PM), Mario Scheuermann (MS); Peter Gago, Jamie Sach, Andrew Caillard.

> *Whereas Grange uses American oak, RWT definitely shows off French oak like a Parisian lady showing off her mink coat.* — Neal Martin.

1997–2010

Very well crafted wines with beautiful fruit definition, plenty of sweet fruit flavours and well-balanced French oak. With further evolution 'this could become known as one of the very great wines of the South Hemisphere'. (MS) The early vintages show more obvious granular tannins but the style settles down during the 2000s with better balance of fruit, oak and structure. This is a Barossa Shiraz with power, elegance and refinement. The French oak element gives the wine a savoury complexity and finesse. All of the wines are still within their drinking window. 'One can revel in the primacy of these wines, although they certainly benefit from ageing and step into their stride with five or six years in bottle age.' (NB) Highlights included 1998, 1999, 2002, 2004, 2006, 2008 and 2010.

'A grade' shiraz grapes are sourced from mature vineyards located in the west and north-west Barossa around Kalimna, Koonunga, Moppa and Ebenezer. Typically the vines are aged around twenty to a hundred years old. Other top vineyards are located in the central west district around Stonewell, Marananga and Seppeltsfield. The fruit is small-berried, thick-skinned and concentrated with balanced acidity and even tannin ripeness.

RWT Shiraz is batch-vinified in headed-down open stainless steel fermenters and then racked into new tightly grained French oak to complete fermentation. The wine is matured in new and seasoned French oak, with periodic 'rack and returns', for around twelve to fifteen months before bottling. The overall winemaking is identical to Grange; the result differs because of its regional identity and maturation in French oak. Peter Gago says: 'The first vintages are works in progress. Since the 2000s, the French oak is better matched to the character of the fruit in a quest to create understated power, finesse and balance.'

Bin 169 Cabernet Sauvignon

FIRST VINTAGE	2008.
VARIETY	Cabernet sauvignon.
ORIGIN	Prime Coonawarra vineyards, including Blocks 10 and 20. All located on classic *terra rossa* soils.
FERMENTATION	Static fermenters with wooden header boards at Nuriootpa. Some components vinified in open fermenters with wax-lined header boards at Magill Estate. All components complete fermentation in barrel.
MATURATION	12–15 months in mostly new French oak hogsheads.
COMMENTS	Approximately 1000 cases. A contemporary alternative to Bin 707 with a strong regional imprint. A sibling to RWT.

Bin 169 Cabernet Sauvignon is one of Penfolds newest regional Coonawarra wines. This gorgeously scented and richly concentrated wine is steeped in the origins of Penfolds modern winemaking tradition.

In 1960 Penfolds purchased Sharam's Block from the Redman family. Today the company owns over 100 hectares of prime Coonawarra land. The region is flat and largely unprotected but the unique confluence of cool maritime and warm continental climates is ideal for cabernet sauvignon. With the added benefit of natural aquifers and groundwater, vineyards can be tuned almost perfectly throughout the growing season. The vines are mostly trained on high single wire trellises. Yields are approximately 4–5 tonnes per hectare.

Bin 169 Cabernet Sauvignon is a contemporary showcase of cabernet and Coonawarra. Peter Gago says: 'This wine symbolically guards Bin 707 in the same way that RWT shields Grange.' Typically, the wine shows classic blackcurrant/sage aromas, supple fine-grained tannins and beautifully matched French oak.

In 2008 the cabernet fruit was batch-vinified at Nuriootpa. After draining and pressing off, fermentation was completed in 100 per cent new French oak hogsheads. This was followed by maturation in the same oak for fourteen months. Senior winemaker Steve Lienert explains: 'We fermented the wine at cool temperatures to enhance the aromatic profile of the wine. The savoury nuances of French oak knit in nicely with the regional expression of cabernet sauvignon. This is opposite to Bin 707, where the dark fruit profile is better suited to American oak!'

During the maturation period, the wine was regularly 'splashed in and out' of oak. This technique – known as 'rack and return' – has the effect of micro-exposing the wine to air. It promotes more fruit complexity, harmonises the oak and softens the tannins. The wine also 'toughens up' for medium- or long-term cellaring.

Peter Gago says: 'Bin 169 Coonawarra Cabernet Sauvignon reflects the increasing consistency and availability of A grade Coonawarra fruit. However, it's not only about improved viticulture and winemaking techniques. This is an archetypal style that reflects the personality of Coonawarra and the identity of Penfolds. We believe this has all the hallmarks of an emerging classic.'

 *Panel Review*

Magill Estate, Adelaide – Penfolds Bin 169 Cabernet Sauvignon
Jo Burzynska (JB), Jane Faulkner (JF), James Halliday (JH), Huon Hooke (HH), Ralph Kyte-Powell (RKP), Tony Love (TL), Tyson Stelzer (TS); Peter Gago, Steve Lienert, Andrew Caillard.

These are complete expressions of Cabernet Sauvignon.
There is nothing else quite like this in Australia. — James Halliday.

2008–2010
A contemporary alternative to Bin 707 with French rather than American oak. A Penfolds style, rather than overtly regional in character. The wines possess 'such purity and concentration' and 'plenty of cassis interwoven with new oak' (RKP). Lovely ageing potential. A trio of lovely wines.

Magill Estate Shiraz

FIRST VINTAGE	1983.
VARIETY	Shiraz.
ORIGIN	Single vineyard wine – using selected parcels of fruit from Blocks 1, 2 and 3 of the 5.2 hectare Magill Estate, Adelaide, South Australia.
FERMENTATION	Wax-lined, open concrete fermenters with wooden header boards. After basket pressing, components complete fermentation in barrel.
MATURATION	12–15 months in new French (approximately 65%) and American (approximately 35%) oak hogsheads.
COMMENTS	Approximately 1500–3000 cases. Although the wine was only first made in 1983, the vineyard has been in production since around 1847 (it was planted in 1844).

The historic and heritage-listed 5.2 hectare Penfolds Magill Estate is one of the few single vineyards in the world located within city boundaries. At its peak in 1949 the vineyard – planted to several different grape varieties on rich chocolatey red-brown soils – covered 120 hectares of gentle north-west facing slopes. Regarded as one of the choicest sites in colonial South Australia, the vineyard has gradually diminished because of urban development. Its legacy as the 'Grange Vineyard' ensures its place in Australian wine history.

Magill Estate Shiraz is batch-vinified in open wax-lined concrete tanks. After completion of fermentation, the wine is matured in a combination of two-thirds new French and one-third new American oak for around twelve to fifteen months. This 'soft of nature, yet structured and substantial' wine, is distinctly different to the mainstream Penfolds style, showing intense floral/blackberry/aniseed aromas, smooth richness and a 'long spine without being overtly firm'.

Our aim is to make one of the finest single vineyard or "monopole" wines in Australia. We know we can do this, with history on our side and one or two winemaking tricks up our sleeve.
— Peter Gago, Penfolds Chief Winemaker.

During the 1950s Magill became a centre of winemaking experimentation using Grange Vineyard fruit and drawing more material from surrounding vineyards. This includes the experimental bottlings of Grange Hermitage and the celebrated 1956 Bin 136/Bin S56 Magill Burgundy. The world-famous 1962 Bin 60A Coonawarra Cabernet Sauvignon Kalimna Shiraz was also made and blended at Magill Estate.

In 1972, Penfolds sold off a part of the vineyard for subdivision in the hope of staving off and saving the greater vineyard. However, in 1975 the South Australian Land Commission compulsorily acquired 65 hectares of vine area. The remaining dry-grown vineyard is planted on fertile red-brown earth over limestone. It comprises three blocks: the upper and lower cottage blocks, which are now known as Block 1, were partially replanted in 1951; Block 2, opposite the Grange Cottage, formerly a grenache vineyard, was completely replanted to shiraz in 1967; Block 3, on the north-eastern side of the winery and originally a cabernet vineyard, was replanted with shiraz in 1986–87. All of the vines are planted on their own roots.

None of the original Magill vines used for the production of the very first few Granges survive. The original Grange Cottage, built in 1845, still remains among the vines and Penfolds turn-of-the-century bluestone winery and cellars is once again a working

Chateau Magill Wine

1. Concept

To make a French Chateau-style red wine, distinctly different to the Grange Hermitage style, in that body weight and colour would be approximately half that of Grange, whilst aroma, flavour and character would be individual and pronounced, extractives would be less, resulting in more elegance consistent with lighter body. As such it would be different to Grange Hermitage.

2. Material

Derived from approximately 16 acres Shiraz or Hermitage grapes remaining as per subdivisional plans Adelaide Development Co.

Hermitage in itself would not be sufficient to give the character, breeding and complexity to make the wine as designated, and would require additional 20% minimum involvement Hermitage from selected vineyard Eden Valley area, and Hermitage or Cabernet Sauvignon from our own Coonawarra Vineyard.

Grapes would be hand picked.

3. Yield

Based on average 3 tons per acre, return from Chateau Magill vineyard would be 50 tons. Supplementary tonnage would be 5 tons Eden Valley Hermitage and 5 tons Coonawarra Cabernet Sauvignon. This would represent 20% outside involvement. Total tonnage for processing would be 60 tons.

A handwritten proposal by Max Schubert given to Barry Woodward, Penfolds NSW Sales Manager.
Source: Chris Shanahan and David Farmer.

winery. Magill Estate, components of Grange and St Henri and some of the Penfolds Cellar Reserve wines are vinified in the original cellars.

The Penfolds winemaking team were greatly affected by the loss of the prized Grange Vineyard and, in 1982, a plan was hatched by Don Ditter and Max Schubert to save the remnants of the historic site. The development of a 'château-style red wine distinctly different to the Grange Hermitage' was seen as 'our way of justifying its existence', John Bird recalled.

In the end a more realistic approach to vineyard sourcing was taken, reflecting the ambitions of the winemaking and marketing teams to make a single vineyard wine. Over twenty-three different trial wines were made 'to work out a method of attack'. Don Ditter said, 'We had to design the wine according to the character and constraints of the vineyard. At this level you don't just let it all happen.' The ambition was to make an elegantly styled wine along the lines of a classic claret that would reflect the character of vineyard site rather than something that could go on for a hundred years. At first the specifications required the fruit to be picked at around 11.5–12 Baume, almost identical to the early experimental Granges but well below the levels of contemporary vintages.

Although the original proposal suggested outside vineyard sourcing and a blend of varieties, the wine was released as a single vineyard Shiraz. Nonetheless the first vintages of Magill Estate (1983–1989) were works in progress rather than a clear vineyard style. The older wines are holding, but some vintages have not lasted the distance. Don Ditter suggests that the first vintages were probably left too long before bottling. Substantial vineyard investment, vine maturity and winemaking refinements have put the style on track. During the 1990s and 2000s the Magill Estate Shiraz style has reached a new level of richness and fruit complexity. The wines are more exciting than ever before.

 *Panel Review*

Magill Estate, Adelaide – Penfolds Magill Estate Shiraz
Jo Burzynska (JB), Jane Faulkner (JF), James Halliday (JH), Huon Hooke (HH), Ralph Kyte-Powell (RKP), Tony Love (TL), Tyson Stelzer (TS); Peter Gago, Steve Lienert, John Bird, Andrew Caillard.

Wines of great integrity but still standing apart from the mainstream Penfolds style.
— Jane Faulkner, *The Age*, Melbourne.

1983–1989

The early vintages are variable, reflecting a search for a suitable single vineyard style. 1983 and 1986 have lasted the passage of time but both are nearing the end.

1990–1999

A strong decade. The early vintages are elegant and stylish with 'exceptionally even fruit' (JB) and structured tannins. After 1994 the wines possess more density, fruit sweetness, tannin presence and oak integration. 1991, 1994, 1996 and 1998 are top vintages.

2000–2010

A 'glorious decade'. (TL) 'Almost without exception sweet fruit characters ran through the wine from the start to the very last aftertaste.' (JH) These wines stand apart from mainstream Penfolds wines because of their graphite, earthy nuances, muscular tannin and strong voice of place. 'The highlights of this flight broke some of the trends of other Penfolds brackets, making me wonder if this monopole is deserving of its own appellation.' (TS) Peter Gago: 'An open weave, texturally different wine thanks to vineyard character, open fermenters, basket pressing and barrel fermentation.' A suite of great vintages: 2002, 2004, 2006, 2008, 2009 and 2010.

St Henri

FIRST VINTAGE	1953–1956 experimental; 1957 first commercial release.
VARIETIES	Shiraz and cabernet sauvignon.
ORIGIN	Multi-district blend, South Australia. Significant contributions of shiraz from Barossa Valley, Eden Valley, Clare Valley, McLaren Vale, Langhorne Creek, Robe and Bordertown; cabernet sauvignon from Coonawarra and Barossa Valley. Increasing components of Adelaide Hills fruit in recent vintages.
FERMENTATION	Stainless steel tanks with wax-lined/wooden header boards at Nuriootpa. Some components are vinified at Magill Estate.
MATURATION	18 months in large (1460 litres) old, oak vats.
COMMENTS	Labelled 'Claret' until 1989 vintage. Cabernet sauvignon plays a secondary role. Packaged in laser-etched bottles since the 1996 vintage. Released under screwcap in some markets since 2005.

Penfolds St Henri is a great Australian red with a unique stature, emblematic story and enduring quality. Steeped in South Australia's colonial heritage and inextricably linked to the development of Grange, this beautifully scented, gorgeously seductive and finely structured wine, resurrected by legendary winemaker John Davoren, lies at the heart of Australia's ultra-fine wine scene. It is a sentimental favourite among collectors, a reference point for winemakers and a philosophical football for wine writers. It has been loved and argued about for its entire history.

Over the last twenty-five years of Rewards of Patience tastings, St Henri has been described as old-fashioned, contemporary, modern and exciting. Intrinsically it has remained true to its original blueprint, while the ever-changing perspectives are a reflection of a fast-moving world and altering vogues. After six decades of contiguous vintages under the Penfolds St Henri moniker, it has a proven cellaring track record and enormous reputation among devotees and collectors.

A unique retrospective of Penfolds St Henri was held on a balmy early summer's day in Berlin, for the European tasting panel. With its bewildering diversity and troubled past, Germany's capital city has risen from the ashes of post-war gloom to become the most exciting cultural centre in Western Europe. When Penfolds Auldana St Henri Claret was first made in 1953, an uprising in communist East Germany began events that would lead to the construction of the Berlin Wall (in 1961) and a prolonged period of fear and social despair. Today Berlin is once again enjoying a golden age of art, music, literature and architecture, not seen since the 1920s. It is where the traditional and the avant garde collide, the perfect milieu for a wine that was inspired by the past and now represents an expression of the future.

Auldana Cellars and Vineyards, adjacent to the Grange Vineyards at Magill, were established in 1853 by Patrick Auld (1811–86). He arrived in Adelaide with his wife Eliza, son William Patrick and daughters Agnes and Georgiana, on 6 April 1842 on the *Fortitude*.

Soon after arrival he purchased two sections comprising 'each of 230 acres at £1 per acre', the usual price charged by the Crown. Auld began life in South Australia as a publican and then became a wine and spirits merchant in Hindley Street, Adelaide. He planted a small vineyard at first, but impressed by the quality of the fruit on return from a brief stint in England he started commercial winegrowing in 1853. In 1861 he floated the South Auldana Vineyard Association with a market capitalisation of £12 000. The Association's first vintage – in 1862 – produced 3000 gallons of white and red wine. Patrick Auld was described as 'one of the most persevering and painstaking winemakers in the colony'. His meticulous attention to detail was legendary. *The South Australian Weekly Chronicle* wrote: 'Mr Auld has been taught by experience the necessity of ascertaining by careful study and observation the particular kinds of grapes adapted to produce the best kinds of wine and the exact proportion in which they must be respectively used for that purpose and then of classifying his vineyard accordingly.'

The relatively young Auldana Vineyard quickly became one of the most important and well known vineyards in South Australia. In *The Vineyards and Orchards of South Australia* (Adelaide, 1862), journalist Ebenezer Ward wrote:

> Entering the south vineyard on its northernmost side, the visitor finds himself at the foot of what is known as Verdeilho Hill [sic]. This hill and those beyond it to the south are admirably situated for the growth of the vine, inasmuch as they form a perfect natural basin and the slopes on which the vines are planted shelter each other from all winds, especially from the destructive wind which blows periodically from the south west.

The vineyard was planted to a fruit salad of varieties, including tokay, muscat of Alexandria, grenache, verdelho, 'Carbonet [sic] – grafted on Carignan', mataro, malbec and shiraz. The high price of labour and shortages of manpower restricted vineyard expansion. Nonetheless, by 1875 Auldana Vineyard was one of the colony's largest. A promising market in Victoria was hindered by substantial import duties but Auld, who had opened an office in London in 1871, 'had succeeded in opening a good market in England whither all his wine is shipped as soon as it is of sufficient age', according to a 1875 vintage report in *The South Australian Register*. The Auld family was producing two mainstream commercial wines: 'Auldana White' and 'Auldana Ruby'. By this time around 40 000–50 000 gallons of wine were stored in 500 and 800 gallon casks at Auldana. In failing health Patrick Auld moved to New Zealand in 1881 with his daughter Georgiana. He died on 21 January 1886.

> *Mr. Auld's persevering efforts in bringing our wines into notice in Europe had a most beneficial effect on the colonial wine trade. As a vigneron he was very successful, and his wines have been recognised as being of a high class.*
> — *The South Australian Register*, 1886.

In 1882 Hally Burton of Auld, Burton and Co., which imported and distributed Auldana Vineyard wines in England, declared himself bankrupt. The business was renamed the Australian Wine Company, with an emu as its logo. In 1885 it was acquired and renamed The Emu Wine Company, becoming one of the most successful Australian export businesses at the turn of the century, with markets in the United Kingdom and Canada. However, prolonged financial difficulties eventually led to Auldana being mortgaged, and finally Josiah Symon, the mortgagee, gained ownership of the property in 1888.

Symon – a prominent Adelaide identity – was a vocal advocate of Federation. The fledgling South Australian wine industry had much to benefit from the colony becoming a part of the new Australian Commonwealth nation. Until 1901 tariffs between the colonies had created artificial trade barriers resulting in localised wine markets. After Federation, South Australia experienced a substantial increase in vineyard plantings. The Auld family continued to be involved with the South Australian wine industry and pressed for legislation to prevent the introduction into the colony of *Phylloxera vastatrix*, which had caused widespread damage in New South Wales and Victoria. William Patrick Auld (1840–1912), son of Patrick, accompanied John McDouall Stuart on an expedition which crossed the continent in 1861–62, and became the first secretary of the Provisional Phylloxera Board after the introduction of the South Australian *Phylloxera Act 1899*.

Léon Edmond Mazure (1860–1939), the original creator of St Henri Claret, started at Auldana Cellars around 1887, after working at George Burney Young's St George Cellars at Kanmantoo, near Mount Barker in the Adelaide Hills. His *Australian Dictionary of Biography*'s entry includes:

> Mazure was among the first vignerons in South Australia to make champagne [sic] on a large scale (in 1896), to preserve olives and to introduce *levures* [selected yeasts] into the making of wine. In 1887–1912, while at Auldana, he was awarded eighty-three first prizes, seventy-one seconds and twelve thirds by the Royal Agricultural and Horticultural Society at the Adelaide wine shows. For three years in succession Auldana hock, chablis and sherry gained the champion ten-guinea cup against all Australia. A councillor of the South Australian Vignerons' Association, Mazure became a wine and pruning judge, initiated a pruning competition for boys under 18, and took out several patents for ideas including the Mazure corkscrew, a corking machine [which filled bottles consistently to a high level] and a windmill bird scarer. He was a member of the Adelaide Stock Exchange and was appointed a justice of the peace in 1901.

Léon Edmond Mazure is also credited as the creator of the famous and unique style of Australian Sparkling Burgundy!

The original St Henri Claret was probably named after Mazure's son Henri or wife Henrietta. First produced around 1890, it is surprising how few bottles have surfaced from this era. One bottle of 1896 Auldana Cellars St Henri Claret – found in the cellar of a Tasmanian collector – was discovered in the late 1980s, while a bottle from 1911 appeared at a Penfolds Red Wine Re-corking Clinic in Hobart in 1996.

An early vintage of St Henri Claret was submitted to the Bordeaux Exposition of 1895, an annual wine fair held by the *Société Philomatique*. In a letter to *The Advertiser* in October 1895, Edmond Mazure was furious with the newspaper's omission of Auldana Cellars' successes:

> It was too bad of your London correspondent to quite ignore the Auldana vineyard when telegraphing the awards to South Australian exhibitors at this important international exhibition. His telegram makes no mention of the fact that St Henri Claret won a silver medal. I am not boastful, but I could not believe that in the home of claret and with French judges, who know what genuine claret is, our St Henri had taken a place and a high one. So I was disappointed to read your correspondent's telegram as were our friends. But now I am content if you will kindly do us the justice of making public this correction so that it may be known that even Bordeaux thinks highly of Auldana St Henri.

Edmond Mazure and Josiah Symon were also interested in the effects of shipping high-quality wine to faraway markets. At the end of 1894 a shipment of St Henri Claret, in bottles especially sealed by Edmond Mazure's corking machine, was sent to England 'to test the effect of a double voyage'. Although the wine had 'twice stood the test of the tropics' it was pronounced by experts – and reported in *The South Australian Register*, as being 'absolute sound, delicious of flavour and possessed of a remarkably fine bouquet'. At this time Australian wine exports were plagued by 'sickness', or sweet wine disease, which often ruined shipments. Auldana Vineyards received an order for 1000 cases from a London agent shortly afterwards.

After Federation in 1901, Auldana Vineyards continued to prosper throughout the Australian Commonwealth. A Special Commissioner from the *Adelaide Chronicle* (1909) wrote:

> From a spectacular point of view there are few places more impressive than Auldana in full working order during the crushing season. From the delivery of grapes in the Old World style per bullock-dray, through the various stages of fermentation and treatment, to a walk below ground along the bottle-lined corridors hewn out of limestone, until one emerges upon the scene of champagne-making [sic] in its many processes, and winds up with a banquet in the hall of a thousand casks, no better effect could be stage managed to afford a glimpse of another world.

Turn-of-the-century Auldana Cellars St Henri Clarets were beautifully packaged in Bordeaux-type bottles with unique tamper-proof riveted capsules. Historian and vigneron John Wilson of Wilson Vineyard in the Clare Valley believes that St Henri was the first 'trade-named' wine in Australia.

By 1912, *The Mercury* in Hobart reported, 'Gold and silver medals and championship cups have been awarded so repeatedly and constantly that the Auldana wines may well be described as the undisputed champions throughout the Commonwealth.' The 'red-capsuled' Auldana Special St Henri Claret, 'a wine which won the championship cup against all Australia' was establishing itself as the newly formed state's most celebrated wine.

Edmond Mazure left Auldana around 1914, purchasing the adjacent Auld family's Home Park Vineyard, but remained an important figure in South Australia's wine industry until his death in 1939. Jack Lang, who had worked at Auldana prior to World War I, returned in 1918 and was the winemaker there until 1942. He then worked as the cellar supervisor, with winemakers John Davoren and Newton Harris, until his retirement in 1963.

The Auldana vineyard was acquired by Penfolds in 1943, after the business ran into financial difficulties. The assets offered by the liquidator included 303 acres of freehold property (including vineyard and grazing land), cellar plant, wood storage and wine stocks including sweet wines, dry wines and sparkling white. John Davoren, who was managing Penfolds Kalimna property in the Barossa Valley, was transferred to Magill in 1947 to take charge of the newly acquired winery. Ray Beckwith said, 'Of course John Davoren was eminently suited to that. He'd grown up in the vineyards on the Hunter River, with his father as manager. He'd grown up with vines.' Davoren had also worked at Penfolds Minchinbury vineyard as a sparkling winemaker before the war. Once installed as Auldana Cellars' new manager all of the table wines from Magill were bottled at Auldana and 'John carried on with building up the sparkling wine' and developing St Henri. During the 1950s and 1960s about half of Magill's table wine production, with the exception of Grange later on, was shipped up in bulk to Tempe Cellars in New South Wales for bottling and distribution on Australia's eastern seaboard.

The wine stocks at Auldana were in a terrible state, having been left unattended for several months. Over three years Ray Beckwith and John Davoren saved 50 000 gallons of flor sherry and sorted out the whites 'by blending it with some good wine and going through the sparkling process'. The red table wines, riddled with acetic bacteria, were blended away gradually into commercial blends, mostly sparkling burgundy. Ray Beckwith said, 'We had a mammoth task to straighten out the wines that were bequeathed to us, and we did so without loss.' After that 'John was able to work with sound wines and put his own stamp on the resulting blends – and how well he did it!'

Gordon Colquist, a second-generation family employee (his father Herman had joined the firm in 1928 to look after the boilers; the Nuriootpa winery machines at the time were steam-driven) joined the Magill cellars in 1938 and celebrated fifty vintages at Penfolds in 1988. He observed, 'First of all, John was a viticulturalist. He grew peas between the vines to get more nitrogen. He used deep shears on a plough to aerate the ground and ensure that water supply was equally spread. Then they put him on to winemaking and with Ray Beckwith they developed a yeast that made beautiful controlled champagne. Later he became a dry wine specialist.'

The St Henri label was revived by John Davoren (1915–91), whose family had enjoyed a long association with Penfolds. His father and grandfather had worked at Dalwood, an historic nineteenth-century Hunter Valley vineyard near Branxton. Originally owned by the Wyndham family, it was subsequently split in half, with one portion of 52 hectares sold to Penfolds in 1904. John Davoren's father John, a legendary Hunter winemaker, became manager. It was his 1930 Hunter River Cabernet, bottled for the UK market, that inspired Max Lake, a great supporter and collector of Penfolds, to establish Lake's Folly, one of Australia's earliest boutique wineries, in 1963.

John Davoren the younger – known as 'Jack' to distinguish him from his father – started making wine at Penfolds Dalwood during the 1930s and also worked at Penfolds Griffith, alongside his much older brother Harold, the winery manager, where he handled fortified wine production and distillation. Soon afterwards he was appointed manager of the now defunct Penfolds Minchinbury Vineyards, specialising in sparkling wine production, at Rooty Hill in Sydney. He enlisted with the Royal Australian Air Force in 1942 and spent the rest of the Pacific war 'flying typewriters' in New Guinea. After a brief stint managing the newly purchased Kalimna Vineyard in the Barossa Valley, he was appointed manager of Penfolds Auldana Cellars in 1947.

The revival and development of St Henri Claret mirrored the story of Grange, except that Davoren deliberately looked at the heritage of Auldana Cellars and his own family winemaking traditions for inspiration. John Davoren was keen to establish a wine based on the original work of Léon Edmond Mazure. His objective was to make 'a genuine claret style' wine that could be compared favourably with the greatest *grand cru classé* wines of Bordeaux.

The first experimental Penfolds St Henri vintage was made in 1953, on the hundredth anniversary of Auldana Cellars; the bottles were labelled with cast-off St Henri labels, found lying in the storage loft at Auldana. While the 1957 vintage is generally recognised as the first commercial release, John Davoren was still calling the early St Henris 'trials' until 1960. The wines were shown to visitors and were regularly mentioned by wine writers and observers. He considered the 1958 St Henri as one of his best early vintages.

Penfolds were old when your Grandparents were young.
— Advertising slogan, 1950s.

Davoren's vision was to make an 'homage' rather than to develop a completely new style of wine. His gentle personality and modest ambition will partly explain why Grange overshadowed St Henri. He did not enjoy being the focus of attention, despite 'his friendly welcome, that twinkle in his eye and sense of fun', as Ray Beckwith remembers. He was a very spiritual man who saw God as a lived dimension within his life and family. His son, Rob, recalls that his father had a 'beautiful custom that combined his family and winemaking. On major celebrations he would blend and bottle wine in honour of a wedding, ordination, birth, etc.' He was very much a winemaker's winemaker, steeped in the love of wine and with an intuitive sixth sense for growing things. It is

probably this feeling of place and inherent knowledge of fruit quality, through gardening and viticulture, that made him a cut above other winemakers of his generation.

Although Penfolds St Henri is now a shiraz-dominant blend, the first trials were centred on the Barossa Valley's Kalimna cabernet sauvignon, Auldana's Quarry Block mataro and later Paracombe Vineyard's shiraz. Subsequent commercial releases in the 1950s and 1960s sourced fruit from Auldana, Magill, Morphett Vale, Modbury, Paracombe and Adelaide Hills. By the early 1970s St Henri was predominantly a shiraz–cabernet blend.

The first vintages were reportedly foot-stomped in open-ended hogsheads. A relatively high percentage of stalks was also retained in the vinification. Davoren once explained this practice: 'we add stalks deliberately to keep the skins apart for the plunging cap, and to get colour as quickly as possible'. The St Henri style, to this day, is a highly perfumed, elegantly structured wine based on fruit clarity and maturation in older oak. For many years St Henri was partially aged in two-year-old American oak hogsheads, first used for Grange.

Sandie Coff, Max Schubert's daughter, says, 'I remember Dad giving me my first taste of St Henri. He introduced it by saying it was a very special wine – the number two wine the company made. He had great respect for St Henri and for John Davoren. If anyone ever asked Dad to sign a bottle of St Henri he would always refuse. All credit for this wine should go to John Davoren, he would say.'

St Henri Claret – An Australian red to make the French turn green.
— Advertising slogan, 1980s.

Both St Henri and Grange were regarded as classic Penfolds wines and distinguished Australian reds within a decade of first release. In 1966 Dan Murphy, a respected Melbourne wine merchant, described these 'special bottlings of claret' as 'the best firm styles Australia makes'.

Through his work at Auldana, John Davoren established a reputation as one of Australia's great winemakers. Robin Moody, who began work as a sparkling winemaker at Auldana during the 1970s, described Davoren as 'very outgoing with a really nice sort of personality, a really good palate and this enthusiasm to keep developing, finding out things and testing the boundaries.' The late Murray Tyrrell said, 'there is no doubt the greatest Australian winemaker is John (Jack) Davoren. He always fully

understood his subject and is the most interesting, innovative winemaker in Australia. As a winemaker, Jack, responsible for Penfolds St Henri, showed great anticipation and observation during fermentation and maturation of his wines, which resulted in some of the best reds we've seen in this country.'

The success of Grange was very much enhanced by the contrasting St Henri style. The two wines began life together in a climate of intense excitement, experimentation and research. Veteran Penfolds senior winemaker John Bird, who worked alongside both Max Schubert and John Davoren, said:

Much has been spoken about the intense competitive relationship between the two men. The strong personal rivalry made good copy, but in fact Davoren reported to Schubert. Penfolds worked under a veil of secrecy throughout the 1950s and 1960s; winemakers were not allowed to talk about their work to outsiders. Robust arguments and strongly held views were aired between the two men, but always within the framework of a common purpose. They were friends. You can see by the comparative styles that St Henri and Grange come from the same stable. The wines can look remarkably similar to each other – especially between ten and fifteen years of age.

In fact, without Max Schubert's support St Henri might not have been released. Sandie Coff reflects:

Dad and I were very close. He travelled a lot when I was a child in the '50s and '60s, particularly to the Barossa, Riverland, Coonawarra, Griffith and Sydney. When he was in Adelaide I would be with him no matter what he was doing. This included visits to Penfolds and colleagues before and after school, at weekends and holidays. I knew all of Dad's colleagues and the cellar hands, gardener, secretaries and management. Each person I referred to as Mr, Mrs or Miss – except one. Uncle John Davoren. Dad would take me to Auldana with him and Uncle John would always greet us with a beaming smile. I have a photo of Dad pinning a long-service pin on him and the look of pure affection on John's face explains the relationship they had.

Nonetheless, the development of Grange and St Henri was like a mini–space race, with a friendly competition for available resources and wine industry accolades. These wines were regularly presented to Beef and Burgundy clubs, social gatherings and industry events to build recognition. With their origins so inexorably linked, St Henri and Grange were like Siamese twins in the market.

St Henri Claret. It's not cheap, it's not easy to find. But it is one of the finest red wines in Australia.
— Advertising slogan, circa 1980.

Initially St Henri achieved greater commercial success than Grange. It was a more elegant, approachable style whereas the revolutionary Grange was something of a blockbuster with a richness and fullness 'that few people cared for'. Reports from the critics of the 1960s refer to St Henri as 'one of the only true claret [sic] styles in Australia'. Don Ditter – former Chief Winemaker – says, 'There was a strong following for St Henri from the very outset. Initially both Grange and St Henri were priced at the same level. The demand for each of the wines however was soon quite similar, some preferring the lightly wooded maturation style of St Henri over the more strongly flavoured, barrel-fermented and new-oak matured style of Grange.'

During the 1960s, the main Auldana winery became a sparkling wine production centre for the South Australian and Victorian markets. John Bird remembers: 'It was doing the same sort of thing that Minchinbury was doing in Sydney, but it also was a major bottler of table wines.'

Urban encroachment and the subsequent wholesale clearance of prime vineyards within the Adelaide city boundaries led to the eventual 'pulling-up' of the Quarry Paddock, a highly prized mataro vineyard lying between Auldana and Magill. Its loss was greatly felt by Penfolds. Along with other great old Adelaide vineyards it now lies under a housing estate. The Auldana Vineyards and winery ceased production in 1975, much to John Davoren's great sadness. He never really recovered from this disappointment and retired from Penfolds soon after. Auldana's long-serving winemaker Newton Harris was also broken-hearted and died three years later. In 1980 Auldana was sold for housing estate development, to be followed by neighbouring Modbury Vineyard and most of Magill Estate – except for the front blocks – in 1984.

John Davoren died in 1991. His son, Rob Davoren, SJ, gave the eulogy: 'With his earthly vintages at an end, he was returned to the soil from which he came; soil that was so much a part of his life and love, whether the vineyards of the Hunter Valley, Griffith, Barossa, Magill or his own garden at home.'

Nowadays St Henri is a multi-district blend drawing shiraz from the Barossa Valley, Eden Valley, McLaren Vale, Clare Valley, Langhorne Creek and Limestone Coast. The Adelaide foothills – once a primary source for St Henri – are again making significant contributions to the blend: Penfolds now draws about 10–15 per cent of its fruit for St Henri from Paracombe, Rostrevor, Stonyfell, Waterfall Gully and Williamstown. Cabernet sauvignon, which adds both firmness and structure to the St Henri style, is sourced from the Barossa Valley and Coonawarra. Fruit quality has improved substantially over the last decade through vigilant vineyard management and optimum pick dates.

The rumoured practice of stalk retention – which in theory adds perfume and structure to the wine – is no longer used; these attributes have already been achieved through vineyard management and selection. Peter Gago says, 'We can identify the desired aromatics, concentration and structure of potential St Henri fruit in the vineyard. The best parcels are typified by strong praline/chocolate characters, obvious fruit sweetness, very intense flavour development and supple tannins.'

St Henri Shiraz is batch fermented in headed-down stainless steel tanks. Each component is classified according to fruit profile and structure. A bucket of stalks, in addition to a stainless steel cage, is often placed near the bottom valve of the open fermenters to aid drainage for the daily 'rack and returns'. The young wines which make 'the cut' are then matured for between fifteen and eighteen months in large, 1460 litre old oak casks. Over this period, the fruit builds up further complexity and richness, without any oak pick up, while the tannins soften and develop. St Henri has a lacy firm-grained palate texture that distinguishes it from other Penfolds shirazes.

Peter Gago comments, 'John Davoren's ambition to make a great Australian Claret style is still relevant. The work in the vineyards is never finished, our reference points are ever changing, and our impression of great wine is continually challenged. Yet St Henri symbolises our approach to winemaking. St Henri was our least fashionable wine during the late 1960s and 1970s. Consumers wanted fresh wines with new oak. Penfolds calmly ignored mainstream fashions, fads, and remained true to John Davoren's original vision.'

The fruit was allowed to speak unimpeded and the vintage variation was there to be seen and savoured. In addition, the vertical attested to the longevity of St Henri. I would not broach a bottle for fifteen years, whilst the finest vintages are easily capable of lasting forty.
— Neal Martin.

Berlin, Germany – Penfolds St Henri

Neil Beckett (NB), Frank Kämmer (FK), Peter Keller (PK), Andreas Larsson (AL), Neal Martin (NM), Peter Moser (PM), Mario Scheuermann (MS); Peter Gago, Andrew Caillard.

> *St Henri has remained true to what it stood for when John Davoren re-introduced the wine in the 1950s. The style has altered, but its character remains intact. A masterclass in the art of blending.* — Neil Martin.

1957–1963

'Expressive wines of potency, vigour and style. Vaguely *claret* like, but with a viscosity, fat and gloss rarely found in European wines of this age.' (NB) Fully mature and alive with impressive complexity and condition. 1957, 1959 and 1962 were stand-outs.

1964–1969

'The evolution continues with glorious refinement.' (NB) The slight stemmy notes and silky tannins give an almost Burgundian character. A beautiful consistency across vintages with a 'certain rusticity' (AL). 1964, 1965, 1966 and 1969 'the Doppelganger' being the strongest performers.

1970–1979

Still very lively, elegant with finesse and complexity. The early vintages are quite hedonistic with lovely freshness and richness. The stemmy notes are less evident and the fruit is more nuanced. The shiraz component is more obvious. A variable decade reflecting experimentation, evolution and engagement of new technology and ideas. Peter Gago: 'The decade reflects a growing interest in the management of our vineyards. The term "controlled neglect" was invented by Penfolds to describe the effort in growing optimum fruit: low yield, some stress and not trying too hard.'

1980–1989

Consistent and modern, with more clarity and definition on the palate; nothing obtrusive. The tannin quality and overall fruit definition emancipates St Henri from European reference points. An attractive balance between fruit and maturity is present in many vintages. 1982, 1983, 1986, 1988 and the 'perfectly drinkable' 1989 were highlights. 'These wines have matured *without* the crutch of new oak. *If you added new oak, it would ruin it.*' (NM)

1990–1999

A strong Australian identity is emerging during this decade. Youthful colours are still present in the mid to late 1990s. There's richness without heaviness, a firm tannin backbone, and opulent fruit. Mature characters are still to emerge, but this will happen over the next 15–20 years. The wines possess lovely freshness, purity and tannin presence. 1990, 1991, 1994, 1996, 1998 and 1999 were strong performers.

2000–2010

Big, fat and opulent! 'A drippingly vivid quality with primary fruit intensity and nothing jutting out. Monumental in scale but seamless integration between fruit and tannins top to bottom.' (NM) 'Complexity, power and length are achieved completely by fruit selection.' (AL) 2002, 2004, 2006, 2008, 2009 and the 'exciting' 2010 were strong vintages.

St Henri vintages from the 1950s and 1960s still regularly appear at auction. While Grange has soared in value over the last fifty years, St Henri has remained an auction staple, attracting solid reliable demand at comparatively affordable prices. Nonetheless, rare vintages at auction now command prices in excess of $2000 while a bottle of 1957 Penfolds St Henri, even rarer than its Grange counterpart, recently fetched $8110. The 1971 St Henri has achieved higher prices at auction than the more famous 1971 Grange, reflecting the rarity of older vintages. With proven cellaring potential and a genuinely fascinating narrative, St Henri enjoys a unique status in the ultra-fine wine market. It is classified as 'Outstanding' by the authoritative Langton's Classification of Australian Wine, in recognition of its proven track record of quality, identity and cellaring potential.

St Henri is typified by fresh mulberry/blueberry/dark chocolate/liquorice aromas and flavours, mid-palate richness, fruit sweetness and fine chocolatey tannin structure. Cabernet sauvignon provides aromatic top notes of violets and cassis and firmness at the finish. Maturation in older oak brings these components together into a harmonious whole. With age these wines further develop, gaining more complexity, generosity, velvety texture and weight.

The Berlin panel found the St Henri retrospective tasting to be 'a beautiful experience' (AL). Although opinions varied, there was a feeling that St Henri is a unique Australian wine with a unique personality. The absence of new oak, the emphasis on fruit complexity, its timeless quality and ageing potential were greatly appreciated.

Reserve Bin A Chardonnay

FIRST VINTAGE	1994.
VARIETY	Chardonnay.
ORIGIN	Adelaide Hills – extending from the Piccadilly Valley to Birdwood.
FERMENTATION	100% barrel fermented/100% malolactic fermentation.
MATURATION	Matured for 8–9 months in specially selected tightly grained French oak barriques of varying ages. Lees retention and stirring (*battonage*).
COMMENTS	The early vintages include some Tumbarumba fruit. Vintages prior to 2001: drink now/past. Released entirely under screwcap since the 2003 vintage. Not made in 1996, 1997, 1999, 2001, 2002 and 2011.

Reserve Bin A Chardonnay has established a strong following since it was first released in 1994. It is a fresh, minerally style that lies at the vanguard of the Chardonnay genre in Australia. With its distinctive flinty aromatics, creamy flavours and razor-sharp acidity, it perfectly articulates the fruit complexity and mouth-watering attributes of cool-climate Adelaide Hills fruit.

The impressive successes at wine shows of Reserve Bin A Chardonnay, both in Australia and the UK, has brought significant worldwide attention to this hallmark Penfolds style. The celebrated 2009 Reserve Bin 09A Chardonnay blitzed the Australian wine show circuit and it also won the International Wine Challenge Champion White Wine award. Senior white winemaker Kym Schroeter's imaginative winemaking

skills landed him a nomination as *Gourmet Traveller Wine*'s prestigious Winemaker of the Year in 2012 and recognition as 'one of the top three white winemakers in the world'.

The Reserve Bin Chardonnay style is a culmination of many years of experimentation and refinements. Its origins are inextricably linked with the development of Penfolds ultra-cuvée Bin 144 Yattarna Chardonnay. During the late 1980s the winemaking team were asked to trial and develop white winemaking techniques that would mirror the success of Penfolds red wines in mainstream markets.

The first trials were based on three main varietals: semillon, sauvignon blanc and chardonnay. At a very early stage it was believed that the wine should be relatively taut and linear, with plenty of ageing potential.

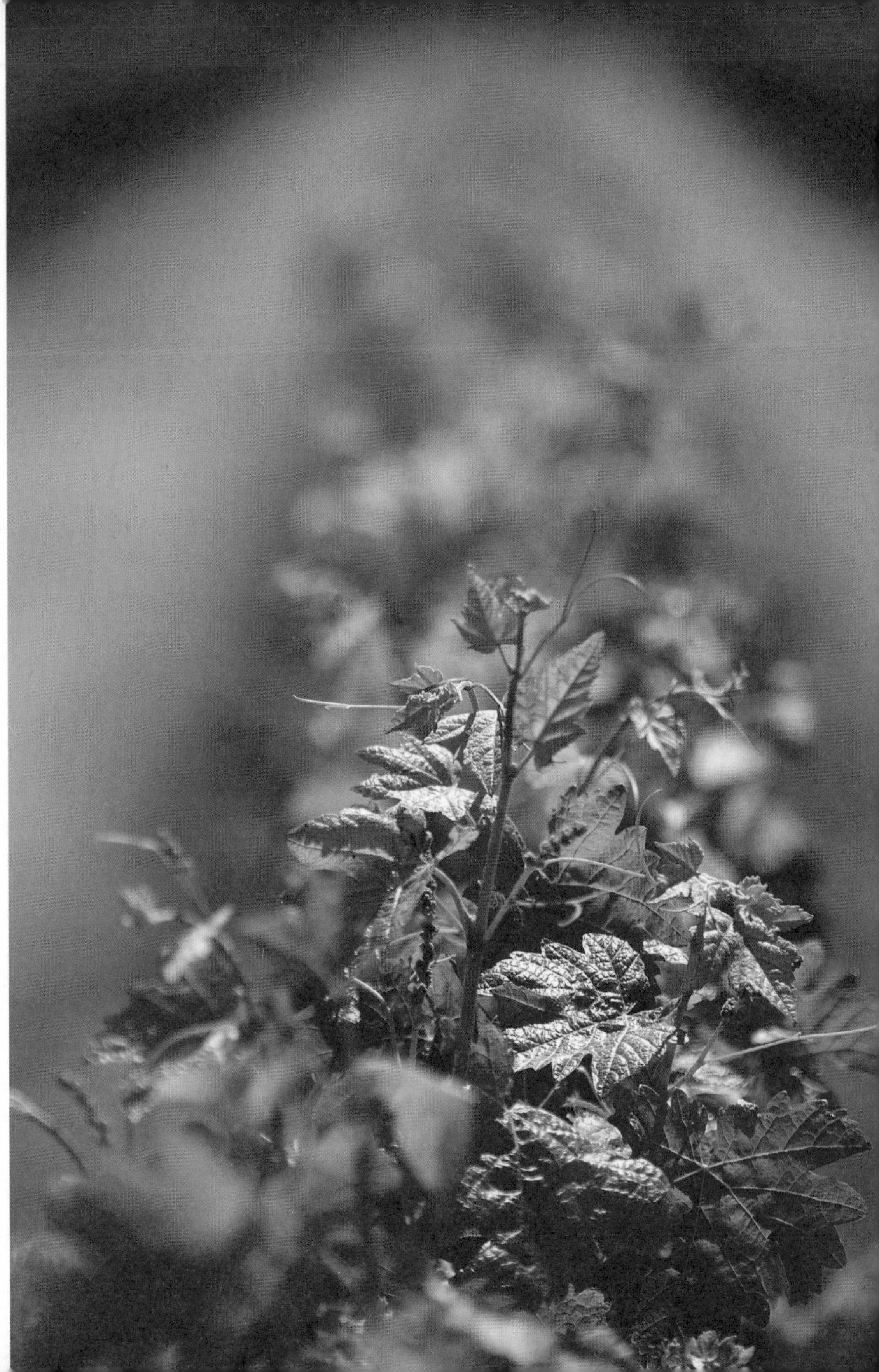

Cool-climate chardonnay from Adelaide Hills and Tumbarumba was chosen as the main source material; the pristine fruit and balanced acidities were perfect for the style. A touch of sauvignon blanc was added to the first wines because it gave an appealing lift and freshness but they developed too quickly and before too long Reserve Bin A was made as a 100 per cent Adelaide Hills Chardonnay.

The grapes for Penfolds Reserve Bin A are from up to 80 vineyards, mostly located around Birdwood, Balhannah, Piccadilly, Woodside, Morialta and Gumeracha – a long stretch of varied, cool microclimates at higher elevations of the Adelaide Hills. The wine centres specifically on ten key vineyards, all owned by loyal independent growers. Yields are kept low, at around three to four tonnes per hectare, to maximise flavour development and balance.

The fruit is picked at night to optimise aromatics, flavour and acidity. On delivery to the winery, the whole bunches are gently transferred into a press, naturally drained and then lightly pressed. The juice is sequentially filled direct from the press into barriques.

Typically the wine is naturally fermented in new and one-year-old tightly grained French oak barriques and then allowed to go through 100 per cent malolactic fermentation to achieve further integration of fruit and oak. Some parcels are purposely sulphide-influenced or fermented on solids with wild yeasts. At the end of fermentation the wines are regularly stirred on their lees to bring more richness and depth of flavour. Once regarded as extreme winemaking practices, these techniques are now considered quite standard in achieving heightened complexity, flavours and texture. Reserve Bin A has profoundly influenced Australia's Chardonnay landscape and further adds to Penfolds long and proud winemaking history.

 *Panel Review*

Magill Estate, Adelaide – Penfolds Reserve Bin A Chardonnay
Jo Burzynska (JB), Jane Faulkner (JF), James Halliday (JH), Huon Hooke (HH), Ralph Kyte-Powell (RKP), Tony Love (TL), Tyson Stelzer (TS); Peter Gago, Steve Lienert, John Bird, Kym Schroeter, Andrew Caillard.

> *All finesse and elegance. They reflect the exciting and ongoing changes in Australian style.*
> — James Halliday, *The Australian.*

2000, 2003–2010

A classic modern style with clear fruit aromas combined with flinty complexity, beautiful volume and richness. The most recent vintages reflect the extraordinary progress taking place in Australian ultrafine Chardonnay. These are stylish wines with underplaying fruit complexity, but great presence and sustained flavours. There is a sense of daring in their execution, making this Reserve Bin A series real leaders in the field. Highlights included 2004, 2006, 2008, 2009 and 2010.

' I believe [the 1962 Bin 60A Coonawarra Cabernet
Sauvignon Kalimna Shiraz] to be the greatest
Australian red wine ever made.'

JAMES HALLIDAY

SPECIAL BINS
Chapter six

The declaration of a Penfolds Special Bin wine is not taken lightly. Decades later collectors remain judgemental.
— Peter Gago, Penfolds Chief Winemaker

Penfolds Special Bin wines hold a special significance and place among generations of wine collectors. Some of these wines are preliminary sketches for masterpieces such as Grange or Bin 707. Others are exploratory compositions of nature's colour and the vineyard landscape. Since Max Schubert's first experimental bottlings of Bin 1 Grange Hermitage, Penfolds has been at the forefront of fine winemaking in Australia and the Penfolds red wine-making style has been developed and improved through its ongoing limited production of Special Bin vintages.

The story of Penfolds Special Bins is largely about Max Schubert's frustration and triumph with cabernet sauvignon, while Coonawarra, Penfolds Kalimna Vineyard and shiraz, as a blending component, are consistent themes. The success of 1962 Bin 60A, 1966 Bin 620 and 1967 Bin 7, a trinity of great and lasting cabernet–shiraz 'claret' styles, not only, according to Len Evans, 'proved forever that the two varieties can blend beautifully together', but also contributed to Coonawarra's reputation as one of Australia's greatest wine regions. The Penfolds Special Bins are extremely rare, reflecting the meticulous standards and expectations of the winemaking team.

These storied wines are dramatic, expressive and evocative. One moment I am spirited away into the dusty Australian outback; another I am in a magical field of violets.
— Tony Love, *The Advertiser*, Adelaide.

Block 42 Kalimna Cabernet Sauvignon

FIRST VINTAGE	Only ever released in 1953 (Grange Cabernet), 1961, 1963, 1964 (Bin 707), 1996 and 2004.
VARIETY	Cabernet sauvignon.
ORIGIN	Block 42 – Kalimna Vineyard, Barossa Valley.
FERMENTATION	Stainless steel tanks with wooden header boards to submerge cap. Daily rack and returns.
MATURATION	Fermentation completed in new American oak hogsheads. Approximately 13–18 months' maturation in oak (300 litres).
COMMENTS	Block 42 played a critical role in the development of Grange and particularly Bin 707. There are probably no examples left of the 1948 Kalimna Cabernet Sauvignon. The 2004 vintage was selected for the Penfolds Ampoule.

Planted in the mid-1880s, Block 42 is located at the edge of the Kalimna Vineyard in the Moppa sub-region of the Barossa Valley and comprises cabernet sauvignon vines of ancient genetic origins. It belongs to a national heritage of great old nineteenth-century pre-phylloxera Australian vineyards that include Henschke's Hill of Grace – in the Eden Valley sub-region of the Barossa – and the central Victorian shiraz vineyards planted by Bests and Tahbilk.

The 10 acre (4.05 ha) Block 42 was planted only thirty years after the great 1855 Bordeaux Classification, and comprises the oldest plantings of cabernet sauvignon in the world. These original, low-yielding, contorted, knotted and serpentine vines have deep penetrating roots which extend through alluvial sands, rich brown soils and into the fissures and

cracks of the bedrock. For nearly 130 vintages the vines – planted on their own roots – have produced flavour-intense and mineral-rich fruit with exceptional concentration and balance.

Max Schubert experimented with parcels of Block 42 Cabernet during the development phase of Penfolds Grange. He recognised the extraordinary potential of this vineyard site through the special bottlings of the late 1940s. Those early but very limited wines – made from pressings and matured in old oak puncheons – were deeply concentrated, with immense flavour and classic fine-grained tannins.

Schubert initially hoped that he could use Block 42 cabernet sauvignon as raw material for his 'different and lasting' Grange. The experimental 1952 and 1953 Grange Cabernets were Block 42 wines, but the low-yielding vines could not provide enough fruit nor the required consistency every vintage. Nonetheless, Schubert regularly returned to this historic vineyard for inspiration. Providing 'lift and structure', Block 42 cabernet sauvignon was occasionally blended into the early Penfolds Grange vintages.

'One-off' Block 42 bottlings were periodically released. 1961 Bin 58 Cabernet Sauvignon was followed by 1963 Bin 64 Cabernet Sauvignon, which won the prestigious Jimmy Watson Memorial Trophy for a one-year-old dry red at the 1964 Royal Melbourne Wine Show. The inaugural 1964 Bin 707 Cabernet Sauvignon is also one of the classic Block 42 wines of the 1960s. The valuable and historic Block 42 vineyard continues to provide important backbone and substance to Bin 707 and occasionally Grange.

Block 42 remained a relatively anonymous patch of vines through the 1970s and 1980s. Since the 1960s, Penfolds has only released two single-vineyard Barossa wines: the 1996 and 2004 Block 42 Cabernet Sauvignon, years in which mild dry weather during the final stages of fruit development led to perfect ripening conditions. After vinification in stainless steel, the 1996 Block 42 was 100 per cent barrel fermented and then matured in American oak hogsheads for approximately eighteen months.

These rare Block 42 Cabernet Sauvignons are an important fragment of the Penfolds story. They are magical wines, steeped in nineteenth-century Barossa Valley heritage, and possess 'an ethereal dimension and saturated blackness on the palate'. The next release will be dependent on a bit of luck, ideal weather conditions and the right parcels of fruit.

Bin 60A Cabernet Sauvignon Shiraz

FIRST VINTAGE	Only ever released in 1962 and 2004.
VARIETIES	Cabernet sauvignon and shiraz.
ORIGIN	**1962** Two-thirds Coonawarra cabernet sauvignon (Sharam's Block and Block 20), one-third Barossa Valley shiraz (Kalimna Vineyard, original shiraz blocks). **2004** 56% Coonawarra cabernet sauvignon (Block 20), 44% Barossa Valley shiraz (Kalimna Vineyard, Blocks 4 and 14; Koonunga Hill, Block 53G).
FERMENTATION	Stainless steel tanks with wooden header boards to submerge cap. Daily rack and returns. The 2004 was vinified at Magill and Nuriootpa. The 1962 was made in its entirety at Magill.
MATURATION	Fermentation completed in new oak. Blending of components took place at Magill. Approximately 15 months' maturation in American oak hogsheads (300 litres).
COMMENTS	The legendary 1962 Bin 60A is the most famous wine ever produced by Penfolds. Similar vintage conditions inspired Penfolds to make 2004 Bin 60A, an homage wine.

All who have tasted it regard [1962 Bin 60A] as the single greatest Australian red wine of the twentieth century.
— James Halliday, *The Weekend Australian*, 2009.

1962 Penfolds Bin 60A Coonawarra Cabernet Kalimna Shiraz is a legendary Australian wine. In a world where egos readily clash, it unified wine critics and show judges. 1962 Bin 60A is Penfolds most successful show wine, winning nineteen trophies and thirty-three Gold medals in a relatively short timespan. It was a profound oenological, physical and philosophical achievement for its time. Still fresh and alive, it is a lasting model of Max Schubert's ground-breaking winemaking practices and ideas of multi-regional and cross-varietal blending.

James Halliday, Australia's leading wine author, gave 1962 Bin 60A the ultimate tasting note: 'an utterly superb wine, a glorious freak of nature and Man; ethereal and beguiling, yet the palate is virtually endless, with a peacock's tail stolen from the greatest of Burgundies; the fruit sweetness perfectly offset by acidity rather than VA. The 100 point dry red? Why not!'

The grapes were partially foot-crushed. Max Schubert fermented the wine in the classic Penfolds winemaking style, using header boards and rack and returns. Towards completion of fermentation the wine was basket pressed and barrel fermented.

The fame of Bin 60A reached all corners of the globe. Max Schubert's direct contemporary, André Tchelistcheff (1901–94), the founding father of the modern Californian wine industry, once demanded of a room of startled Napa Valley vignerons: 'Gentlemen, you will all stand in the presence of this wine!' Len Evans (1930–2006), who apparently brought that bottle to California, once described the wine as 'one of the great reds I cut my palate on, and proved [to me] forever that the two varieties can blend beautifully together'.

1962 Bin 60A is the only Australian (and non-French) wine to reach *Decanter* magazine's Top 10 ranking 'wines to try before you die'. Harvey Steiman, *Wine Spectator*'s veteran editor-at-large, described it as 'one of the greatest wines I have tasted anywhere'.

The 1962-like vintage conditions in 2004 inspired the Penfolds winemakers to recreate Max Schubert's great wine. 2004 Penfolds Bin 60A Coonawarra Cabernet Sauvignon Barossa Shiraz is a 'select parcel' blend of Block 20 Coonawarra cabernet sauvignon, and Koonunga Hill and Kalimna Vineyard Barossa shiraz. It was vinified at Nuriootpa and Magill in open and closed headed-down fermenters. The components completed fermentation in 100 per cent new oak barriques before assemblage and bottling. 2004 Bin 60A, described as a wine 'wrapped in a starless night sky' as a youngster, was initially released as a *primeur*, or futures offer. It is still saturated in colour, powerful, dense and expressive with a long future ahead of it. In Adelaide's *The Advertiser* Tony Love described it as 'holding on to its inner truth'.

Bin 60A Story

To: Peter Gago
Date: 24 December 2005

Joy Lake and I used to taste the first vintages of Grange with Max Schubert and Geoffrey Penfold Hyland in the lab at Penfolds central office at Magill. I can't clearly recall how the connection began, but it had something to do with the fact that I had started writing about wine, and they liked the independence of an articulate surgeon giving opinions on controversial wines they were making. There is no truth in the fable that I got a lot of those early treasures, tops at half a dozen bottles.

Anyhow the friendship flourished, what with rare fine wines and private medical advice to Max. Which is how one morning soon after my own vineyard was producing some pretty handy wines in the Lower Hunter Valley I sat at Max's desk in Adelaide confronted by a row of new Penfold cleanskins, produced for an opinion free of any company influence. This was quite a privilege as I regarded the company as the best consistent maker of red wines in Australia. I still do.

One of the wines on the table was a stand-out. I asked what was the chance of getting some. He looked hesitantly at the assistant. I knew instantly the problem was with show quantities. The Australian wine show system specifies a minimum quantity of an entry to be available for sale in each class. Otherwise some genius would make a spoonful of an ambrosial drop never to be seen by anyone else.

Penfolds had had an inspection in the recent past and were rather particular about meeting these minimal requirements. When it looked as though I was going to miss out, I offered to refrain from opening a single bottle until they gave me permission, however long that might be. And thus their total count would remain intact, to be called on if necessary. As they hadn't definitely decided to make this a show entry, and my guarantee covered all the bases, Max agreed to let me have some of the wine.

Somehow we managed to pack fourteen cases of 1962 Bin 60A Coonawarra Cabernet Kalimna Shiraz into our car. There was just enough room for me to drive, but otherwise it was stacked to the gunnels. And that is how I drove slowly and cautiously by the shortest main road to Sydney, the vehicle flat on its springs, the wheels fortunately mostly clear.

The wine remained coolly maturing in my excellent cellar at home until that wonderful day when Max notified me it was out of embargo and OK to drink.

Over the next few years Joy and I got through about five cases of this magical potion, usually feeling compelled to share the experience with others of whom Leonard Paul Evans was one. The wine just got better and quickly entered my pantheon, shared with the Maurice O'Shea/Roger Warren Kings Paddock B76 1945, Colin Preece's J34 Cabernet blend ('53 or '57), 1945 Ch. Latour, and Penfolds own 1930 Dalwood Cabernet/Petit Verdot. The 60A led the field.

In the early weeks of this first phase of '60A-dom' I realised I had never paid for it, and after a few letters and phone calls from me, Max confessed they had no idea what to charge for it, and would a book entry (by their accountant) of $1.67 per bottle be satisfactory? There was never a speedier delivery of a cheque. Fortune's child, that's me. And thank you Max. Wherever you are.

Max Lake,
Sydney
Christmas Eve, 2005

Bin 620 Cabernet Shiraz

FIRST VINTAGE	Only ever released in 1966 and 2008.
VARIETIES	Cabernet sauvignon and shiraz.
ORIGIN	**1966** Sharam's Block and Block 20 (cabernet sauvignon), Coonawarra. **2008** Blocks 5, 10 and 20, Coonawarra.
FERMENTATION	Stainless steel tanks with wooden header boards to submerge cap. Temperature maintained at less than 22°C. Daily rack and returns.
MATURATION	Fermentation completed and then maturation for 12 months in new American and French oak hogsheads (300 litres).
COMMENTS	1966 Bin 620 Cabernet Shiraz is one of the most famous wines of the 1960s. Similar vintage conditions inspired Penfolds to make 2008 Bin 620 Cabernet Shiraz, an homage wine. It won a trophy for the best wine over $50 at the National Wine Show in Canberra in 2010.

The first vines were planted in Coonawarra around 1891, a year after the foundation of the Coonawarra Fruit Colony by wealthy pastoralist John Riddoch in 1890. One of the region's earliest 'clarets' made quite an impression on the colonial journalist W. Catton Grasby, who wrote in 1899: 'It promises to have a very high and wide reputation – indeed, there is no doubt but that it will be a beautiful wine of good body, fine colour, delicate bouquet and low alcoholic strength.'

Over half a century ago, in 1960, when Bill Redman approached Max Schubert and general manager Jeffrey Penfold Hyland to buy the Sharam's Block in Coonawarra, the table-wine market was still in its infancy and Coonawarra was barely known as a wine region. However, Schubert hoped that the potential fruit profiles of cool-climate Coonawarra and warm-climate Barossa were complementary and would lead to an 'all-round and balanced wine'.

The first Penfolds Coonawarra wines were crushed at Redman's Rouge Homme Winery and then tankered up to Magill and Kalimna for fermentation and maturation. This was – and still is today – the procedure when Max Schubert reprised the great 1962 Bin 60A with his 1966 Penfolds Bin 620 Coonawarra Cabernet Shiraz (and then the following year with 1967 Bin 7). After vinification in open fermenters it finished off fermentation and maturation in 'French casks', although it is more than likely that a large proportion of American oak featured in the wine. It never reached the legendary status of 1962 Bin 60A, but was greatly admired by wine show judges for its superb regional integrity and longevity.

2008 Penfolds Bin 620 Cabernet Shiraz, which derives from Blocks 5, 10 and 20 within the original boundary of the Coonawarra Fruit Colony in the 'hundred of Comaum', is the latest Special Bin release. It is poignant that veteran winemaker John Bird, who was involved in the original Bin 620 as a young laboratory assistant, played a part in the selection of the fruit. With its deep purple, blackstrap liquorice colour, intense blackcurrant/elderberry/cedar aromas and flavours, silky textured, superbly buoyant palate and gorgeous tannin plume, it fully articulates the potential and potency of Coonawarra. This is the epitome of Australian First Growth Claret. After vinification in headed-down fermenters, it completed barrel fermentation in 100 per cent new French and American oak before further maturation in the same oak hogsheads for twelve months. It has all the substance, richness, balance and vinosity for long-term ageing. Although it can be enjoyed early because of its inherent fruit sweetness and ripe tannin structure, this is a wine that will benefit from further cellaring.

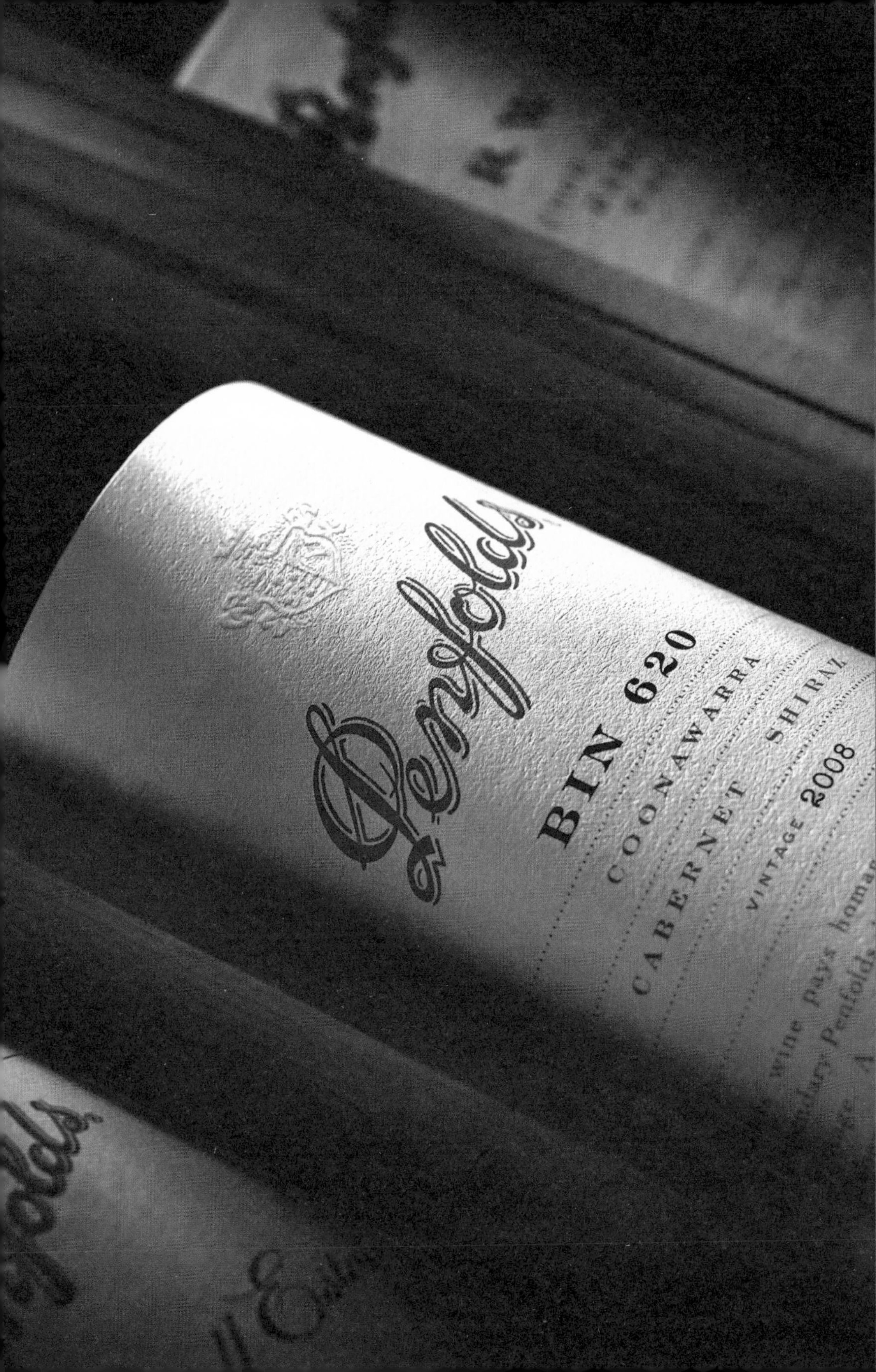
Penfolds
BIN 620
COONAWARRA
CABERNET SHIRAZ
VINTAGE 2008

Other Special Bin wines

FIRST VINTAGE

1948 Kalimna Cabernet Sauvignon is the first known Special Bin–type wine. There have been an astonishing number of 'one-off' releases – often limited to a single barrel or micro-blend.

VARIETIES

Generally based on shiraz and cabernet sauvignon.

ORIGIN

Multi-regional sourcing. Block 20, Coonawarra cabernet sauvignon, and Kalimna and Koonunga Hill shiraz feature regularly.

FERMENTATION

Stainless steel tanks with wooden header boards to submerge cap. Fermentation completed in new oak.

MATURATION

Approximately 15 months in new American or French oak hogsheads (300 litres).

COMMENTS

Special Bins are exceptional parcels of limited release wines especially made to highlight winemaking advances, experiments of interest or vintages of significant importance.

During the 1950s, 1960s and 1970s, Max Schubert experimented with several different combinations of vineyards and grape varieties. He was assisted by a team of winemakers and laboratory technicians: Ray Beckwith, Penfolds chemist, provided him with the crucial science to develop his principles and philosophies of winemaking, while some special bottlings were made by John Davoren at Auldana and Ivan Combet at Minchinbury, near Sydney. The most successful Bin wines of the 1960s were centred on the Kalimna Vineyard in the Barossa and the newly acquired Sharam's Block in Coonawarra. Vineyards have always been the tangible yet enigmatic wildcard of Penfolds Special Bins.

1956 Bin 136 Magill Burgundy was originally sent up from Magill to Sydney as blending material for a special Qantas bottling of Shiraz Mataro. However, Don Ditter was so impressed by the wine that he decided to bottle it separately. Penfolds Bin S56 was its sister wine, made from a similar parcel but bottled at Magill.

The rare 1957 Bin 14 Minchinbury Dry Red was made by veteran winemaker Ivan Combet to commemorate the last red wine vintage at the historic Minchinbury Estate. This nineteenth-century property had been granted to Captain Minchin, a veteran of Wellington's Peninsular War, by Governor Lachlan Macquarie in 1812. It was purchased by Penfolds in 1912 and became the site of sparkling white and red production. Leo Buring – one of the great pioneer winemakers of the early twentieth century – was

instrumental in establishing the reputation of the Minchinbury vineyard, nowadays a western suburb of Sydney.

In addition to 1962 Bin 60A, Max Schubert bottled an experimental one-third Coonawarra cabernet sauvignon and two-thirds Kalimna shiraz blend. 1962 Bin 60, the 'off-blend', is largely forgotten; the wine was not widely released. Schubert said, 'It was never up to the Bin 60A, but a bloody good wine.'

Not much is known about the limited release 1962 Bin 434 Coonawarra Cabernet Shiraz except that it was bottled at Auldana. It is quite possible that it belonged to the same trials as Bin 60 and Bin 60A. John Bird says it was also 'a forerunner to the 1966 Bin 620 Coonawarra Cabernet Shiraz'. Coonawarra, long resisted by Max Schubert as a source of premium fruit, was in some respects a fait accompli, as during the early 1960s Penfolds came under increasing pressure to sell its vineyard holdings in and around Adelaide. The company simply needed to plan for continuity of supply. Coonawarra's ascendency was assured.

1967 Bin 7, also a two-thirds Coonawarra cabernet sauvignon, one-third Kalimna shiraz blend, was a famous show wine of its day, but it never reached the legendary status of Bin 60A. All the same, it still commands plenty of interest in the secondary market; it remains a compelling and enduring follow up to its older sibling.

1973 Bin 169 Coonawarra Cabernet Sauvignon and Bin 170 Kalimna Shiraz, which were both released as single-region wines, were originally earmarked as blending companions – a 1970s version of Bin 60A. While 1973 was not considered as a particularly great vintage the components, roughly 750 gallons each, were 'pretty good'. Nonetheless Chris Hancock and John Bird were instructed not to blend the wine.

Bin 80A, Bin 820, Bin 90A and Bin 920 are homage wines, remembering Max Schubert's two most famous experimental vintages. Using special parcels of fruit, these wines were largely vinified, matured and blended to Schubert's original specifications. Bin 80A and 820 would never have been released without his blessing; both vintages are indelibly marked with Schubert's fingerprints. 1990 Bin 90A and 1990 Bin 920 are John Duval's reprise of Schubert's famous experimental work. These wines also reflect the amazing advances in viticulture and winemaking. 1990 is generally regarded as one of the greatest Penfolds vintages in memory.

2008 Bin 620 is a rendition of the original 1966 Bin 620. This extraordinary wine foreshadows a new golden era for Penfolds in Coonawarra.

2010 Bin 170 Kalimna Shiraz, a single-vineyard wine from a spectacular Barossa vintage, shows the remarkable first-growth quality of this nineteenth-century vineyard and remembers the Penfolds experimental bottlings of the 1960s and 1970s, including the 1973 Bin 170 Shiraz.

The Penfolds Special Bins were once described as 'a remarkable body of work comprising the good, the great, the avant garde, the post-moderns and the curious'. These wines belong to Penfolds great tradition of experimentation, 'pushing boundaries and challenging norms'.

 *Panel Review*

Magill Estate, Adelaide – Penfolds Special Bins
Jo Burzynska (JB), Jane Faulkner (JF), James Halliday (JH), Huon Hooke (HH), Ralph Kyte-Powell (RKP), Tony Love (TL), Tyson Stelzer (TS); Peter Gago, Steve Lienert, Andrew Caillard.

Spellbinding, dramatic and expressive; the story of Penfolds wines in a single tasting. — Tony Love.
The epitome of what old wines should be. — Jo Burzynska.

1950s
The evocative and rare 1953 Bin 2 Grange Hermitage, 1953 Bin 9 Grange Cabernet Sauvignon and 1957 Bin 136 Magill Burgundy were 'gentle and delicate' (JB).

1960s
'One hit after another', the wines are steeped in classicism with an enduring and moving quality. 1962 Bin 60A, 1964 Bin 707 and 1967 Bin 7 were highlights.

1970s
Represented by 1973 Bin 169 and 1973 Bin 170. They were 'fascinating yet show a period of uncertainty' (HH).

1980s
Both 1980 Bin 80A and 1982 Bin 820 show concentration, power and fruit complexity.

1990s
1990 Bin 90A, 1990 Bin 920 and 1996 Block 42 are 'starting to open up their petals' (JF).

2000s
2004 Bin 60A, 2004 Block 42, 2008 Bin 620 and 2010 Bin 170 are plush, opulent and velvety with great integrity of style and superb ageing potential.

The Ampoule

The Penfolds Ampoule Project is a collaboration of designer-maker Hendrik Forster, furniture craftsman Andrew Bartlett, glass artist Nick Mount and scientific glassblower Ray Leake. At its core is the Kalimna Vineyard's celebrated 2004 Block 42 Cabernet Sauvignon.

Peter Gago explained the concept: 'The Ampoule Project is a provocative statement about the art and science of wine. When it was first brought to the world's attention in 2012 it became a media sensation. The limited-release Ampoule – only twelve were crafted – is the ultimate wine curiosity, an experience combining Penfolds heritage and South Australian ingenuity and identity.

'Each Penfolds Ampoule was filled with the exceptional 2004 Penfolds Block 42 Cabernet Sauvignon, before being encased by a hand-blown glass plumb-bob shaped vessel and suspended in a beautifully fashioned jarrah timber cabinet. Without any possibility of air reaching it, this wine can slowly evolve for generations.

'The wine should age very well, there's adequate headspace in the ampoule, it's like ageing under the almost hermetic seal of a screwcap, but using glass only; a substance of proven durability and safety.'

When the time comes for the wine to be enjoyed, there will be a unique opening ceremony. A Penfolds winemaker will travel to the destination of choice, anywhere in the world. The ampoule will be carefully removed from its casing and snapped open, using a specially designed, tungsten-tipped, sterling silver scribe-snap. The winemaker will then prepare the wine using a beautifully crafted sterling silver tastevin.

Each Ampoule comprises six essential elements:

- The scientific-grade ampoule, designed to store the wine in ideal conditions, created by glassblower Ray Leake.
- 2004 Block 42 Cabernet Sauvignon, transferred into the ampoule without exposure to air by Chief Winemaker Peter Gago, and then precision sealed by Ray Leake.
- The glass sculpture, crafted by glass artist Nick Mount. To encase the ampoule, Nick has designed and hand-blown a conical, elongated plumb-bob of transparent grey glass with a ruby red 'cotton-reel' top.
- The precious metal detailing, including tastevin and scribe-snap, prepared by designer-maker Hendrik Forster.
- The jarrah timber cabinet, in which the ampoule is suspended and protected, custom-made by furniture craftsman Andrew Bartlett.
- A unique 'making of' booklet and certificate of ownership and authenticity, signed by Chief Winemaker Peter Gago and all contributing artists.

The Ampoule is a work of art, and at its heart is a great wine. It pushes boundaries yet it also furthers the Penfolds story of tradition and culture. Provenance, collaboration, craftsmanship, meticulous execution and planning are all inextricably linked in this ambitious design.

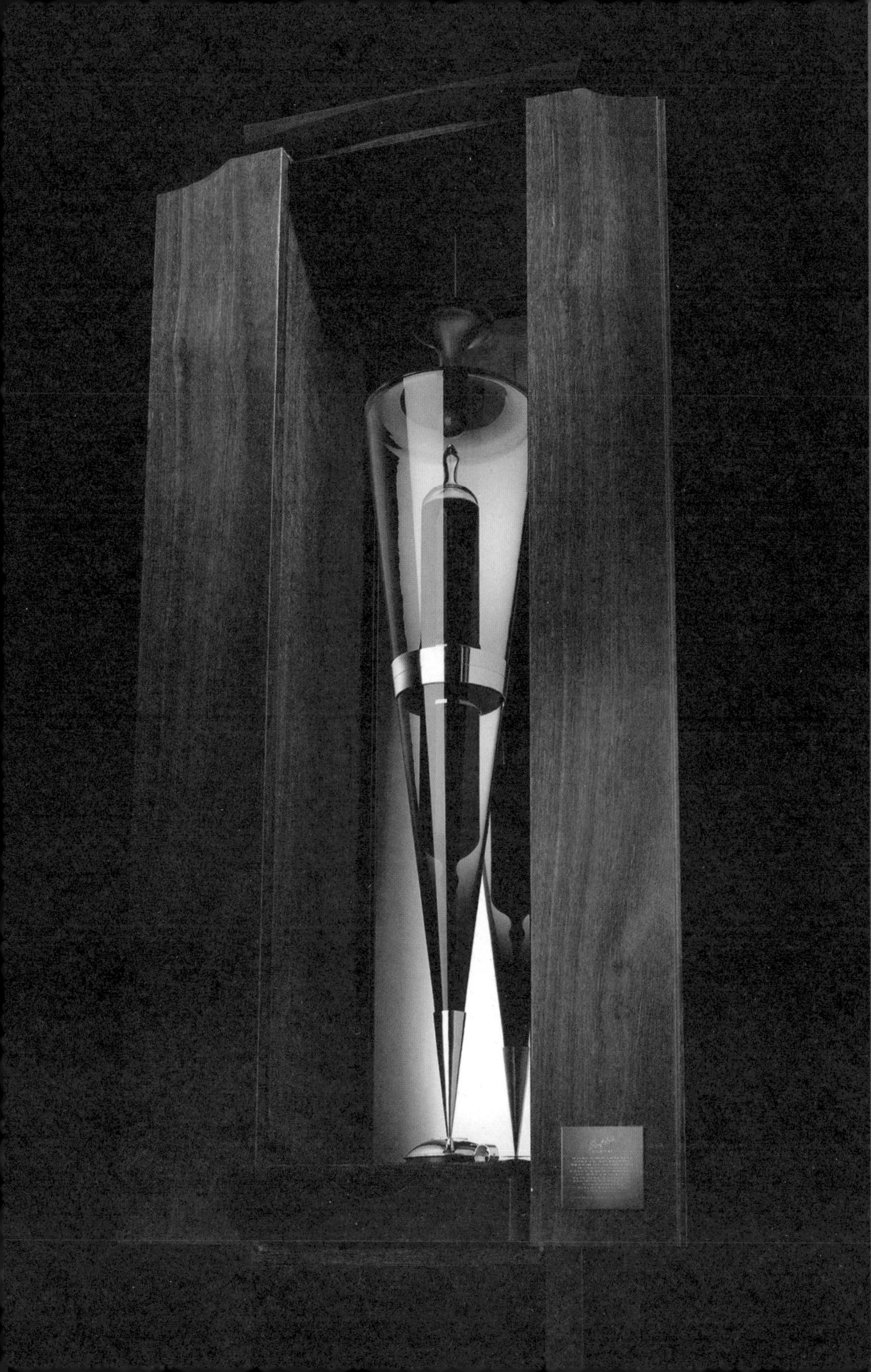

'We must not be afraid to put into effect the strength of
our own convictions, continue to use our imagination in
winemaking generally, and be prepared to experiment
in order to gain something extra, different and unique
in the world of wine.'

MAX SCHUBERT

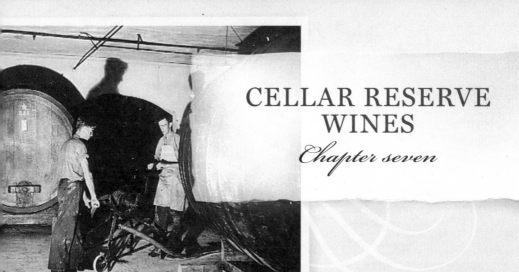

CELLAR RESERVE WINES

Chapter seven

Cellar Reserve wines are alternative, experimental or 'one-off' works in progress for collectors, cellar door visitors and fine wine specialists with an interest in quirky labels. These expressive and non-conforming wines have been developed without the constraints of the mainstream wine market. They reflect the ever-changing viticultural landscape and the spirit of making something meaningful and different.

First introduced during the 1995 vintage by Peter Gago at Magill, with an experimental Pinot Noir, these wines have become a valuable addition to the Penfolds portfolio. Cellar Reserve and Reserve Bins, comprising both whites (not reviewed, although gewürztraminer, pinot gris, chardonnay and a dessert-style viognier have been released under the Cellar Reserve label) and reds, are alternative wine styles for the traditional Penfolds drinker.

Cellar Reserve Adelaide Hills Pinot Noir

FIRST VINTAGE	1995 experimental; 1997 commercial release.
VARIETY	Pinot noir.
ORIGIN	Adelaide Hills and sometimes a very small percentage of Eden Valley grapes.
FERMENTATION	Cold soaked for 3–7 days and batch-vinified (up to 30% whole-bunch) with natural yeasts in open fermenters. No pressings ever included.
MATURATION	10 months in 1-year-old and up to 70% new French oak barriques on gross lees. No fining or filtration.

The 1997 Cellar Reserve Pinot Noir was first released after a series of winemaking trials. The fruit is sourced from various cool-climate vineyard sites along the spine of the Adelaide Hills and neighbouring Eden Valley. The style is typically richly concentrated, with musky black cherry aromas and chocolatey tannins.

The fruit is cold soaked for up to seven days in open fermenters before partial whole-bunch vinification with natural yeasts and around-the-clock hand-plunging (*pigeage*). Varying percentages of stalks remain in the ferment to achieve more complexity and tannin structure. After completion of fermentation the wine is run off into new and one-year-old French Dargaud & Jaegle oak barriques for around ten months' maturation.

Magill Estate, Adelaide – Penfolds Cellar Reserve Pinot Noir
Huon Hooke (HH); Peter Gago, Steve Lienert, John Bird, Andrew Caillard.

Yes, a Penfolds Pinot Noir! — Peter Gago.

A Penfolds style with an emphasis on structure and savoury characters. Not pretty or fruit accentuated.
The tannins tend to dominate older vintages. The younger wines are more vibrant and red fruited, with
richness and strong tannin presence. Highlights are 2008 and 2010.

Cellar Reserve Barossa Valley Sangiovese

FIRST VINTAGE	1998.
VARIETY	Sangiovese, an Italian grape variety best known for the wines of Tuscany, especially Chianti.
ORIGIN	First made from 1982 experimental plantings in the Penfolds Kalimna Vineyard, Barossa Valley. Since 2003 also sourced from an independent grower in the Marananga sub-region of the Barossa Valley. Average age of vines is 25–30 years.
FERMENTATION	Vinified in open fermenters with natural yeasts. Up to 5 weeks post-fermentation skins.
MATURATION	11 months in up to 6-year-old French oak barriques – 'as old as you can get them!' Bottled unfined and unfiltered.

Early trials from eight rows of vines planted at Kalimna in 1982, in conjunction with the South Australian Department of Agriculture, were followed by several unreleased experimental wines. Encouraged by the results, Barossa grower Paul Georgiadis, a former Penfolds Grower Liaison Manager, planted his own block of sangiovese during the mid 1990s. These wines are typically deep in colour with fresh black cherry aromas and velvety tannins. Winemaking is laissez faire, with no added yeast, no pH correction, no new oak, no addition of any other variety, no fining, no filtration. The wines are vinified in open headed-down fermenters at Magill Estate and then allowed to macerate for five weeks on skins before being basket-pressed to barrique.

Panel Review

Magill Estate, Adelaide – Penfolds Cellar Reserve Sangiovese
Jo Burzynska (JB), Jane Faulkner (JF), James Halliday (JH), Huon Hooke (HH), Ralph Kyte-Powell (RKP), Tony Love (TL), Tyson Stelzer (TS); Peter Gago, Steve Lienert, John Bird, Andrew Caillard.

Good solid style with the stamp of Penfolds all over it. — Tyson Stelzer.

1998–2010

A work in progress, with the most recent vintage reaching a new level of fruit complexity and balance. The wines are not overly exuberant and possess strong muscular tannins and 'flowing' acidity. The oldest vintages are now drying out but the younger wines are great drinking. 'Older vines and better viticulture will lead to greater things.' (JH) 2004, 2006, 2008 and 2010 are best vintages.

Cellar Reserve McLaren Vale Tempranillo

VINTAGES	2008, 2009, 2010.
VARIETY	Tempranillo, a Spanish grape variety best known for the wines of Rioja and Ribera del Duero.
ORIGIN	McLaren Vale – single vineyard.
FERMENTATION	Vinified in 5 tonne open fermenters at Magill Estate with natural yeasts. Basket-pressed and filled direct to barrique.
MATURATION	11 months in up to 6-year-old French oak barriques. Bottled unfined and unfiltered.

Tempranillo is one of the latest alternative varieties to be 'cellar reserved' by Penfolds. It was probably introduced into Australia during the nineteenth century although records are sketchy. New clones were imported during the 1960s, 1970s and 1990s. This has gradually resulted in a surprising diversity of plantings around Australia. McLaren Vale has shown much promise for this variety. Typically the Cellar Reserve Tempranillo, unencumbered by new oak, is intensely fruited with ripe blackberry fruit, black olive flavours and ripe round tannins.

 *Panel Review*

Magill Estate, Adelaide – Penfolds Cellar Reserve Tempranillo
Jo Burzynska (JB), Jane Faulkner (JF), James Halliday (JH), Huon Hooke (HH), Ralph Kyte-Powell
(RKP), Tony Love (TL), Tyson Stelzer (TS); Peter Gago, Steve Lienert, John Bird, Jamie Sach, Andrew
Caillard.

> *I am more and more impressed with the potential of this variety.* — Ralph Kyte-Powell.

2008–2010

Powerfully black-fruited wines, with fabulous violet, tobacco and flesh. Not definitively tempranillo but
a work in progress. 2008 and 2010 are highlights.

Cellar Reserve Barossa Valley Cabernet Sauvignon

VINTAGES	2005, 2006, 2008, 2010.
VARIETY	Cabernet sauvignon.
ORIGIN	One-off cuvées based on parcels of selected Barossa Valley and Eden Valley fruit. Includes Block 42 and Block 41 Kalimna fruit.
FERMENTATION	Open headed-down stainless steel fermenters. Fermentation finishes in barrel.
MATURATION	All matured in new and old oak for up to 16 months.

Cellar Reserve Cabernet Sauvignon is sourced from select parcels of A1 Barossa Valley and Eden Valley fruit. This includes 1880s planted Block 42 and 1955 planted Block 41 cabernet sauvignon. The style is very close to the Special Bin Block 42 style, but draws fruit from a wider spectrum of vineyards and soil profiles. Typically it is opulent in structure with ripe cassis, dark chocolate aromas, chocolatey tannins and underlying savoury new oak. The wines, vinified using classic Penfolds techniques, have a superb lasting quality.

 Panel Review

Magill Estate, Adelaide – Penfolds Cellar Reserve Cabernet Sauvignon

Jo Burzynska (JB), Jane Faulkner (JB), James Halliday (JH), Huon Hooke (HH), Ralph Kyte-Powell (RKP), Tony Love (TL), Tyson Stelzer (TS); Peter Gago, Steve Lienert, John Bird, Jamie Sach, Andrew Caillard.

2006, 2008 and 2010 are a stunning trilogy. — Tyson Stelzer.

2005, 2006, 2008 and 2010

A magnificent line up comprising wines of hedonistic richness and volume. Black fruits with profound depth and power. 2006, 2008 and 2010 are top performers.

Cellar Reserve Coonawarra Cabernet Barossa Valley Shiraz

VINTAGES	1993, 2005.
VARIETIES	Cabernet sauvignon and/or shiraz.
ORIGIN	Almost 'one-off' cuvées based on parcels of selected Barossa Valley, Coonawarra, Clare Valley and Padthaway fruit.
FERMENTATION	Open headed-down fermenters and small stainless steel static fermenters.
MATURATION	All matured in new and old oak for up to 16 months.
COMMENTS	'Sweet fruit, density and supple tannins', Jo Burzynska.

Cellar Reserve Barossa Valley Grenache

FIRST VINTAGE	2002.
VARIETY	Grenache.
ORIGIN	Mature bush vines, Marananga, Barossa Valley.
FERMENTATION	Open headed-down fermenters and small stainless steel static fermenters.
MATURATION	11 months in older French hogsheads.
COMMENTS	'The greatest Grenache vintage in the Barossa Valley for a decade. We are still waiting for the next!' Peter Gago.

Cellar Reserve Kalimna Block 25 Mataro

FIRST VINTAGE	2010.
VARIETY	Mataro.
ORIGIN	Block 25, Kalimna Vineyard, Barossa Valley – planted in 1964.
FERMENTATION	Small stainless steel static fermenters.
MATURATION	11 months in older French hogsheads.
COMMENTS	'Extraordinary', Tony Love. The wine won 'Best Red Wine' at the London Wine Trade Fair in May 2012.

Cellar Reserve Adelaide Hills Merlot

FIRST VINTAGE	2010.
VARIETY	Merlot.
ORIGIN	Adelaide Hills.
FERMENTATION	Open headed-down fermenters.
MATURATION	11 months in older French hogsheads.
COMMENTS	'Medium-bodied and floral with a strong tannin presence', Jane Faulkner.

'I have been involved with making Bin 389 for over thirty years. For me this is what Penfolds is all about; an inheritance and tradition for future generations.'

STEVE LIENERT

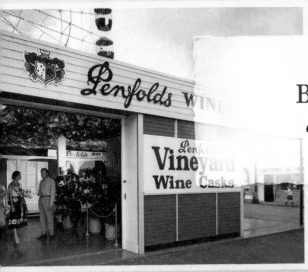

BIN WINES
Chapter eight

Bin 389 Cabernet Shiraz

FIRST VINTAGE	1960.
VARIETIES	Cabernet sauvignon and shiraz.
ORIGIN	A multi-district blend, South Australia. Barossa Valley, Coonawarra, Padthaway, Robe, McLaren Vale, Langhorne Creek and Clare Valley.
FERMENTATION	Stainless steel tanks with wax-lined/wooden header boards. Some components complete fermentation in barrel.
MATURATION	18 months in American oak hogsheads (300 litres); 20–30% new, 70–80% 1- and 2-year-old oak, including barrels used for the previous vintage of Grange and Bin 707.

Bin 389 Cabernet Shiraz, affectionately known as 'poor man's Grange' or 'baby Grange', is one of Australia's great cellaring red wines. First produced in 1960, its history is connected with the development of Grange and Max Schubert's ambition of creating 'a dynasty of wines which all bear an unmistakable resemblance and relationship to each other'. This classic, much-loved and beautifully consistent Australian style epitomises Penfolds winemaking philosophy and the benefits of cross-varietal and multi-regional blending. Bin 389 is the quintessential expression of the Penfolds red wine style. Typically it is fresh, generous and buoyant with ripe dark chocolate/dark berry fruit, beautifully extracted flavours, fine-grained tannins and underlying new oak characters. The best vintages can develop and improve for decades.

Bin 389 is the most popular wine in the Australian secondary wine market because of its heritage, consistency and reputation. Its 'Outstanding' rating in Langton's Classification of Australian Wine underscores its currency as an Australian 'Super Second'.

Bin 389 is an Australian Grand Vin; a selection and interpretation of the best parcels, from the best vineyard sites, by a winemaking team steeped in the ethos and tradition of a great Australian wine style.

While Max Schubert experimented with Grange during the 1950s, Penfolds had been producing claret- and burgundy-style red table wines for around a century. These wines had won prizes at Australian wine shows, and at various exhibitions in Paris and London before and after Federation. At first the light dry table wines were made from the Grange Vineyards at Magill. From the 1920s and 1930s, however, the demands of an ascendant fortified and table wine market led to a Penfolds tradition of blending its barrels of stock, without concern of origin. That the wine was Penfolds was all the consumer really needed to know. Nonetheless, Penfolds No. 1, Royal Reserve and Blue Label clarets continued to win prizes and accolades.

As Penfolds newly appointed senior winemaker in 1948, Max Schubert was keenly interested in modernising the range. Aside from the significant work to be done with fortified wine production and sparkling wine, there was a growing belief that red table wine would become increasingly important in post-war Australia. Improvement was certainly needed: in 1951, Penfolds clarets and burgundies comprised Private Bin labels (priced at 4s 9d) and Extra Special Royal Reserve Claret (at 3s 6d). There were other bottlings, taken from barrel, for various Beef and Burgundy clubs, but the range was minimal.

On Max Schubert's return from his European trip, a new energy and competitive atmosphere quickly appeared at Penfolds. He had observed barrel fermentation in Jerez, new oak maturation in Bordeaux and myriad other vinification techniques. His vision of making a claret that would last at least twenty years manifested in the development of Grange and a unique way of making red wine to suit Australian conditions. At the same time John Davoren, a senior red winemaker, was experimenting with red wine fermentation at Auldana. The development of St Henri Claret was not only happening simultaneously with Grange, but also the observations and experience were shared. Ray Beckwith, the Penfolds chemist (he would now be known as the oenologist) also played a key role, by providing technical support. Not surprisingly, both of these wines were presented to the market as sibling clarets during the 1960s.

The 1950s was an exciting decade that saw the birth of Bin 28, as well as Grange and St Henri. Experimentation with grape varieties, vineyard sites and blending combinations yielded new experimental wines and 'one-offs', sometimes as little as a few bottles. Max Schubert was already tinkering with cabernet shiraz blends, using combinations of Kalimna, Grange and Auldana fruit. The experimental St Henri Clarets, initially cabernet mataro blends, became increasingly reliant on shiraz. The technical advances, experimental winemaking and planning during the 1950s provided Penfolds with a genuine competitive and commercial edge. The release of the 'claret-style' 1960 Bin 389 Cabernet Shiraz – named after its original binning compartment, and both vinified and matured at Magill cellars – heralded Penfolds new, signature method of winemaking, just as signs of Australia's fine red wine boom were underway.

Following Max Schubert's observations in France, he set about producing the definitive standard in Australian premium red winemaking by selecting suitable grape material (by variety and location) and finishing the fermentation in oak hogsheads and storing the wine therein for a protracted period of 1½ to 2 years.
— Ray Beckwith, unpublished memoirs, 2001.

Ray Beckwith described the Penfolds winemaking process in 1967:

The reds are fermented at a maximum of about 85 degrees Fahrenheit … allowing good colour extraction. Excessive temperatures result in loss of flavour and spoilage by bacteria. As fermentation proceeds the winemaker watches the development of colour, and when he judges that it is sufficient, the fermenting juice is removed from the skins, and the process completed in other tanks, with a constant check being made in temperature. For certain special red wines (including batches of Bin 389), fermentation is completed in new oak hogsheads of about 65 gallons capacity to allow extraction of the oak tannin.

From the very start, Max Schubert favoured warm-climate fruit as a source for Bin 389; reflecting his strong preference for richly concentrated 'buoyant' fruit. He wasn't really looking for varietal character; rather, he was looking for maturity and complexity of aromas, flavour development and structure. Schubert fully understood the potential of cabernet and shiraz as blending companions. He believed that each component could bring something extra and different to a wine. The philosophy holds true today. Warm-climate cabernet sauvignon is beautifully perfumed and intensely flavoured, with firm chocolatey tannins. Shiraz is more opulent and fleshy, with power and generosity of fruit. Combined, these two elements bring together extraordinary volume of fruit, mid-palate richness and finesse.

Bin 389 just can't work with cool-climate fruit.
— James Halliday.

No controversy accompanied Bin 389. Doug Crittenden, a Melbourne wine merchant, was an early supporter and used to transport the wine over from Adelaide by barrel. Gradually the pressure of a growing market and Adelaide's urban encroachment resulted in the wine becoming a multi-district blend. The early vintages were made from vineyards around the Adelaide foothills, including Auldana and Magill,

then almost entirely from Barossa Valley fruit and over the last twenty years from distinguished vineyards around South Australia including the Barossa Valley, Coonawarra, Padthaway, McLaren Vale, Langhorne Creek, Clare Valley and Robe. Although the best of regional characters are blended together, it is ultimately the personality and the structure of the fruit that really matters.

The introduction of refrigeration and stainless steel saw major progress in style and control; the wine could be fermented at lower temperatures, resulting in intense clear fruit aromas and palate freshness. Vineyard management and sourcing of grapes, new membrane presses, improved seasoning of new American oak, meticulous barrel selection, rigorous classification and 'a lot more walking around the rows' are further refinements during the last forty years. The overall winemaking practices have not changed: the classical heading down in open fermenters and partial barrel fermentation remain key Penfolds techniques. The original blueprint remains the reference.

I have been involved with making Bin 389 for over thirty years. For me this is what Penfolds is all about; an inheritance and tradition for future generations.
— Steve Lienert, Senior Red Winemaker.

Around 20–30 per cent of Bin 389's components complete fermentation in new A. P. John–coopered American oak hogsheads to enhance complexity, richness and the integration of flavours. The remaining elements are vinified in stainless steel fermenters and then matured for around eighteen months in seasoned (one- and two-year-old) ex-Grange and ex-Bin 707 hogsheads. With its generous ripe dark chocolate/dark berry fruit profile, beautifully extracted flavours, fine-grained tannins and underlying new oak, Bin 389 is a classic Penfolds red wine with a great track record for cellaring.

 *Panel Review*

Magill Estate, Adelaide – Penfolds Bin 389 Cabernet Shiraz
Jo Burzynska (JB), Jane Faulkner (JF), James Halliday (JH), Ralph Kyte-Powell (RKP), Tony Love (TL), Tyson Stelzer (TS); Peter Gago, Steve Lienert, John Bird, Andrew Caillard.

A world-class flight of wines; the strongest vintages encapsulating the ageability of Australian Cabernet Shiraz.
— Tyson Stelzer.

1961–1969

Although a flying start with the 1961, this decade is quite variable. The 1964 and 1967 were attractive and supple but the wines are unlikely to improve. 1968 and 1969 not tasted.

1970–1979

1971 and 1976 mirrored the reputation of Grange, with plenty of fruit complexity, mushroom notes and restrained power, but ultimately this decade is unlikely to improve with further cellaring.

1980–1989

A decade showing more concentration, density and finesse. The tannin quality, as expected, has improved greatly. A succession of consistent wines, most possessing lovely fruit richness, integrated tannins and oak. 'Shows the compelling harmony between cabernet sauvignon and shiraz. A real progression of building on the past.' (TS) Highlights included 1982, 1986, 1988 and 1989.

1990–1999

'A pretty serious bracket of big wines looking youthful and bright.' (JB) 'Bolshie wines that unfurl in the glass.' (JF) Relatively tanninic wines with beautiful fruit, richness and power; 1991, 1994, 1996 and 1998 are highlights.

2000–2010

A shift in style occurs around 2001 with more mouth flow and softer textures. The wines are more concentrated, with superb freshness, fruit concentration, integrated tannins and oak; all reflecting older vines, new vineyards on great soils and better vineyard and tannin management. Highlights include 2002, 2004, 2006, 2008, 2009 and 2010.

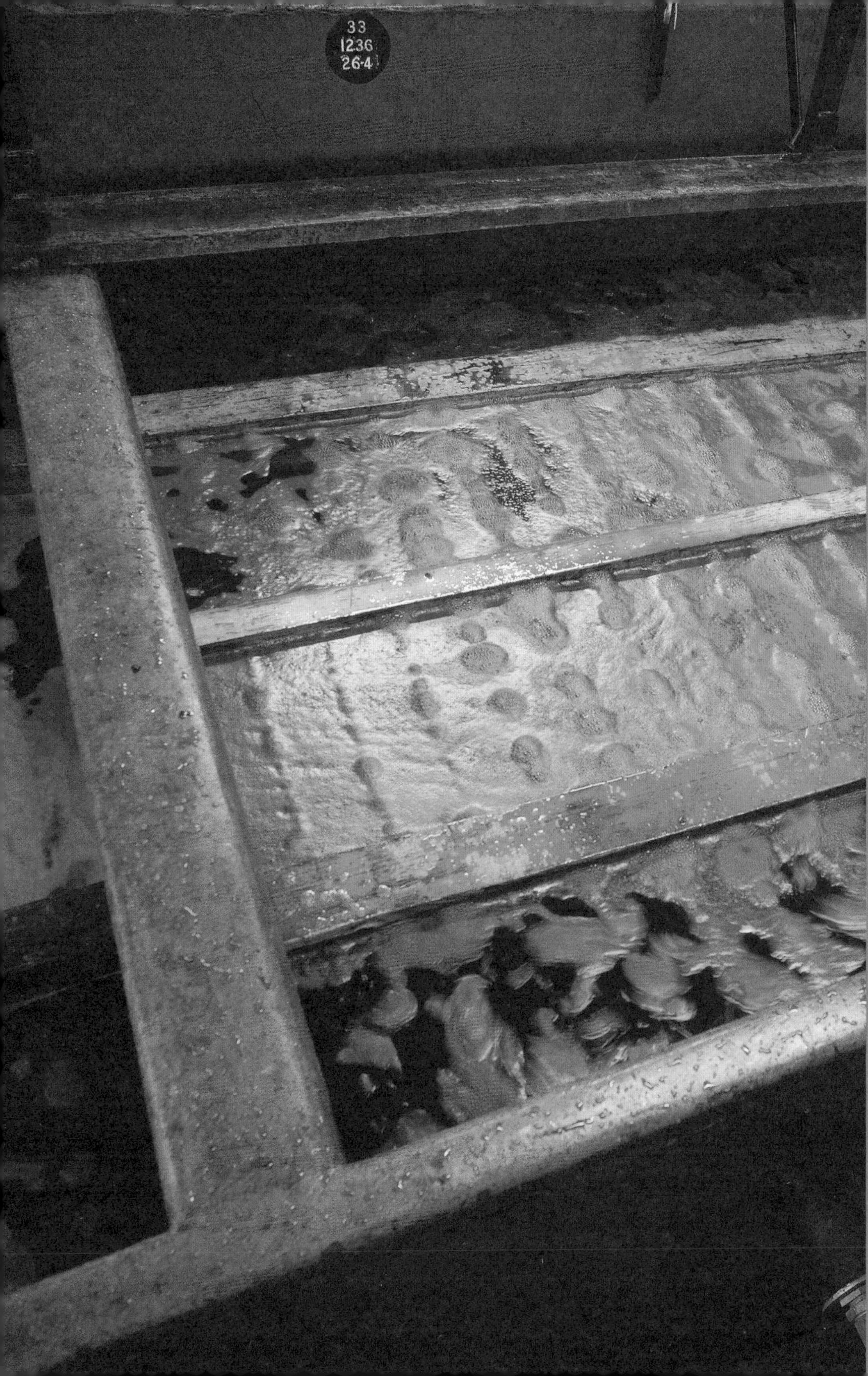
33
1236
26-4

Bin 150 Marananga Shiraz

FIRST VINTAGE	2008.
VARIETY	Shiraz.
ORIGIN	Walton's, Zilm and grower vineyards at Marananga in the central Barossa Valley.
FERMENTATION	Static fermenters with wooden header boards. After basket pressing, components complete fermentation in barrel.
MATURATION	16-18 months in new (50%) and seasoned American and French oak hogsheads and puncheons (500 litres).
COMMENTS	A new, classic Bin with superb sub-regional Barossa Valley provenance.

Bin 150 Marananga Shiraz is a heartland Barossa style with a clear Penfolds winemaking signature. It was first introduced in response to increasing availability of exceptional fruit, a growing fascination about Barossa *terroir* and the expectations of new and exciting wines by the market. It is a superb addition to Penfolds Bin series, with its saturated colours, fruit density, generous tannin structure and potential longevity.

The low-yielding vineyard blocks are located on sandy loams and red-brown soils. Small berries, deep colour, intense aromas and juicy concentrated flavours are typical characteristics of Marananga Shiraz. Senior red winemaker Steve Lienert says, 'The fruit quality from some of these blocks is about as good as you can possibly get around here.'

The wine is made in the traditional Penfolds method. After vinification in headed-down static stainless steel fermenters the wine completes fermentation in a combination of new (50 per cent) and seasoned American and French oak hogsheads (and increasingly more puncheons). It is then matured in the same barrels for around sixteen to eighteen months, before bottling.

Bin 150 Marananga Shiraz is unmistakably Penfolds in character, with seductive, rich concentrated fruit and expressive sub-regional base notes. The style has a beautiful lasting quality and will repay medium- to long-term cellaring.

 *Panel Review*

Magill Estate, Adelaide – Penfolds Bin 150 Marananga Shiraz
Jo Burzynska (JB), Jane Faulkner (JF), James Halliday (JH), Ralph Kyte-Powell (RKP), Tony Love (TL), Tyson Stelzer (TS); Peter Gago, Steve Lienert, John Bird, Andrew Caillard.

Consistently terrific. — Tony Love.

2008–2010
'Wonderful shiraz fruit; not overblown. 2010 walks away with the prize. A beautiful wine.' (TL) Impressive sub-regional style with terrific potential longevity. Highlights 2008 and 2010.

Bin 407 Cabernet Sauvignon

FIRST VINTAGE	1990.
VARIETY	Cabernet sauvignon.
ORIGIN	Multi-district blend, South Australia/Western Australia including Bordertown, Padthaway, Coonawarra, Robe, McLaren Vale, Clare Valley, Barossa Valley, Wrattonbully and Margaret River, Western Australia.
FERMENTATION	Stainless steel fermenters with wooden header boards. Some components complete fermentation in barrel.
MATURATION	12 months in new French and American (approximately 30%) oak. Also 1- and 2-year-old hogsheads – some of which were used for previous vintages of Bin 707.
COMMENTS	From 2004 bottled under screwcap in some markets.

Bin 407 Cabernet Sauvignon, as its Bin number suggests, is related in style to the rich and opulent Bin 707. The wine was developed in response to the increasing availability of high-quality cabernet sauvignon fruit and mounting pressure from the market for a versatile wine that could be enjoyed for immediate drinking or could develop well in the cellar.

The wine has achieved great success in the market as a classic Penfolds wine. In 2005 it was rated

'Distinguished' by Langton's, in recognition of its growing importance in the secondary market. It has become one of Australia's most popular collectible wines, and in China it enjoys an almost cult status.

Bin 407 is a fresh elegant style with clear varietal blackcurrant/cassis aromas, fine-grained firm tannins and underlying cedary/savoury oak. Further cellaring for around four to six years allows the wine to develop more richness, volume and bottle complexity. The best vintages have tremendous lasting power, as indicated in this edition's panel tasting (the 1990, now in its third decade, is drinking wonderfully).

First produced in 1990, it is an important Penfolds House Style wine based on rigorous selection of multi-regional fruit and the flexible use of French and American oak maturation. The fruit is primarily sourced from the Limestone Coast, including Coonawarra, Robe and Bordertown. Selected parcels of fruit from the Clare Valley, Barossa Valley and McLaren Vale are sometimes used to achieve consistency.

The overall winemaking philosophy follows similar lines to Bin 389, down to the maturation of Bin 407 in second-use Bin 707 hogsheads. The wine is vinified in open stainless steel tanks with wax-lined wooden header boards, a traditional Penfolds technique. Some components are barrel fermented in new French and American oak to increase complexity and blending options. After fermentation the wine is matured for twelve months in a combination of new French (15 per cent) and American (15 per cent) oak; the remainder is aged in one- and two-year-old French and American oak.

Bin 407 is a textbook Cabernet Sauvignon. While it reflects the growing season of each vintage, the selection and classification of parcels, followed by meticulous attention to detail, gives the wine a reputation for outstanding consistency and reliability. Bin 407 and its sister wine Bin 389 exemplify the ambitions of Penfolds winemakers and the core qualities of the Bin Series.

Panel Review

Magill Estate, Adelaide – Penfolds Bin 407 Cabernet Sauvignon
Jo Burzynska (JB), Jane Faulkner (JF), James Halliday (JH), Ralph Kyte-Powell (RKP), Tony Love (TL), Tyson Stelzer (TS); Peter Gago, Steve Lienert, John Bird, Andrew Caillard.

Once again, as we have seen in other lines, Bin 407 shows remarkable consistency of style and character; the hallmark of Penfolds. — Ralph Kyte-Powell.

1990–2010

Full-flavoured and robust during the 1990s. Balanced and modulated during the 2000s with more fruit definition, varietal integrity and superbly integrated tannins. 'A very close family resemblance across two decades' (TS) despite being multi-regionally sourced. Lovely hedonistic wines with restrained power and opulence. Highlights include 1990, 1991, 1996, 1998, 2002, 2004, 2008, 2009 and 2010.

Kalimna Bin 28 Shiraz

FIRST VINTAGE	1959.
VARIETY	Shiraz.
ORIGIN	Multi-district blend, South Australia. Significant contributions from Barossa Valley, McLaren Vale, Clare Valley and Langhorne Creek.
FERMENTATION	Stainless steel fermenters with wooden header boards.
MATURATION	15 months in older American hogsheads.
COMMENTS	Screwcap closures available in most markets since 2004. Available in magnum format.

Bin 28 Shiraz is an Australian classic, valued for its integrity as a maturation-style fine wine and suitability as a fresh, early drinking style for many different occasions.

Penfolds Kalimna Bin 28 Shiraz, with its signature characters of abundant fruit, ripe fleshy palate and generosity of flavour, is one of the most popular wines in Penfolds portfolio. First produced in 1959, Bin 28 is Penfolds earliest Bin-range wine. The first releases were single-vineyard wines made from the renowned Kalimna Vineyard in the Barossa Valley. Kalimna is now a trademarked name used for this wine, but the vineyard's history is steeped in the beginnings of South Australia's colonial wine industry.

The Kalimna property, on Moppa scrub country on the western side of the Barossa Valley, was first planted around 1884. In 1904, the 364 acre vineyard and cellars, with 250 000 gallons of storage, sold its entire production of wine – mostly 'Burgundy' – for export. After trading difficulties during World War II, the Kalimna Vineyard and Cellars was sold to Penfolds. Along with Auldana and Magill, it became a mother vineyard of Penfolds red wine revolution during the 1950s and early 1960s.

Kalimna Bin 28 Shiraz was first made while Grange was still being made in secrecy. Many of the new vinification techniques developed by Ray Beckwith and Max Schubert were employed in the development of this commercial 'claret-style' wine, including the use of open headed-down fermenters and barrel fermentation. It is understood the very first vintages were matured, by default, in a small proportion of new oak. Nowadays, the wine is matured in one- and two-year-old American oak barrels, to preserve freshness and improve fruit complexity and tannin structure. The additional use of screwcaps at bottling, since 2004, further enhances cellaring potential and longevity.

Many of Max Schubert's early wines were drawn off primary ferments or out of barrel, lab-blended and aged in glass flagons. Sometimes he may have hand-bottled and labelled up only a few examples for friends or visiting dignitaries. The appearance of a bottle of 1959 Bin 28 shows that standardisation and [bookkeeping] control was still a long way off at Penfolds during that time.
— John Bird, senior winemaker, who joined Penfolds as a laboratory technical assistant in 1960.

During the early 1960s Bin 28 quickly established a reputation as an 'authentic Barossa-type red' which would develop 'additional character' with further cellaring. The success of this fresh but classic style led to a change of fruit sourcing during the late 1970s. With its extraordinary portfolio of vineyards across several wine regions, access to independently grown fruit and newly developed grape-quality classification system, Penfolds could now offer the market one of the most reliable and consistent of wines within its class. Hence Kalimna Bin 28 Shiraz can now be found in the swishest restaurants in London, New York, Shanghai and Moscow, as well as in a dusty outback pub in the Northern Territory.

The introduction of membrane, or air bag, presses during the 1980s led to a more gentle extraction of colour, flavour and tannins. The quality of pressings, the last fraction of thick, concentrated, tannin-rich wine, improved greatly during this period. Blended back into the free-run wine, this constituent provides the underlying power and stuffing of the contemporary Bin 28 style.

The wine has received numerous awards at various international wine shows around the world, including the International Wine Challenge in London. In 2005, it was classified 'Distinguished' by Langton's Classification of Australian Wine, in recognition of its visibility in the collectible ultra-fine wine market.

The real point of difference between today and the past – aside from better presses and overall technology – is the relationship between the winemaker and the grower. Winemakers and grower liaison viticulturalists are spending more time in the field explaining exactly what their needs are. Most growers – including our own vineyard managers – have a real interest in the final destination of their fruit. This goes beyond the quality incentives provided by fruit grading and price premiums. Penfolds growers are second-to-none when it comes to attention to detail and teamwork.
— Steve Lienert, senior red winemaker.

2008 Kalimna Bin 28 Shiraz marked the fiftieth vintage of this line, an important milestone in the history of Penfolds winemaking. The wine characterises Penfolds contemporary winemaking philosophy. Bin 28, typically sourced from premium vineyards in the Barossa Valley, McLaren Vale, Clare Valley, Langhorne Creek and Padthaway, is a vinous jigsaw puzzle where the winemaker's craft and empathy allows each parcel to blend together harmoniously.

Magill Estate, Adelaide – Penfolds Kalimna Bin 28 Shiraz

Jo Burzynska (JB), Jane Faulkner (JF), James Halliday (JH), Huon Hooke (HH), Ralph Kyte-Powell (RKP), Tony Love (TL), Tyson Stelzer (TS); Peter Gago, Steve Lienert, John Bird, Jamie Sach, Andrew Caillard.

A remarkable line-up of vintages. An early to medium drinking style that punches above its weight in the market. The best vintages, especially under screwcap, have an impressive lasting quality.

1963–1969

Faded old grandparents that have developed beyond their optimum drinking window. 'The waves of fruit have left the rocky shore!' (TL) 1964 was still in good condition, but at full maturity.

1970–1989

More consistent and 'joyous' but still variable. 'These wines show the inevitable variation of old bottles under cork, and the highlights and lowlights may even exchange places from different bottles on a different day.' (JH) 1971, 1977, 1978, 1986 and 1989 were highlights.

1990–1999

'A very strong line-up of wines. Youthful, vibrant and stonkingly good.' (TS) 'After fifteen years of bottle age, Bin 28 is at the height of its powers.' (JF) Best vintages 1990, 1991, 1994 and 1999.

2000–2010

'A very consistent decade; a function of youth, fruit quality and amplitude. Star anise, blackberry, chocolate and spicy notes frequently encountered. Plush voluptuous fruit with lovely tannin finesse.' (HH) Highlights include 2002, 2006, 2008 and 2010.

Bin 128 Coonawarra Shiraz

FIRST VINTAGE	1962.
VARIETY	Shiraz.
ORIGIN	Coonawarra.
FERMENTATION	Stainless steel tanks with wooden header boards.
MATURATION	12 months in new (20%) and aged (80%) French hogsheads. The changeover from American to French oak was progressive; it began in 1979 and was completed by the 1983 vintage.
COMMENTS	Screwcaps are available in selected markets. In 1981 the wine's name changed from Bin 128 Coonawarra Claret to Bin 128 Coonawarra Shiraz.

Bin 128 Coonawarra Shiraz is one of Australia's most recognised wines, with a popular profile in both the primary and secondary wine markets. The wine was first made in 1962, just a few years after Penfolds invested in the region. The style embodies the evolutionary vineyard management and winemaking practices over the last fifty years. Consistent and reliable, Bin 128 exemplifies the perfume, transparency and seductive nature of cool-climate Coonawarra shiraz. Although it is made for early to medium drinking, the best vintages can last for decades.

Bin 128 counterpoints the more opulent and richly concentrated warm-climate Kalimna Bin 28 Shiraz style. Collectors love the contrast in style, often buying a case of each!
— Peter Gago, Chief Winemaker.

The Coonawarra Fruit Colony was first established in 1890 when the region was settled by 'blockers' who planted orchards and vines. With constant labour shortages and poor access to markets, the early vignerons struggled to survive. Bill Redman, one of Coonawarra's great wine men, observed, 'from 1890 to 1945 you can write failure across the face of Coonawarra'.

Although still isolated, the region began attracting investment capital after the release of the famed Woodley's Treasure Chest Series (1949–56) and the 'freak' 1955 Wynns Coonawarra Estate Michael Shiraz. Penfolds was a late Coonawarra entrant, because Max Schubert preferred more richly concentrated, warm-climate fruit, and was unsure of the region's potential. It was Bill Redman who eventually persuaded him to purchase the Redman family's Sharam's Block.

Throughout the 1960s Max Schubert experimented with Coonawarra fruit. His trials included 'on' and 'off' blends, comprising various percentages of Coonawarra grape varieties. The wine show successes of 1962 Bin 60A, 1967 Bin 7 and 1966 Bin 620, and other cabernet-dominant wines, inevitably led to an increase in plantings of cabernet sauvignon. Nonetheless, Bin 128 Coonawarra Shiraz still found favour with wine drinkers and critics. Max Lake, a leading wine author of the time, used to tell people that Coonawarra Shiraz had many similarities to Cabernet, with its cassis-like aromas and cedar-like tannins.

One wine critic who was particularly impressed with the initial vintages was Len Evans, who once described the 1966 Bin 128, in its early years, as:

one of the top young reds of the last couple of years; the nose is very full and quite extraordinary for the straight Shiraz that it is. The fruit is so very big yet full of that delightfully austere Coonawarra character. There is evidence of small oak maturation; this is probably being with American oak. There is little doubt that Penfolds are really masters of this kind of treatment. Unlike many reds being made today for the demanding market, there is no suggestion of thinness before the finish asserts. I feel this wine will go on for many years. It could be eight to ten years before it shows its best.

Although past its optimum drinking window, this same wine still possessed good depth and substantial grip in the most recent review of Bin 128.

In the early years of Penfolds move to Coonawarra, systems were primitive. During the 1960s, the grapes were crushed with the aid of an old car engine in a tin shed and the wine pumps were operated with steam. In the 1970s mechanisation overcame labour shortages and brought new efficiencies. The vineyards were mechanised and minimal pruning was introduced. The results were outstanding initially, but the architecture of the vines was eventually compromised. Over the last few decades Penfolds has rectified this by investing heavily and reconfiguring its Coonawarra vineyards: initiating better canopy control, improving block management and adopting new data-gathering technologies. The winemakers and viticulturalists are also walking more regularly through the vineyards to monitor ripening.

By the 1970s, Penfolds gradually adjusted its barrel maturation programme to match the pastille-like fruit profile of Coonawarra shiraz and to further differentiate Bin 128 from Bin 28 and Bin 389. A gradual changeover from American to French oak between 1979 and 1983 saw a marked improvement in its style.

Coonawarra is enjoying a modern renaissance. This unique *terroir*, with its cigar-shaped stretch of *terra rossa* soils, is now realising its true potential. The small Coonawarra community of winemakers, through trial and error over the last fifty years, has nurtured and created one of Australia's great wine regions. Bin 128 Shiraz is just one small part of that journey. In 2005 it was classified 'Distinguished' by Langton's Classification of Australian Wine, for its enduring quality and popularity with wine collectors.

Bin 128 has never been better. The fruit coming out of Coonawarra is sensational. Our aim is to preserve the character of the vintage by minimal interference and matching the oak maturation to the style.
— Andrew Baldwin, Penfolds red winemaker.

After vinification in stainless steel, the wine is transferred into new and used French oak hogsheads to finish fermentation. It is then matured in approximately 20 per cent new, the remainder one- and two-year-old tightly grained French oak for about twelve months prior to bottling. Typically, Bin 128 Shiraz is deep in colour with intense liquorice, red fruit aromas and sage/herb garden notes. The style is buoyant and juicy with loose-knit tannins. A small measure of oak adds roundness and complexity. Although Bin 128 is best to drink when relatively young, recognised vintages can develop for fifteen years and beyond.

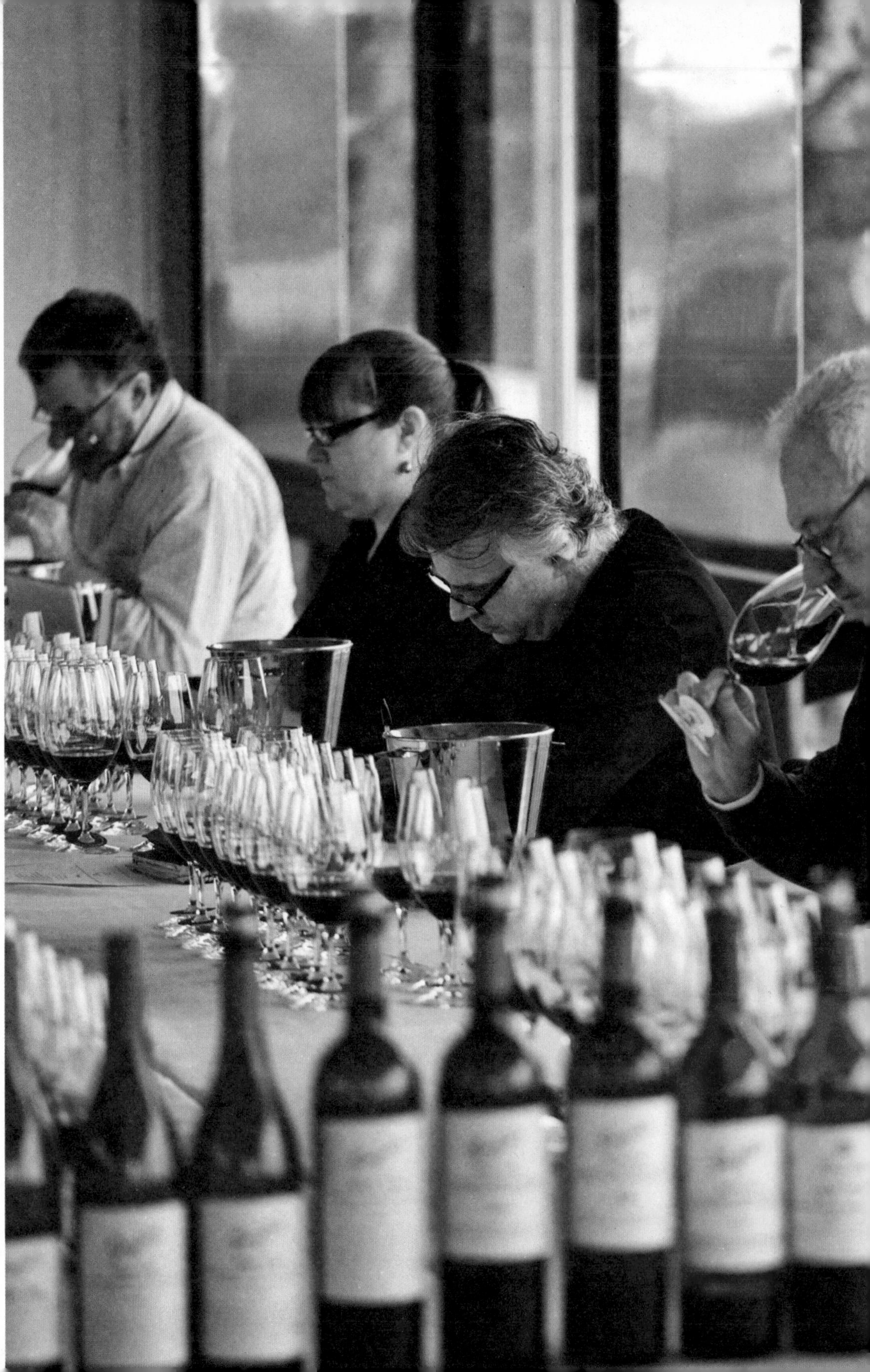

Magill Estate, Adelaide – Penfolds Bin 128 Coonawarra Shiraz
Huon Hooke (HH); Peter Gago, Steve Lienert, John Bird, Andrew Caillard.

Bin 128 is a rich, fuller-bodied style that reflects our focus on vineyard management, selection and oak handling. This is Coonawarra Shiraz with the full Penfolds attention to detail. — Peter Gago.

1963–1978

The older wines are looking fresh, but experience suggests that most bottles are past. The wines from the 1970s are beginning to dry out.

1980–1989

The ups and downs during this decade are mostly attributed to evolving vineyard management and philosophies of minimal pruning and shading of fruit. The style was also changed with the gradual introduction of French oak. Up to the late 1970s the wines were stored in large old American oak.

1990–1999

The wines have more body and flesh. The fruit rarely has any minty or peppery notes. The structure has improved and there is no obvious oakiness. Best years are 1994, 1996 and 1998.

2000–2010

The wines show more freshness, buoyancy, fruit concentration and tannin presence. This is one of the few Bin wines where the primary fruit sourcing has not changed since first vintage. 'The wines reach a new level of richness and power in 2004 and consistently perform.' (HH) The wines generally reach their optimum around 5–12 years of age. Highlights are 2002, 2004, 2006, 2008 and 2010. 'One of the surprises of this tasting was the perceived longevity of the Bin 128 style,' Peter Gago.

Note: 1962, 1964, 1965, 1967, 1968, 1969, 1977, 1979, 1984 and 1987 were not tasted.

Bin 138 Barossa Valley Grenache Shiraz Mataro

FIRST VINTAGE	1992.
VARIETIES	Grenache, shiraz and mourvèdre (mataro), in varying proportions depending on vintage conditions.
ORIGIN	Barossa Valley.
FERMENTATION	Stainless steel tanks with wooden header boards.
MATURATION	12–15 months in older (often more than 6 years) seasoned oak hogsheads.
COMMENTS	Sourced from several vineyards. Vine age varies, from 40 to 110 years. Since 1998 the vintage elevated to Bin range from 'Barossa Valley Old Vine'. Since 2010 vintage, released as Grenache Shiraz Mataro.

The only Penfolds red (other than the pinot noirs) left in component form during barrel maturation, largely because of picking dates and the fickle nature of grenache.
— Peter Gago.

Penfolds Bin 138 Barossa Valley Grenache Shiraz Mataro (Mourvèdre) is a traditional regional blend based on old vines planted as far back as 1895. It was first released in response to an increasing international interest in regional wines and the resurgence in popularity of the Barossa during the early 1990s. The first vintages (1992–1997) were released as 'Old Vine Grenache Shiraz Mourvèdre'. The heady perfume, juicy texture and abundant fruit of grenache is tempered by the inky-coloured, chocolatey smooth texture of shiraz and the spicy top notes and savoury tannins of mourvèdre, also known as mataro. Bin 138 is particularly versatile because it can be enjoyed as a youngster or as a medium-term wine.

The contribution of each variety is sotto voce *yet the wine evokes the richness and generosity of the Barossa landscape.*
— Peter Gago, Chief Winemaker.

Bin 138 fruit is sourced from low-yielding vineyards on the north-western fringe of the Barossa Valley, around Greenock, Kalimna, Moppa, Ebenezer, Marananga and Stonewell. The age of the vines varies from around 40 to 110 years old. This includes venerated dry-grown low-yielding bush vine material. At the turn of the twentieth century, many of these old vineyards provided fruit for fortified wine production or bulk table wine export markets. It is a combination of colonial vision and happenstance that grenache, shiraz and mataro (as mourvèdre was then called) were planted in the Barossa. These varieties form the backbone of the region's vinicultural heritage.

Grenache, shiraz and mourvèdre ripen at different stages during vintage and picking dates between the first batches of shiraz and the last parcels of mourvèdre can be as much as two months. Batch vinification takes place in open stainless steel fermenters followed by maturation for around twelve to fifteen months in seasoned (as old as six years) American and French hogsheads. The various components are blended together just prior to bottling.

If I was able to sleep in a Penfolds Bin, it would be Bin 138!
— Tony Love, wine writer, *The Advertiser*, Adelaide.

Bin 138 is typically fragrant and minerally, with musky plum, dark chocolate, star anise aromas, juicy raspberry pastille, plum, liquorice flavours, lacy fine tannins and underlying spicy notes. Its youthful exuberance, generosity of flavours and soft structure make it ideal to drink now or soon. With further cellaring it can develop more fruit complexity and interesting maturation characters.

Panel Review

Magill Estate, Adelaide – Penfolds Bin 138 Grenache Shiraz Mourvèdre
Jo Burzynska (JB), Jane Faulkner (JF), James Halliday (JH), Huon Hooke (HH), Ralph Kyte-Powell (RKP), Tony Love (TL), Tyson Stelzer (TS); Peter Gago, Steve Lienert, John Bird, Andrew Caillard.

'Gastronomic guzzling wines' best enjoyed in the short to medium term.
Develops savoury characters with age.

1992–2010

An attractive early drinking style, best enjoyed before it reaches its first decade. A stylistic jump in 2005 results in brighter fruit and better-integrated tannins. The more recent vintages are generous, immediate and hedonistic. Highlights are 2006, 2008, 2009 and 2010.

Bin 23 Adelaide Hills Pinot Noir

FIRST VINTAGE	2009.
VARIETY	Pinot noir.
ORIGIN	Adelaide Hills.
FERMENTATION	Cold soaked and naturally fermented in open fermenters.
MATURATION	12 months in seasoned French oak barriques.
COMMENTS	Named after Maturation Cellar 23 at Magill Estate. A barrel by barrel selection.

Penfolds Bin 23 Adelaide Hills Pinot Noir is sourced from around twenty vineyards scattered around the elevated slopes of the Adelaide Hills. The Bin number is derived from Maturation Cellar 23 at the historic Penfolds Magill Estate, where the wine is vinified. Chief Winemaker Peter Gago says, 'This is unashamedly a Penfolds wine. No attempt has been made to create a Burgundian look-alike.' The wine is made in a rich, voluminous style with intense aromatics, plenty of fruit density and tannins. Although it is made for early drinking, it does have medium-term cellaring potential.

After light crushing, the must is cold soaked and batch-vinified in small open fermenters. The free-run wine is then racked into a combination of new (around 35 per cent) and seasoned French oak barriques for around ten months before bottling. No pressings are added to the wine because optimum colour and flavour extraction have already been achieved. Some components finish fermentation in barrel to optimise the blending options.

The wine is a distinctive style especially made for Penfolds drinkers who enjoy rich, mouth-filling wines with plenty of fruit generosity and flavour length. Typically it is saturated in colour, with cherry/brambly aromas, plenty of juicy fruit and chocolatey tannins, and underlying new oak.

 *Panel Review*

Magill Estate, Adelaide – Penfolds Bin 23 Adelaide Hills Pinot Noir
Huon Hooke (HH); Peter Gago, Steve Lienert, John Bird, Andrew Caillard.

This wine is not for the fainthearted! — Peter Gago.

2009–2011

A new addition to the Penfolds Bin range. Freshness, ripe red fruits, buoyancy and noticeable tannin presence are a common thread. Best to drink early.

Bin 2 Shiraz Mourvèdre

FIRST VINTAGE	1960.
VARIETIES	Shiraz and mourvèdre (mataro).
ORIGIN	Multi-district blend, South Australia/Victoria, including Barossa Valley, Clare Valley, McLaren Vale and Langhorne Creek.
FERMENTATION	Stainless steel fermenters.
MATURATION	12 months' maturation in 5- or 6-year-old oak hogsheads and larger vats.
COMMENTS	Originally labelled Shiraz Mataro. Reintroduced to the Australian market in 2004.

Bin 2 Shiraz Mourvèdre is a classic Penfolds wine, steeped in nineteenth-century tradition. Many of the early Penfolds 'Burgundy' dry reds were shiraz mataro (mourvèdre) blends and won medals at colonial wine shows. 1956 Bin 136 Magill Burgundy, also bottled as Bin S56, is perhaps the most famous post-war bottling; it was originally designated as blending material for a special Qantas bottling of shiraz mataro.

Bin 2 was developed during the 1950s and first released in 1960. It was discontinued during the 1970s at the height of the white wine boom and reintroduced to the Australian wine market in the early 1980s for a brief period. For many years it was an export-only wine. Since the release of 2004 Bin 2 Shiraz Mourvèdre, the style has once again found a ready market. Typically it is a generous medium-bodied wine based on shiraz, but shot with the aromatics and muscular strength of mourvèdre. It's best to drink early, but like many of Penfolds reds can last the distance in the cellar.

The contemporary Bin 2 style is based on warmer vineyard sites including the Barossa Valley (together with the Kalimna Vineyard), Clare Valley, McLaren Vale and Langhorne Creek. The wine is vinified in stainless steel and aged in five- to six-year-old hogsheads and larger vats. This allows the maturation effect to further evolve and soften the wine.

Panel Review

Magill Estate, Adelaide – Penfolds Bin 2 Shiraz Mourvèdre
Huon Hooke (HH); Peter Gago, Steve Lienert, John Bird, Andrew Caillard.

Dark fruits, spices and tannin backbone. — Huon Hooke.

1966, 2004–2010

A medium-bodied style with a stronger emphasis on Barossa Valley, McLaren Vale and Clare Valley fruit in recent vintages. Fruit character, fullness, richness and generosity together with an assertive tannin backbone are hallmarks. 2008 and 2010 are top vintages.

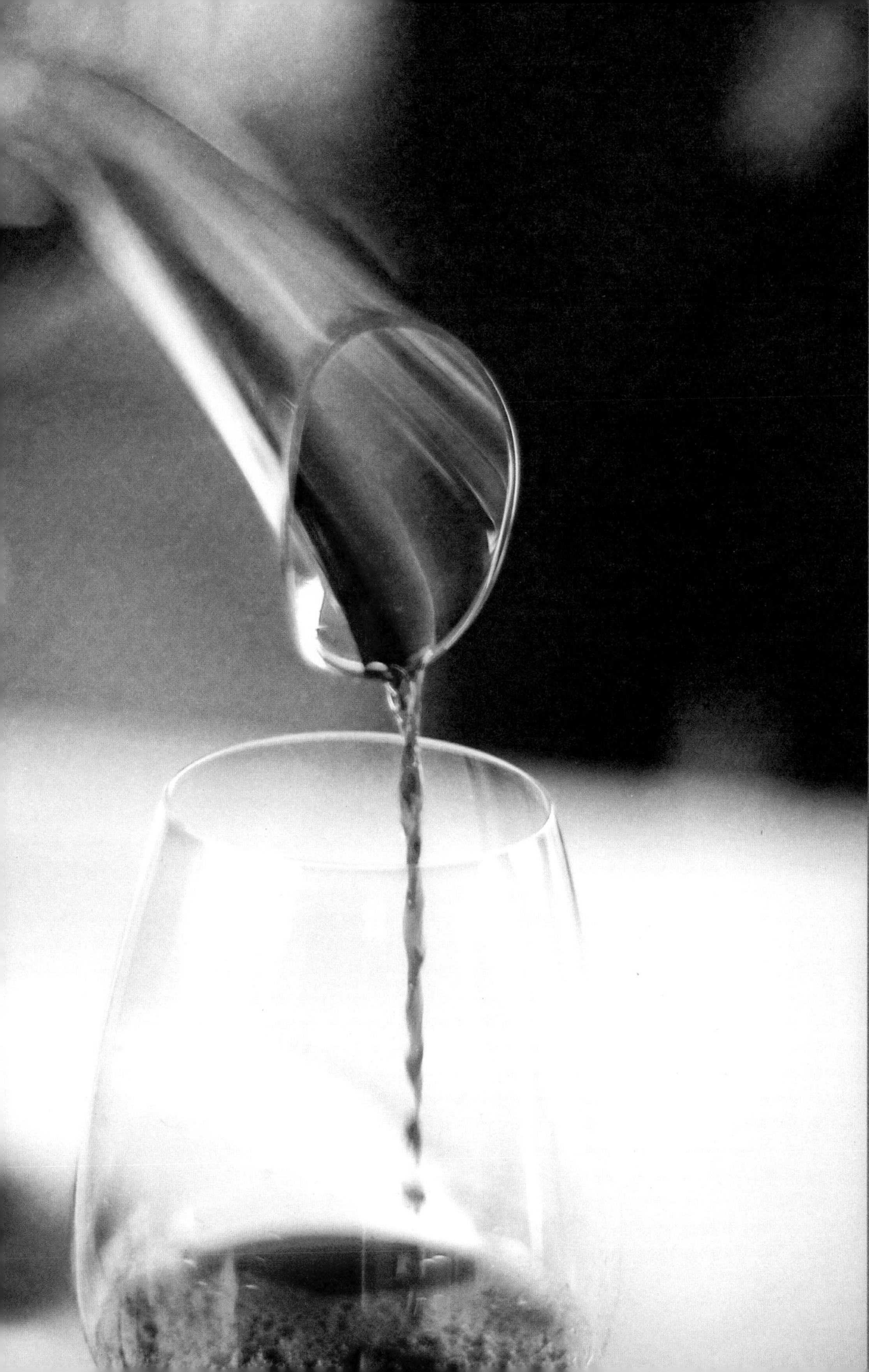

Bin 8 Cabernet Shiraz

FIRST VINTAGE	2003.
VARIETIES	Cabernet sauvignon and shiraz.
ORIGIN	A multi-district blend drawing fruit from vineyards throughout South Australia including Robe, Bordertown, Coonawarra, Barossa Valley and Riverland.
FERMENTATION	Stainless steel fermenters with wooden header boards.
MATURATION	12 months in a combination of 2- and 3-year-old French and American oak hogsheads.
COMMENTS	An early drinking style which will keep for a few years but best soon after release.

Penfolds Bin 8 Cabernet Shiraz is an early drinking style. Typically it is fresh and vivacious with ripe black fruits, plenty of richness and supple velvety tannins. The magical combination of cabernet sauvignon and shiraz brings extra fruit complexity, buoyancy, smoothness and minerally freshness.

Bin 8 is crafted in the traditional Penfolds style. The wine is vinified in headed-down open stainless steel fermenters and then aged for around twelve months in two- and three-year-old French and American oak hogsheads, previously used for Bin 389, Bin 128 and Bin 28.

Panel Review

Penfolds Bin 8 Cabernet Shiraz

An early drinking style and best consumed within a few years of release.

Ripe, balanced fruit flavours are supported by softly integrated oak, resulting in a silky, smooth red wine. If stored in a cool cellar or cabinet, it may keep for a longer time. Not reviewed by tasting panel.

Bin 311 Chardonnay

FIRST VINTAGE	2005.
VARIETY	Chardonnay.
ORIGIN	A cool-climate single-region style usually from Tumbarumba, NSW, with the exception of Orange, NSW (2007) and Henty, Victoria (2011).
FERMENTATION	100% barrel fermented in seasoned old French oak barriques.
MATURATION	Matured for around 7–9 months in seasoned old French oak barriques.
COMMENTS	Sealed with screwcap in all markets.

Bin 311 Chardonnay is a cutting-edge style that reflects the ongoing experimentation and developmental work of Yattarna and the Bin A series. This offshoot wine comprises declassified parcels of A1 graded chardonnay grapes and seasoned oak previously used for Yattarna. As a consequence Bin 311 has established a reputation similar to Bin 389 and represents one of the very best value ultra-fine Australian Chardonnays available. Typically it is fresh and crunchy with fragrant white peach, pear skin aromas, flinty complexity, creamy textures and long indelible acidity.

High altitude cool-climate fruit from Tumbarumba, on the southern slopes of the Snowy Mountains in New South Wales, is the main source material for Bin 311. Orange, NSW, and Henty, Victoria, have also featured in some vintages, reflecting the winemakers' mantra 'we always go where the fruit grows best and where it best suits style'.

Penfolds has developed a hallmark chardonnay style that is as easily recognisable as its famed red wines. Kym Schroeter, senior white winemaker, has introduced exquisite refinements to Bin 311 over the last five years, bringing extra fruit complexity, freshness and vinosity. His exacting standards and empathy with chardonnay are widely recognised.

Bin 311 Chardonnay is barrel fermented and then matured for around seven to nine months in seasoned French oak barriques, previously used for the maturation of Yattarna. *Battonage*, or lees stirring, is regularly carried out to build up overall palate richness and flavour complexity. Kym Schroeter says, 'This is a fresh flinty style with zippy freshness and creamy textures.' Sealed under screwcap, to preserve freshness, this wine is delicious to drink now, or it can be cellared for a while.

Panel Review

Magill Estate, Adelaide – Penfolds Bin 311 Chardonnay
Jo Burzynska (JB), Jane Faulkner (JF), James Halliday (JH), Huon Hooke (HH), Ralph Kyte-Powell (RKP), Tony Love (TL), Tyson Stelzer (TS); Peter Gago, Steve Lienert, John Bird, Kym Schroeter, Andrew Caillard.

Wines of flinty freshness and mineral intensity. — Jo Burzynska.

2005–2011

'An accurate translation of Tumbarumba's cool-climate environment with its minerally notes, chalky structure and textural presence.' (TS) A light- to medium-bodied style with an emphasis on fruit complexity, energy, persistence, 'line and length'. Highlights included 2008, 2009 and 2010.

Bin 51 Eden Valley Riesling

FIRST VINTAGE	1999 – originally labelled Eden Valley Reserve Bin Riesling.
VARIETY	Riesling.
ORIGIN	Eden Valley, South Australia.
FERMENTATION	Stainless steel tanks. Bottled immediately post-vintage to retain freshness and vibrancy.
MATURATION	Bottled early and released towards the end of vintage year.
COMMENTS	Bin 51 from 2006 vintage onwards. Screwcaps introduced in 2002.

The Bin 51 Eden Valley Riesling, with its pristine fruit quality, natural energy and fine acid backbone, is a classic regional style with the structure and concentration to age for the medium to long term. The fruit is primarily sourced from prime Penfolds-managed Woodbury and High Eden vineyards in the elevated Eden Valley, a sub-region of the Barossa.

Bin 51 is typically fragrant with intense lime/rose petal aromas, pure fruit flavours and fine cutting acidity. It develops complexity and richness with age, without losing clarity, freshness or linear acid structure. The paradox of bone dryness and fruit sweetness brings an extra dimension and appeal. This is one of Eden Valley's most beautiful rieslings.

Eden Valley's soils are derived from schistic and sedimentary rock and are mostly red clay and sandy, silty loams interspersed with gravels. The fruit quality from this region is exceptional. Vineyard parcels are batch-vinified in steel tanks and bottled early under screwcaps to preserve the pristine fruit characters, freshness and natural mineral quality. Since the introduction of screwcap in 2002, Bin 51 has become a much better cellaring option. The wine ages gracefully and evenly, without losing freshness or vigour.

 Panel Review

Magill Estate, Adelaide – Penfolds Bin 51 Eden Valley Riesling
Huon Hooke (HH); Peter Gago, Steve Lienert, John Bird, Andrew Caillard.

Power, delicacy and longevity. — Huon Hooke.

1999–2012

The youngest wines possess an extra degree of citrus blossom/floral notes. A recurring theme of lemon/lime aromas and flavours with dryness, softness and finesse. Stylish and consistent. Highlights include 2002, 2004, 2005, 2008 and 2010.

'No guilt, no apologies. Koonunga Hill Shiraz Cabernet
is a real Penfolds red wine.'

PETER GAGO

KOONUNGA HILL
Chapter nine

Koonunga Hill Shiraz Cabernet

FIRST VINTAGE	1976.
VARIETIES	Shiraz and cabernet sauvignon. The proportions vary but shiraz is always the dominant variety.
ORIGIN	Multi-regional blend, South Australia, from Barossa Valley, McLaren Vale, Clare Valley, Coonawarra, Padthaway, Langhorne Creek and Bordertown. May vary considerably depending on vintage conditions.
FERMENTATION	Stainless steel fermenters.
MATURATION	12 months' maturation in 3- to 4-year-old oak hogsheads.
COMMENTS	Released and labelled as Koonunga Hill Claret until 1991 vintage. A drink-now style with a reputation for ageing. Screwcap was introduced in 2003 in some markets. The 1999 was voted Great Value Red Wine of the Year at the London International Wine Challenge (2001) and the 2002 was listed as one of the top 50 New World wines by *Decanter* magazine.

Cast it over Broadway, spread it over New York City, that John Webber would not leave a wine cellar!
— American actor/comedian John Webber, on his reluctance to leave Penfolds cellars at Magill in 1908.

The Penfolds wine experience begins with Koonunga Hill. This entry-level brand has established a world-wide reputation for value, quality and consistency. When the 1976 Koonunga Hill Shiraz Cabernet was first released, it created a tradition and expectations for the future well beyond its modest price point – it was first offered to customers at around $1.69 a bottle. Nowadays this vintage can fetch over $250 at auction, illustrating the strength of its reputation with wine collectors and drinkers. The Koonunga Hill range is all about freshness, generosity of flavour, richness and vivacity of fruit. These are affordable and great every-day drinking styles that reflect Penfolds remarkable vineyard resources and outstanding red and white winemaking skills.

The Koonunga Hill brand is linked to its South Australian origins. The Koonunga District is on the northern edge of the Barossa Valley and 'Koonunga Station', established in 1841, was one of the region's earliest colonial properties. 'Koonunga' probably means 'good shelter' in the local Aboriginal dialect.

During the early 1960s an automatic telephone exchange was established in the area. A 'Hill' was added to avoid confusion with other place names. Penfolds Koonunga Hill Vineyard was planted on local grazing land in 1973.

No guilt, no apologies. Koonunga Hill Shiraz Cabernet is a real Penfolds red wine.
— Peter Gago, New York, June 2012.

When the legendary 1976 Koonunga Hill Claret was first released into the Australian market it created a sensation. With its hallmarks of 'elegance and power' and strong Penfolds winemaking stamp, it offered buyers a 'real Penfolds red wine' at a working man's prices. In a depressed wine market, it sold like hotcakes. The savviest wine collectors, recognising its superb richness, balance and lasting quality, bought dozens to keep as an everyday wine and to cellar for the future. Over the last thirty years this 'very expressive' claret-style wine has endured well beyond expectations. Made for early drinking, it has outlasted and outgunned some of the world's most expensive wines. Although the rare 1976 Koonunga Hill Claret nowadays achieves over fifty times its original release price at auction, new releases still offer a relatively inexpensive but hugely engaging Penfolds red wine experience.

Koonunga Hill is a bellwether wine. It is where the art of winemaking and the Penfolds journey truly begins. Peter Gago says, 'The emotional investment in Koonunga Hill brings credibility and momentum to the Penfolds brand. We measure ourselves as winemakers by the overall success of this wine.' Its fruit sourcing, vinification and maturation are given meticulous attention. Even at this level winemakers walk through the vineyards batching fruit for the Koonunga Hill harvest.

With widespread vineyard resources across South Australia it is possible, with extreme vigilance and 'a lot of walking', to find exceptional parcels of premium fruit. The approach to selection and winemaking for the Koonunga Hill range is identical to Penfolds Bin wines. The difference ultimately lies in the scale of production, complexity of fruit sourcing and maturation.

Koonunga Hill is typically a shiraz–cabernet sauvignon blend with deep colour, ripe black fruit aromas, voluminous richness and plentiful tannins. Sourced primarily from the Barossa, Coonawarra, Robe and Clare Valley, it offers a compelling vinosity and freshness, usually associated with ultra-fine Australian wine. Peter Gago says, 'Koonunga Hill

and Grange are bookends of the Penfolds experience. Although completely different in style and quality, there is a common signature.'

On a glorious early summer's morning on the rooftop of New York's Mondrian Hotel in the swanky bohemian district of Soho, a group of wine journalists from Canada, the US and South America assembled to taste a complete vertical of Koonunga Hill Shiraz Cabernet. The previous day they had reviewed six decades of Grange, the most significant tasting of this wine ever. The atmosphere was relaxed and possibly resigned. Was this really a penance for yesterday's extraordinary cultural experience?

Penfolds ambassador Jamie Sach dexterously unfoiled, uncorked and checked old bottles while an army of sommeliers and helpers poured samples of Koonunga Hill into vintage-tagged Riedel wine glasses. Australia's most famous wine photographer, Milton Wordley, flitted and stretched across the light-filled room, like a remote-controlled fly. The whirring and chirping of his camera shutter and the shuffling of feet gradually dissipated as the room fell into a rhythm of thought and introspection. Thirty-five vintages is half a lifetime, a generation and five years of teamwork. Each glass with its vermilion crimson robe and ever-changing scent tells a story. It captures a single growing season and all the complexities and beauty of South Australia's ancient landscape. And here in New York the aromas of this faraway and strange countryside waft through the air like released spirits. 'How will they feel about our beloved Koonunga Hill?'

Penfolds senior winemaker Steve Lienert first started making wine when Koonunga Hill was just a fledgling brand. 'When Max Schubert retired in 1976, Don Ditter took over winemaking. At the time the Australian wine market was suffering from a post-red wine boom with downward price pressure on our Bin wines. Koonunga Hill was introduced to protect the integrity of our Bin range and to meet the market head on! As far as we were concerned Koonunga Hill was a Bin wine and we would do everything within the resources provided to make something special and different.' However, within the context of the international market, is it just another commercial wine?

A tasting of this type can seriously backfire, if the wines fail to meet expectations. Fortunately the performance of Koonunga Hill is well documented through previous editions of *Penfolds The Rewards of Patience* and the Penfolds Re-corking Clinics. The mature bracket of Koonunga Hill Shiraz Cabernet (1976–1979) impressed the panel. Prominent

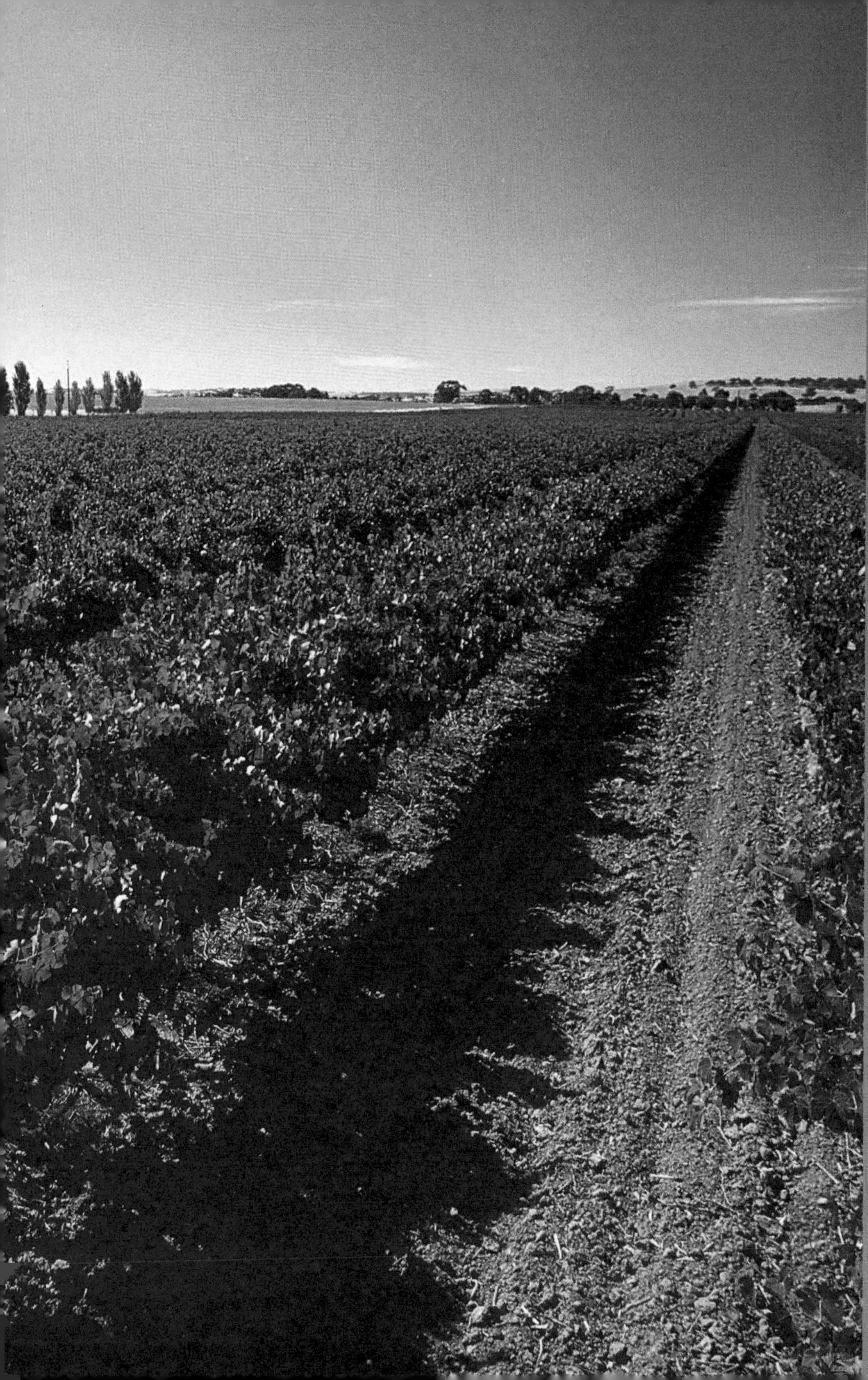

Prominent American wine writer Josh Raynolds was 'pleasantly surprised' by the Médoc-like structure of the best vintages: 'It lives up to its original claret moniker. The 1979, for instance, is very perfumed and expressive.' Canadian writer and broadcaster Tony Gismondi observed, 'These old Koonunga Hills are shockingly fine to drink! Four-square and sturdy, harmonious and generous, they deliver with some oomph! Although the 1976 has reached a tipping point it's still outstanding.'

At vintage time the fruit arrives at Penfolds Nuriootpa winery in South Australia's Barossa Valley on schedule. Penfolds winemakers have already prepared the cellar and fermentation tanks for vinification. The picking times, always weather-dependent, have been finalised. Most of the grapes are harvested by machine overnight to preserve freshness, flavour and acidity. Shiraz arrives earliest as it ripens before cabernet sauvignon.

Vintage time is frenetic and every step is organised to optimise intake and quality. During March and early April Penfolds is like a busy airport, with trucks arriving throughout night and day. The freshly picked grapes from nearby vineyards are unloaded into stainless steel receival bins and then gently moved along by Archimedes screw into destemmers and crushers. Fruit from more far-flung wine regions – such as Robe and Bordertown in South Australia's far eastern corner – are crushed and chilled where they are picked, to produce 'must', then are tankered up overnight to the Barossa. Every batch is matched to vineyard source and analysed by the Penfolds winemaking team.

In the 1980–1989 Koonunga Hill bracket, US wine correspondent Linda Murphy observed, 'a progression of richness and depth'. Brazilian publisher Marcelo Copello said, 'the wines are ageing with dignity. The 1980s are rich and well concentrated with lively tannins.' *Washington Post* wine writer Dave McIntyre found that many vintages possessed 'a silky texture, almost like Egyptian cotton'. Joe Czerwinski found the wines had staying power, while Tony Gismondi described the 1986 as being the best in the group: 'It had a sweet spine with harmonious fruit and grippy tannins at the very end.' The panel all agreed that it would probably go down the same path as the 1976. This of course harks back to Don Ditter's comment, 'there's never been such a thing as a crook Koonunga Hill. It's always a reliable drop, but in top years you can expect something out of the ordinary.'

Koonunga Hill is a winemaker's jigsaw puzzle. Every vintage brings new challenges, but experience and knowledge of vineyard parcels ensures a level of certainty. Although the growing season can throw up unexpected events, such as weather extremes and disease pressure, vineyard management and constant vigilance can overcome potential difficulties. Even the notoriously complicated 2011 vintage, the wettest growing season in forty years, yielded consistent results.

Koonunga Hill Shiraz Cabernet is a classic South Australian blend and reflects the original aims of Max Schubert. He discovered through his experimental winemaking that shiraz and cabernet sauvignon were highly compatible varieties. During the late 1940s he would have observed that some Bordeaux negociants, or winemakers, used syrah – the European name for the shiraz variety – as a blending component before the *appellation contrôlée* rules were tightened up. This was called 'hermitaging'.

Schubert's winemakers also believed that rigorous selection and blending of shiraz and cabernet sauvignon from multi-regional sites could bring consistency and continuity of supply, regardless of most vintage conditions. Koonunga Hill is a product of this theory. In general the shiraz component brings ripeness, volume and sweetness of fruit. Cabernet sauvignon offers fragrance, further fruit complexity and a rigid tannin base. Melded and matured in seasoned old oak hogsheads for a year, the combination has an extra magic and richness.

The introduction of a comprehensive vineyard classification, computerised tracking system and membrane bag presses revolutionised winemaking during the 1990s. Peter Gago says, 'Koonunga Hill is very much a scaled-up boutique wine. The hand of the winemaker is evident at every stage, from the vineyard to final bottling. Batch vinification in stainless steel fermenters, maturation in seasoned oak, meticulous classification and blending all contribute to the overall consistency and quality of Koonunga Hill Shiraz Cabernet.'

The 1990–1999 Koonunga Hill bracket impressed the panel. The wines are still showing primary fruit, a touch of spice and 'underlying cabernet tannin backbone'. Dave McIntyre went further: 'This decade is Koonunga Hill's sweet spot. The flight really impressed me. There is a textural sensation reminiscent of Grange. I am forgetting that these wines are made for early drinking!'

'The wines even smell far more expensive than they really are!' says wine correspondent and columnist Linda Murphy. 'I really appreciated their gorgeous generous quality.' Josh Raynolds described the 1990s decade as 'effortless', and added, 'These wines are so easy to drink with their red fruit flavours, silky textures and refreshing acidity at the finish.'

1991 Koonunga Hill, now in its third decade, was a highlight, with its 'earthy dark fruit, sweet spice and soft tannins'. 1998 was also a star performer with its 'ripe fruit and fresh grilled toast complexity'. Importantly, every vintage from the 1990s had its supporter, illustrating the overall consistency and deliciousness of the wines.

The 2000–2010 bracket illustrated a further evolution of the Koonunga Hill style. The fruit characters are typically 'bright and pure', reflecting both their youth and the significant progress in vineyard sourcing and management. Joe Czerwinski said, 'Most wines in this price category are boring corporate concoctions with jammy fruit and lack of character. These 2000s possess a lovely brightness and immediacy. The wines are easy to drink now, but they have a vinosity and energy to further develop.' Tony Gismondi observed, 'All of these wines over-deliver in quality. Koonunga Hill is a case of smart people making smart wine. There are few other commercial wines in the world that can offer consumers such a black hole of choice.'

Screwcap closures, in this bracket, have clearly played a role in retaining the freshness and potential longevity of vintages. Even wines nearly ten years old showed flamboyancy and richness of primary fruit. There was a feeling of real excitement about the 2000s. The 2004 was singled out as a favourite: 'A big opulent flamboyant wine showcasing plenty of blue fruits and fine savoury tannins', and 'The garrigue/dried herb characters bounce off the sweet fruit beautifully. This classic Penfolds wine will cellar forever.'

2010 Koonunga Hill, with its 'inky violet colour, superb cassis, liquorice aromas and luxurious chocolatey texture' is still elemental and powerful, young and expressive. Is 2010 Koonunga Hill Shiraz Cabernet the best ever produced?

Freshness, balance and concentration are hallmarks of all good wine. Many of the world's greatest vintages are delicious to drink young because all the elements are in harmony. However, with patience, these wines can develop into something quite ethereal and magical. The success of Koonunga Hill Shiraz Cabernet lies in its origins. Although named after the newly planted Koonunga Hill vineyard on the western edge of the Barossa, the fruit source has always been multi-regional. It was initiated to build up sales in a difficult market. These days it would be called a fighting brand. Except with a major difference. Harking back to the early days of Grange, the wine-making team ignored the instructions of head office and looked at making a wine that would reflect the original ambitions of Max Schubert. Koonunga Hill would be a Penfolds wine 'down to its bootstraps'. This was achieved through meticulous vineyard sourcing and blending 'borrowed material earmarked for more illustrious bin numbers including Grange!' Over the last forty years a revolution has taken place in Penfolds vineyards and winery. Precision viticulture, detailed harvest management, selection and classification, oak maturation, technology and intuition have upheld Koonunga Hill's aristocratic origins.

This rare vertical tasting, held in New York's Manhattan district amid the mahogany-coloured skyline, demonstrated a resilience and authenticity of style. Koonunga Hill Shiraz Cabernet is typically exuberant with delicious sweet fruit and an 'aggressively polite' swagger of tannins! The heady perfume, opulence and seductive flavours are surprising to those who have never tasted the wines.

In good years, it can really last the distance. Ultimately it is one of the world's most versatile wines, offering an unexpected value, quality, character and longevity. Much like New York itself.

 *Panel Review*

New York City, NY, USA – Penfolds Koonunga Hill Shiraz Cabernet
Marcelo Copello (MC), Joe Czerwinski (JC), Anthony Gismondi (TG), Ray Isle (RI), Dave McIntyre (DM), Linda Murphy (LM), Josh Raynolds (JR); Peter Gago, Steve Lienert, Andrew Caillard.

> *Is Koonunga Hill as good or age-worthy as Grange? Of course not. But they share that lively, spicy character of Australian red wine, along with one other important trait: they are refreshing.*
> — Dave McIntyre, *The Washington Post.*

> *Inspirational for the modest collector.* — Tony Gismondi, *The Vancouver Sun.*

1976–1979
Four-square and sturdy with harmonious fruit, tannin softness and roundness. All ready to drink. Highlights are 1976 and 1979.

1980–1989
A progression of richness and depth with a couple of exceptions. Reaches a new level of modernity in 1989. The best performing wines were 1982, 1983, 1986, 1988 and 1989.

1990–1999
Koonunga Hill's 'sweet spot'. Plenty of red fruit characters, suppleness and refreshing acidity across the decade. 'These wines taste far more expensive than they are.' 1990, 1991, 1996, 1998 and 1999 are all strong years.

2000–2010
A consistency of quality, approachability, brightness and immediacy. Sumptuous fruit and polished tannins. Plenty of blue fruits, palate richness and silky soft tannins. Standouts were 2004, 2006, 2008, 2009 and 2010.

Koonunga Hill *Seventy-Six* Shiraz Cabernet

FIRST VINTAGE	2006.
VARIETIES	Shiraz and cabernet sauvignon.
ORIGIN	Multi-regional blend, South Australia, from Barossa Valley, McLaren Vale and Coonawarra.
FERMENTATION	Stainless steel tanks.
MATURATION	12 months' maturation in new and seasoned old oak hogsheads.
COMMENTS	Released as an homage to the 1976 Koonunga Hill 'Claret'.

The 2006 Koonunga Hill *Seventy-Six* Shiraz Cabernet, with its retro packaging, was released as an homage wine to the famous 1976 Koonunga Hill 'Claret', a wine that captured the imagination of avid Penfolds wine enthusiasts and collectors; its extraordinary longevity and reputation is well documented. The refined style of the *Seventy-Six*, aged in a proportion of new oak hogsheads, brings another dimension and complexity to the Koonunga Hill story.

Initially the wine was only available through fine restaurants, duty-free stores and Penfolds cellar door at Magill Estate and Nuriootpa. Following a strong debut, the 2007 *Seventy-Six*, the next vintage, was entered in the Royal Sydney Wine Show where it won Gold and Trophy. Subsequent vintages have attracted critical and wine show acclaim. A prominent Australian wine retailer recently described the Koonunga Hill *Seventy-Six* style as 'pound-for-pound the best Pennies red wine'.

The *Seventy-Six* style 'is more of everything', said Josh Raynolds, 'more power, more depth, more mid-palate input, more elegance, more presence, more tannins!'

Lower-yielding fruit from the Barossa Valley, McLaren Vale and Coonawarra, South Australia's classic wine regions, combined with meticulous selection, intuitive blending and maturation in a combination of new and seasoned American oak hogsheads, give an extra level of concentration and power to the *Seventy-Six* Shiraz Cabernet style.

> *Nowadays, my pick of the lot is the Koonunga Hill Seventy-Six Shiraz Cabernet label – the 2010 current release will make a believer out of readers, if nothing else does.*
> — Lisa Perrotti-Brown, MW, erobertparker.com.

Panel Review

New York City, NY, USA – Penfolds Koonunga Hill *Seventy-Six* Shiraz Cabernet
Marcelo Copello (MC), Joe Czerwinski (JC), Anthony Gismondi (TG), Ray Isle (RI), Dave McIntyre (DM), Linda Murphy (LM), Josh Raynolds (JR); Peter Gago, Steve Lienert, Andrew Caillard.

A fresh plush style with expressive fruit, pure rich flavours and plentiful silky tannins.

2006–2010
A very strong and consistent line up of wines, with greater fruit complexity and richness than the standard Koonunga Hill. 2010 is a stunning example. Other highlights included 2008 and 2009.

Koonunga Hill Autumn Riesling

FIRST VINTAGE	2008.
VARIETY	Riesling.
ORIGIN	Barossa and Eden valleys.
FERMENTATION	Stainless steel fermenters.
MATURATION	Bottled soon after completion of fermentation, to preserve freshness and minerality.
COMMENTS	Screwcap. Retro-packaged in homage to the original 1971 Autumn Riesling release.

'Vibrant and electric.'
— Linda Murphy.

Penfolds Koonunga Hill Autumn Riesling is a 'distinctive South Australian style' with a history connected to modern winemaking and Grange. The riesling grape variety was first brought to Australia by colonial administrator James Busby during the 1830s. Cuttings were planted at the Macarthurs' vineyard at Camden in New South Wales, in the pastoral countryside south of Sydney. By 1842 *Riesling of the Rhine* 'properly prepared for planting and packed into bundles of two hundred each' were regularly advertised for sale in Sydney's *The Australian* newspaper and elsewhere. An article in *The South Australian Register* (June 1867) advised readers that 'the riesling and verdelho, when not tortured, yield wines second only to the Bucellas of Lisbon and the sweeter kinds of Madeira.' One writer of the period observed of Riesling: 'in South Australia, nature herself is opposed to the production of thin, high bouquet wine'.

The standard of dry and sweet riesling wines at the turn of the twentieth century was extremely variable. Many were cloudy, lacked freshness or possessed bitterness. The introduction of pressure fermenters and cool-fermentation techniques during the 1950s allowed winemakers to preserve riesling's natural delicacy, beautiful aromatics, pristine fruit and crisp mouth-watering acidity. Max Schubert and his team experimented with parcels of late-picked riesling (and gewürztraminer) in the hope of introducing 'Lady Grange', as a consort to Grange Hermitage. But in the early 1970s, top management ordered the project to

be stopped. The finished wines, bottled and ready to be released as 'Lady Grange', were sold as Penfolds Autumn Riesling.

The origins of Koonunga Hill Autumn Riesling are steeped in post-war innovation and experimentation. The development of late-picked dry riesling styles during the 1960s established the reputation of Eden Valley and Clare Valley as premium South Australian wine regions. Well ahead of its time, the 1971 Autumn Riesling was sealed with screwcaps to preserve freshness and longevity. Surviving bottles are still drinking very well, but it would take another thirty years before the general market accepted this type of closure in Australia.

Screwcapped, retro-packaged and stylistically true to its origins, today's Penfolds Koonunga Hill Autumn Riesling is the quintessential South Australian Riesling, possessing fresh lemon curd, lime aromas and zesty acidity. Tony Gismondi said, 'These are killer wines with pan-Asian cuisine. Freshness, dryness and texture, combined with a nice progression of age, make them appealing and quite sophisticated.' Joe Czerwinski agreed: 'Collectors would be very surprised how these Autumn Rieslings age. The 2008, for instance, is still lean and tight and still retains some greenish hues. Yet there is a touch of evolution with honey/toasty complexity.'

The American panel observed, 'Koonunga Hill Autumn Riesling is typically dry, austere and crunchy when first bottled, but develops sweetness, volume and richness with age. This is a classic South Australian style; completely different from European Rieslings but just as compelling and delicious.'

New York City, NY, USA – Penfolds Koonunga Hill Autumn Riesling

Marcelo Copello (MC), Joe Czerwinski (JC), Anthony Gismondi (TG), Ray Isle (RI), Dave McIntyre (DM), Linda Murphy (LM), Josh Raynolds (JR); Peter Gago, Steve Lienert, Andrew Caillard.

A delicious late-harvest dry Riesling with spectacular pure fruit and laser-like acidity. —— Linda Murphy.

2008–2011

A sophisticated style with spice, ginger, lime aromas with freshness, zestiness and surprising cellar potential. 'These wines build sweetness, volume and richness with age.' 2008 and 2010 were compelling examples.

'My first contact with Grandfather Port was in 1935 when
Alfred Scholz [the creator of the style] told me that it had
its genesis in the 1915 vintage.'

RAY BECKWITH

FORTIFIEDS
Chapter ten

Penfolds 'Tawny' styles continue a long tradition of fortified wine production. They are a link with yesteryear, evoking memories of summers past, vast stacks of oak barrels filled with slowly maturing tawny fortified wines and a grand wine industry heritage. Many of these beautifully concentrated and complex tawnies are living history, with components of them going back decades. These wines are the epitome of intensity, heat, richness and utter indulgence.

Penfolds first began making fortified wines as early as the 1850s. Sherry, Madeira and Oporto (port) were extremely fashionable in the nineteenth century. Fortification was a practical method of wine production as the addition of grape spirit contributes significantly to wine stability.

Port-style wines were particularly suited to the climate and frontier culture of colonial and post–World War I Australia. They dominated the wine market. Advertising in newspapers and on hoardings was particularly effective for Penfolds and examples of the slogans used include: 'Penfolds Wines are magnificent, says General O'Pinion' (circa 1914); 'Open the port, steward. Penfolds of course sir!' (circa 1920); and 'Penfolds and Sydney – The two best Ports in the world' (1928).

Penfolds Club Tawny began its life in the 1940s but was first released in 1950, as Penfolds Five Star Club Tawny Port. The term 'Port' is no longer used by Penfolds – in keeping with international standards and trade agreements. The Penfolds Tawny range comprises Club Tawny, Club Reserve Tawny, Father Grand Tawny, Grandfather Rare Tawny, Great Grandfather Rare Tawny and 50 Year Old Rare Tawny.

South Australia's Barossa and Riverland regions are the principal source for the Penfolds Tawny range.

Shiraz, mourvèdre/mataro, cabernet sauvignon and grenache, first introduced to South Australia in the very early nineteenth century, are the predominant grape varieties.

Classification, batch vinification, barrel maturation and selection are crucial elements of Penfolds Tawny production. Intensity, fruit richness, the 'maturation effect' derived from ageing in old oak hogshead casks and the selected grape spirit/brandy character are essential to the style. Each brand in the range differs in concentration, richness and complexity. The density and weight of each brand style is proportionate to the evaporation effect – known as the 'Angels' Share' – during cask ageing, but freshness and rich, full-bodied fruit are a common theme.

Batch vinification takes place in small stainless steel fermenters with regular pump overs to extract colour and optimum flavour from the grape skins. Fortification takes place towards the middle stages of fermentation, to preserve fruit intensity and the required sugar sweetness.

The choice of fortification spirit is an important one, as the style and quality of spirit – which adds complexity and mouthweight – also defines in part the Penfolds House Style. Penfolds now uses a distinctive full-flavoured spirit type. It beautifully complements and accentuates the primary fruit qualities at fortification, yet adds further aromatics, balance, freshness and complexity with age.

Penfolds matures all its tawnies for varying periods in seasoned 300 litre oak hogsheads. The fill levels are kept at roughly 280 litres to allow for expansion of volume during the summer months. Maturation time is aligned with each brand style. Some components are aged for three years, where others may age for fifty years or more as reserve age blending stock.

After fourteen years, some of the best tawny material is entered into the 'Grandfather Solera', a six-stage fractional blending system that takes six years to complete. There are three stacks in each stage and 18 individual hogshead cask stacks in total. This type of ageing results in further concentration and fruit complexity, yet establishes amazing continuity and consistency of style.

Club, Club Reserve and Father Grand Tawnies

These tawnies are matured individually in seasoned 300 litre hogshead oak casks (sometimes as old as sixty years) and stacked using a pyramid system. The Club is aged for three years whereas the Club Reserve enjoys an extra two years of barrel ageing. The maturation process deliberately takes place under corrugated iron. At the height of summer the temperatures on the top stack can be as high as 55° Celsius while in winter the temperatures can fall below 8° Celsius. These cyclic swings in temperature are vital to the maturation process; evaporation, concentration and controlled oxidation lead to remarkable fruit complexity and intense aged flavours – the hallmarks of a great tawny style. Father Grand Tawny, previously known as Bluestone, has an average blended age of ten years, including rare old reserve material. Its new moniker confirms its compelling quality and provenance.

Grandfather and Great Grandfather Rare Tawnies

The young newly vintaged fortified wines selected for Grandfather are first barrel-aged in seasoned oak hogsheads for an average of 14 years. The wines are also matured by grape varietal type, because each variety ages differently and will offer different characters. Shiraz, with its overall opulence and fruit density, ages slowly, whereas the complex and sinewy mourvèdre/mataro matures more rapidly. Cabernet sauvignon and grenache, always a small part of any Penfolds tawny blend, contributes mouth-feel, spice and further layers of fruit intensity.

The 14 year-old vintage fortified components, now showing aged tawny characteristics, are blended to match the style and character of the solera. As it is drawn down through the solera, at a rate of one stage a year, it loses its vintage identity and develops further richness and complexity. When it reaches the sixth and final stage Grandfather Rare Tawny is in theory 20 years old, yet it also comprises the elements of even older vintages. This fractional blending system, which takes years to establish, uniquely preserves the consistency and character of the Penfolds Grandfather Rare Tawny style. After two decades of maturation and evaporation it has developed extraordinary complexity and concentration with perfect balance of fruit opulence, weight, sweetness, alcohol, acidity, oak nuances and volatility. Typically, each release of Grandfather Rare Tawny is tawny coloured with a green patina, seductive rich spicy/nutty/panforte aromas and extraordinary intensity and viscosity on the palate. Grandfather Rare Tawny is virtually identical in aromas and taste to its previous release, thanks to the extraordinary method of barrel ageing, the exacting standards of post-maturation blending and the effect of the solera system.

Ray Beckwith said, 'My first contact with Grandfather Port was in 1935 when Alfred Scholz [the creator of the style] told me that it had its genesis in the 1915 vintage. The wine was matured in hogsheads and quarter-casks of American oak and stored on top of the dome tanks in No. 2 Cellar, just underneath the iron saw-tooth roof, thereby being subjected to warm to hot conditions with a consequent high rate of evaporation. Initially Grandfather Port (sic) was not marketed – it was only given away – and in half-gallon (2.25 litre) flagons. As far as I know the wine was confined to Nuriootpa only and it certainly did not appear on stock sheets under the title "Grandfather Port" as there was no such official designation.'

Great Grandfather Rare Tawny was introduced in 1994 to celebrate Penfolds 150th anniversary – and 1994 bottles were produced. Early releases comprised the oldest and most complex wines available; typically exceptional aged solera material and 'other great stuff lying around the cellar'. A more systematic approach was established during the 1990s, followed by the creation of Penfolds three-stage Great Grandfather Solera in 2011. The topping wine is sourced directly from the final stages of the Grandfather Solera. With an average age of over thirty years before release, Great Grandfather Rare Tawny, now in its 13th series, is a remarkably powerful and complex expression of the Penfolds House Style.

50 Year Old Rare Tawny

Penfolds 50 Year Old Rare Tawny is an ultra-rare bottling of living history. Many components have vintage dates of great significance: the oldest material was vintaged in 1915, the same year that Australia became a true nation, through the prism of resolve at Gallipoli. This extraordinarily beautiful wine, also comprising rare tawny parcels from 1940, 1945 and post–World War II decades, characterises resilience, harmony and generosity of spirit. With its deep tawny crimson colour, warm seductive aromas, smooth silky textures and long crystal-clear brandied cut, it is the ultimate Penfolds Tawny. Timeless and classic, it symbolises a century of winemaking and the collaborative legacy of Penfolds winemakers past and present.

Chief Winemaker Peter Gago says: 'This is an extraordinarily expressive and evocative wine. This is our history in a bottle. All our hopes and aspirations, across a hundred years, are somehow distilled through this 50 Year Old Rare Tawny. Although the average age is fifty years, its oldest wine component goes back to 1915, the same year Max Schubert was born. It was made by Alfred Vesey's protégé Alfred Scholz, and represents the beginning of the Penfolds Grandfather story.'

Penfolds 50 Year Old Rare Tawny is hand-filled into unique hand-blown and number-etched bottles, designed by South Australian artist Nick Mount. It is sealed with a precision-turned wooden stopper, made from seasoned oak barrel staves and sourced from Penfolds fortified maturation cellars. The leather label, pewter crest and the unusual hand-twisted muselet wire, all meticulously hand-fixed, complete a retro-modern aesthetic.

'At half a century of average age, it's ready. Not for the cellar, trophy cabinet or mantle-piece. Open, share and indulge,' says Peter Gago.

FIRST RELEASE	2013
VARIETIES	Predominantly shiraz, mataro, cabernet sauvignon and grenache.
ORIGIN	Primarily Barossa Valley fruit with some Riverland material.
FERMENTATION	Wax-lined open slate and stainless steel fermenters depending on vintage and era. Specially selected brandy spirit of various types.
MATURATION	Maturation in seasoned oak hogsheads. Some components were aged further in the original Grandfather Solera. The 1915 component, after barrel maturation was glass-aged in half-gallon flagons. Vintage components also include 1940, 1945 and various vintages and blends from the 1950s, 1960s, 1970s and 1980s, including 1959 to 1962, 1971, 1974 and 1987.
COMMENTS	Presented in unique hand-blown bottles and handcrafted packaging designed by South Australian artist Nick Mount. Sealed with precision-made wood and cork stopper.
SERIES ONE	Only 330 bottles produced.

Panel Review

Magill Estate, Adelaide – Penfolds Tawnies
Jo Burzynska (JB), Jane Faulkner (JF), James Halliday (JH), Huon Hooke (HH), Ralph Kyte-Powell (RKP), Tony Love (TL), Tyson Stelzer (TS); Peter Gago, Steve Lienert, John Bird, Andrew Caillard.

A wonderful archive of classics. — Ralph Kyte-Powell.

A fascinating bracket with different levels of detail. Like enjoying the splendour of sunset. Complexity, richness and interest increase incrementally through the range.

Tasting Notes

Club Tawny   NOW

Medium tawny crimson. Fresh apricot, caramel, nutty aromas with high-pitched aniseed/floral brandy spirit. Fleshy, sweet, mellow palate with nutty, apricot, currant flavours, some chalky textures, fine cutting spirit and plenty of length. A touch sweet. A lovely commercial tawny with hints of rancio. Average blended age of 3 years at release.

'Light apricot, nutty, brandy characters. Easy to drink with balanced sweetness.' (JF) 'A lighter style with apricots, stonefruit, spice, dried fruit notes and subtle savoury nutty undertones.' (JB)

Club Reserve Tawny   NOW

Medium tawny crimson. Earthy, roasted nuts, orange peel aromas. Soft fleshy palate with amontillado, orange peel notes and loose-knit chalky textures with some vanilla notes. A mellow tawny with a peppery/aniseed kick at the finish. An average blended age of 4–5 years at release.

'Barley sugar notes, in addition to nutty, dried fruit flavours bring extra complexity. This is a step up from the Club with more richness/weight and surprising length.' (TL) 'Attractive well-balanced tawny with depth of black fruits, layers of spice and persistence on the finish.' (TS)

Father Grand Tawny   NOW

Medium tawny crimson. Intense fragrant, orange peel, cloves, pomegranate, toasted nuts, chocolate and raisin aromas. Tighter, more voluminous and deep, with orange peel, tangerine flavours, fine chalky textures and clear-spirited brandy notes. Finishes with plenty of sweet fruit, drying textures and tangy freshness. An average age of at least 10 years old on release. Formally known as Bluestone Grand Tawny.

'Spirity nose but fantastic palate, with toasty almond, nutty, rancio complexity, luscious textures and fresh brandy cut.' (RKP) 'Appropriately darker and more complex than Club with its savoury, nutty personality. Finishes dry with taut acidity and lingering richness.' (TS)

Grandfather Rare Tawny   NOW ••• SOON

Deep tawny crimson. Very intense raisin, roasted walnuts, chestnut, liquorice, rancio aromas. Generous, rich and warm with dense sweet fruit, roasted chestnut, raisin flavours and well-balanced brandy spirit cut. Finishes long and sweet with lingering dried fruit, rancio notes. A classic Penfolds tawny. A minimum average blended age of 20 years old at release.

'Richly developed raisin/toffee flavours with genuine aftertaste and lingering complexity.' (TL) 'Dry savoury nose. Elegant and tight with lovely toasted almond, rancio flavours and long pronounced acidity.' (RKP)

Great Grandfather Rare Tawny Series 12 and 13   NOW ••• SOON

Deep tawny crimson. Very intense orange peel, apricot, roasted coffee, hazelnut, spicy frankincense aromas and flavours with a lovely brandy spirit lift. Sweet, supple and voluminous with luscious fruit, plenty of rancio complexity, a touch of wood varnish, lifted floral notes and fine lacy textures. A rich, generous wine with an almost everlasting finish. A minimum average blended age of 30 years at release.

SERIES 12 *'Intense complex hazelnut, fruitcake, spice on nose with chocolatey notes. Layered and richly flavoured with salty/savoury undertones.' (JB) 'Another level of intensity and complexity. Wonderful freshness, balance and length.' (JH)*

SERIES 13 *'Tawny colour. Intense fragrant hazelnut, rancio, grilled walnut, praline, espresso, apricot aromas with a wood varnish/VA lift. Gorgeously seductive sweet apricot, rancio, coffee, wood varnish, grilled nut flavours sustained by lovely brandied spirit cut. A warm after-glow finish. Magnificent.' (AC)*

50 Year Old Rare Tawny   NOW ••• SOON

Deep tawny crimson. Beautiful raisin, apricot, tobacco, rancio aromas with aniseed, shellac notes. Capacious, rich and complex with raisin, dried apricot, prune tobacco, rancio flavours, smooth silky textures and long crystal-clear brandied cut. A gorgeously seductive and evocative aged tawny with superb freshness, fruit

complexity, weight and length. A great unfolding Penfolds experience. A minimum average blended age of 50 years at release.

'The mahogany edges of age, the beautiful rancio characters and complex, yet fresh, expansive palate exquisitely show the special and unique character of this 50 Year Old Rare Tawny.' (PG)

1940 Grandfather Aged Tawny Port   NOW ••• SOON

Deep tawny, brassy. Over-developed but complex wood varnish, shoe polish, camphor, dried herbs, orange peel, rancio aromas. Attractive orange peel, walnut, varnish, dried fruit, rancio flavours, plenty of richness and volume, but the palate is beginning to dry out, with cutting spirit taking over. Bottled 19 August 1965.

'Lovely crème brûlée, orange marmalade, apricot nuances. Silky, unctuous and long. Astonishing!' (JF)
'Potent and complex like a storm brewing in the glass. Some volatility and slight decay.' (JH)

1945 Bin S6 Grandfather Aged Tawny Port   NOW ••• SOON

Deep tawny. Complex roasted coffee, hazelnut, walnut, panforte aromas with nail varnish, dried roses, rancio notes. Beautifully concentrated, rich and supple with roasted coffee, polish, hazelnut, sweet rancio flavours, oily textures and a drying structure. Orange peel and brandy characters emerge at the finish. A curio rather than brilliance now. Best to drink soon. Bottles are varied. Bottled 21 July 1969.

'Elegant molasses, toffee, rancio, almond aromas and flavours. Gorgeous freshness, richness and texture. A lovely long vanilla/toffee finish.' (RKP) *'Remarkably fresh complex aromas of hazelnuts, honey, molasses, butterscotch, ginger nuts and orange glacé. The palate is rich, creamy and sweet, yet finishes dry and spicy. A clear stylistic link to the Grandfathers of today, albeit in a softer and more mellow mould.' (TS)*

The purpose of decanting is to take the wine off its fine film of sediment and to release its aromas and flavours through aeration. Penfolds uses decanters because it creates a sense of occasion and suits the House Style of its wines.

COLLECTING AND DRINKING
Chapter eleven

Cellaring and serving wine

The cellaring and service of fine wine should be an enjoyable experience for everyone. The tradition and culture of fine wine may seem antiquated or intimidating, especially to newcomers, but following these basic guidelines will maximise your Penfolds wine experience.

Store your wine in a cool place
Wine is best stored in cool temperatures, around 13–16° Centigrade, with a relative humidity of 65–75 per cent. These conditions are sometimes difficult to achieve all year round. A constant temperature of 18° Centigrade is better than high temperature fluctuations between winter and summer.

Wine storage cabinets have recently become a popular way of keeping wine, while air-conditioned cellars are losing popularity because they can dry out corks and are expensive to run. People with large collections may build or rent refrigerated cellar space. The best cellars for keeping wine are dark and free of vibrations.

Always lie bottles on their side
Bottles should be stored horizontally, to ensure the cork remains wet. Corks can dry out if a bottle is left standing up and this will lead to ingress of air and thus oxidation. Screwcapped or glass-stoppered bottles are more resilient but it is best to store these bottles lying down as well: if a bottle is damaged you will identify leakage earlier than if it is upright.

Periodically check bottles for any cork movement. It is not unusual to find 'leakers', even in the best cellars. Some collectors retain tissue wrap around all bottles, as it can be a good visual alert for leakage.

Opening the wine
Corkscrews are still needed for opening the majority of wine bottles around the world. Although Penfolds has introduced screwcaps very successfully across its range, the very top wines, especially Grange, are still sealed with carefully selected corks. At the Penfolds Re-corking Clinics we use the long-barrelled standard table model Screwpull corkscrew, which has a Teflon-coated wire screw and a rigid frame that guides the screw into the centre of the cork and pulls it out automatically. A simple 'waiter's friend' will do the job most of the time, especially with younger corks.

How to serve wine
White wines are best served at cool, refrigerated temperatures. However, if the wine is too cold you will find that the aromas and flavours are deadened. Red wine is best served at a comfortable room temperature of around 18–22° Centigrade. In Australia we sometimes cool red wine down a touch if it's a hot day.

How to decant wine
The world is divided into two types of fine wine people: those who like to decant red wine and those who don't. It is unusual for whites to be decanted

unless the wine needs to be aerated or freed of cork dust; Penfolds often decants its Yattarna.

The purpose of decanting is to take the wine off its fine film of sediment and to release its aromas and flavours through aeration. Penfolds uses decanters because it creates a sense of occasion and suits the House Style of its wines, especially old bottles of Grange, Bin 707, RWT, St Henri, Magill Estate and Bin 389.

Penfolds prefers double-decanting as a means of preparing wine for service. Many wine collectors also double-decant to allow easy identification of the bottles on the table during a meal. Follow these steps.

- Bring the red wine out of the cellar or cabinet a good six to eight hours prior to service. Let the bottle stand to allow the wine sediment to settle.
- Unscrew the cap or pull out the cork.
- Pour the wine carefully and steadily into a clean jug or decanter.

  Some people like to use a funnel. Keep observing the wine through the neck and shoulder of the bottle. The wine will be crystal clear until the very end when sediment will appear. At this point stop pouring.

  Some people use a candle or other light source while decanting, to check when to stop pouring. However it can be just as easy to judge this in daylight or in a brightly lit room.

- Rinse out the original bottle with water and then decant the wine back into the bottle using a funnel.

A warning about wine glasses

There are many types of wine glasses available on the market. Choosing a style and shape is very much a personal decision. Some glass manufacturers suggest that 'the shape is responsible for the quality and the intensity of the bouquet and the flow of the wine'. Penfolds prefers simple but decent-sized stemmed, clear glasses that allow a reasonable swirl to aerate the wine and release aromas.

Wine glasses that are stored in most cabinets (especially wooden or antique cupboards) are prone to collecting fine dust, odours and taints, even within a few days. It is always best to wash glasses thoroughly before use. Poorly stored glasses are a common but completely avoidable source of taints in wine. It is always best to wash and polish glasses prior to use unless you are confident how the glasses have been stored.

Smell and taste before you enjoy

The practice of smelling and tasting wine before dinner or at a restaurant is a very practical tradition. It is an opportunity to check the wine is sound and free of fault before serving. The incidence of cork taint (caused by 2,4,6-trichloroanisole, also known as TCA) is – thankfully – on the decline. If the wine smells musty or like wet hessian and tastes horrible it is probably corked or has been served in a poorly stored glass. If it smells flat or stale it has probably oxidised.

Screwcaps have taken much of the risk out of ordering wine, but even so reductive odours may occasionally compromise the wine's freshness and your enjoyment. Sometimes odours will dissipate, but if you are unsure you should ask the sommelier to check the wine for you. If there is no one else around to give a second opinion, ask yourself 'Am I happy to drink this?'

Penfolds Re-corking Clinics

Penfolds Re-corking Clinics celebrated their twentieth anniversary in 2011. The programme was established in 1991 to allow wine collectors to have their bottles assessed and if necessary topped up, re-corked and re-capsuled. The clinics have become an institution, where collectors enjoy talking face-to-face with the Penfolds winemaking team.

Over 120 000 bottles across four continents have been brought to clinics and checked by Penfolds. The idea of holding a re-corking clinic was inspired by Château Lafite Rothschild's practice of re-corking old bottles for its customers. In fact Max Schubert, the creator of Grange, would also re-cork old vintages of Grange for his friends.

Aside from providing collectors with a perfect and free-of-charge after-sales service, the clinics also give Penfolds an insightful understanding of how its wines are ageing under various conditions across the world. Penfolds wines have a reputation for having the character, concentration and balance to age for the long term, even in not-so-perfect conditions. Although the best bottles seen at the clinics are always from great cellars or temperature and humidity controlled storage, Penfolds reds have a natural

RE-CORKING CLINIC 2013

奔富换塞诊所

每瓶佳酿皆有传奇

robustness because of the maturation techniques originally developed by Max Schubert and his wine-making team.

Only wines judged 'in good condition' are re-corked and re-capsuled. A back label signed off by a Penfolds winemaker assures known provenance; if a wine fails the clinic it cannot be traded, while 'clinic-ed bottles' and wines in pristine original condition nowadays achieve similar prices at auction. With fewer bottles of poorly cellared Penfolds wine thus reaching the secondary market, auction buyers and collectors are confidently able to expect a great Penfolds wine experience. The Penfolds Re-corking Clinics also act as an authentication service; a practical and ongoing anti-fraud process, especially for rare and important vintages.

The Penfolds Re-corking Clinics take place yearly in various cities around the world. The service provided at the clinics is not just for Grange, but is available for all Penfolds red wines, fifteen years or older. For wine collectors, a clinic is also the perfect forum for meeting and tasting their wines with Penfolds winemakers.

Following is an outline of how Penfolds assesses wines that are brought to its Re-corking Clinics.

1. A visual check

If the level of the wine has slipped below the shoulder of the bottle, it is a good candidate for topping up and re-corking – especially if the collector intends to keep cellaring the wine. Overall condition of the bottle, age of the wine and reputation of vintage are all factors taken into account.

Volume loss – ullage – occurs because of absorption, leakage and evaporation. Time, cork failure and cellaring conditions can all contribute to varying ullage levels.

2. Opening

Penfolds has developed a system of opening old bottles to overcome the problem of crumbling or welded-in corks. This includes the use of two interlocking long Screwpull corkscrews and other gadgets.

A tasting portion of about 10–15 ml is poured into a tasting glass. The bottle is then immediately gassed with nitrogen, carbon dioxide, a combination of both, or argon, to prevent oxidation of the wine. It is then stoppered temporarily.

3. Assessment

Penfolds winemakers assess the wine against a benchmark of their accumulated knowledge of vintages and perfectly cellared museum examples. The wine is checked for transparency and sheen; aroma; palate freshness and typicity. Any wines deviating from the window of acceptability are failed. Bottles that pass are topped up.

Wines that fail in the assessment, being deemed to be in unacceptable condition, are not topped up. These bottles are sealed with a plain cork and not re-capsuled. This effectively weeds bottles that are in poor condition out of circulation and the mainstream secondary market, thus improving the reliability and reputation of Penfolds.

4. Topping up
The bottle is topped up with a current vintage of the same Penfolds wine. Grange is topped up with Grange, St Henri with St Henri. Like with like.

The bottle is topped up with approximately 10–25 ml, depending on the ullage of the bottle (3 per cent or less of bottle volume), of new wine. Penfolds believes that the new wine will take on the character of the old. It is widely felt – through comparative

tastings over a period of time – that topping up in small amounts does not affect the integrity of the contents.

5. Re-corking
The topped-up bottle is inert-gassed and then re-corked using a reverse pressure (vacuum) re-corking machine creating negative pressure.

The new corks are stamped with 'Penfolds Red Wine Clinic' and dated. Grange bottles are re-corked with stamped 'Grange' corks.

6. Re-capsuling
A 'Penfolds Red Wine Clinic' capsule is moulded onto the bottle with a capsuling machine.

A Clinic back label – numbered, dated and signed by a Penfolds winemaker – is affixed to the bottle, certifying its condition and authenticity. The label is endorsed by specialist wine auctioneers – Langton's Fine Wine Auctions and Christie's – or, if appropriate, by another recognised wine auction house. All re-corked bottle numbers are stored in a data bank.

Penfolds and the secondary wine market

The ageing potential, aesthetic quality and reputation of Penfolds wines have attracted wide appeal for generations of collectors and drinkers. As a result, Penfolds is the leading wine brand in the Australian auction, brokerage and resale markets. Grange leads the market by value while Penfolds dominates Australian wine auctions, by presence and volume. Rarities secure the top ten prices for Australian wine every year.

Penfolds Grange is a cornerstone of the Australian auction market, with a reputation and track record that rivals the great classified growths of Bordeaux and Burgundy. It heads the prestigious Langton's Classification of Australian Wine because of its history and market currency. The fame of the heritage-listed Penfolds Grange has reached far and wide, so it now also regularly appears at wine auctions in the US, UK and Hong Kong.

Rare Grange (1951–1967)
All the 1950 and early 1960s vintages have become rarities. The most valuable Granges are the experimental 1951 and the first release, 1952 Grange Hermitage. 1952, 1953, 1954, 1956, 1957 and 1958 Granges command premium values at auction

because there are few bottles left. 1955 was a show wine and 1959, although a 'hidden Grange', was commercially released. Both the 1955 and 1959, despite stocks being significantly depleted, do not attract the same high level of prices at auction. The early 1960s are also extremely scarce now. A complete collection of Grange, in pristine condition and signed by Max Schubert, once sold for $250 000.

Vintage Grange (1968–present)
Vintage Grange plays a major role in the Australian secondary wine market. The performance of 1971 Grange at auction was used as a market yardstick by economic think-tank Access Economics during the 1980s and 1990s and for many years this wine out-performed other alternative investments, including racehorses, taxi licences and rare coins.

However, all wines have a life – and 1971 has not really kept up with the overall interest rate for some years. At one stage in 2003, though, it enjoyed a 45-fold increase over release price.

1990 Grange is probably the benchmark now. The best performing Grange vintages at auction currently are 1962, 1963, 1966, 1971, 1976, 1986, 1990, 1991, 1996, 1998, 2002, 2004 and 2006. Anniversary vintages also perform well; for instance, 1963, 1973, 1983 and 1992 (21st) may achieve higher than normal prices in 2013.

The market will sometimes pay a premium for bottles with Max Schubert's signature. The Penfolds Red Wine Re-corking Clinics have improved the currency of Penfolds red wines at auction. Buyers will only pay optimum prices for bottles in pristine original condition, or successfully clinic-ed wines.

Recognised 'Penfolds Vintages' are 1952, 1953, 1955, 1962, 1963, 1964 (Bin 707), 1966, 1971, 1976, 1986, 1990, 1991, 1996, 1998, 2002, 2004, 2006, 2008 and 2010.

Other Penfolds wines at auction

Bin 707 Cabernet Sauvignon
A strong secondary wine market performer, because of its history and alignment with Grange. Chinese interest has propelled prices forward in recent years.

RWT Shiraz and Magill Estate Shiraz
A growing interest in regional styles and single vineyards has established a good following for these wines.

St Henri
A sentimental favourite among Australian auction buyers. Experimental and early vintages have achieved stellar results in recent years; 1971 is a good example.

Bin 389 Cabernet Shiraz
The most popular red wine in the auction market, with strong volumes and currency, especially for recognised vintages. It is regarded as an important auction wine.

Bin 407 Cabernet Sauvignon
Market demand has increased in the wake of Bin 707.

Kalimna Bin 28 Shiraz and Bin 128 Coonawarra Shiraz
Both were classified as 'Distinguished' by Langton's in 2005, in acknowledgement of their presence and popularity in the secondary market.

Bin 138 Grenache Shiraz Mourvèdre/Mataro
A modest presence. Not really a collectible.

Bin 150 Marananga Shiraz, Bin 169 Coonawarra Cabernet Sauvignon and Bin 170 Kalimna Shiraz
These have promising potential but no track record at this early stage.

Special Bins
Special Bins traditionally attract strong support from Australian collectors, but achieve mixed success in other markets.

- 1948 Block 42 Kalimna Cabernet – probably depleted.
- 1962 Bin 60A Kalimna Cabernet Coonawarra Shiraz – very rare.
- 1966 Bin 620 Coonawarra Cabernet Shiraz – very rare.
- 1967 Bin 7 Kalimna Cabernet Coonawarra Shiraz – very rare.
- 1980 Bin 80A Coonawarra Cabernet Kalimna Shiraz – rare.
- 1982 Bin 820 Coonawarra Cabernet Shiraz – rare.
- 1990 Bin 90A Coonawarra Cabernet Barossa Valley Shiraz – rare.
- 1990 Bin 920 Coonawarra Cabernet Shiraz – rare.
- 1996 Block 42 Cabernet Sauvignon – rare.
- 2004 Block 42 Cabernet Sauvignon – rare.
- 2004 Bin 60A Coonawarra Cabernet Barossa Valley Shiraz – rare.
- 2008 Bin 620 Coonawarra Cabernet Shiraz – rare.

Yattarna and Reserve Bin A Chardonnays
Moderate performers, reflecting a red wine bias in the secondary collectible markets. Young vintages are mostly sought after.

Auction price data can be found at Langton's website www.langtons.com.au.

PART THREE

Timelines and Lasting Notes 1951–2012

BRANDY

MELBOURNE ... VIC
PERTH .... HOWARD
BRISBANE . 468
LONDON
RUTHERFORD. OSBOR
GREAT TOWER ST

THE
REGISTERED OFFICE OF
DALWOOD VINEYARDS
LTD

Royal Reserve
PORT
BY PENFOLDS WINES LTD
MAGILL, SOUTH AUSTRALIA.

PENFOLDS
WINES

NSW
LH·437

'It's fraught with risk but then, so are many things that are worth doing. Sometimes you have to take a calculated risk.'

PETER GAGO, Penfolds Chief Winemaker

Penfolds *The Rewards of Patience, Seventh Edition* tastings were held, in 2011 and 2012, in Beijing – China, Berlin – Germany, New York – USA and at Penfolds Magill Estate in Adelaide – South Australia. The following tasting notes are a compilation of observations and opinions from many of the world's finest palates.

The Asian, European, Americas and Australasian panels each brought different insights, cultural perspectives, analyses and languages. Almost all of the wines were sourced from Penfolds Museum Cellars at Magill Estate, with the exception of a handful of rare bottles emanating from private cellars (and the collection of Peter Gago). In a few instances, for historical or reference purposes, extra tasting notes have been added. Underperforming bottles from the various international panel tastings were retasted and reviewed.

The tasting notes are compiled by vintage year and contextualised with vintage reports, vintage ratings and timelines. Although these reviews offer a fascinating insight, no bottle develops or matures exactly the same. The enjoyment of wine is ultimately a personal one and linked to the shared experience. The title of this book, *The Rewards of Patience*, suggests something magical might happen if you are prepared to wait; it's a compelling argument.

'How will the wines travel? Will the corks hold up? How will non-Australian palates respond?'

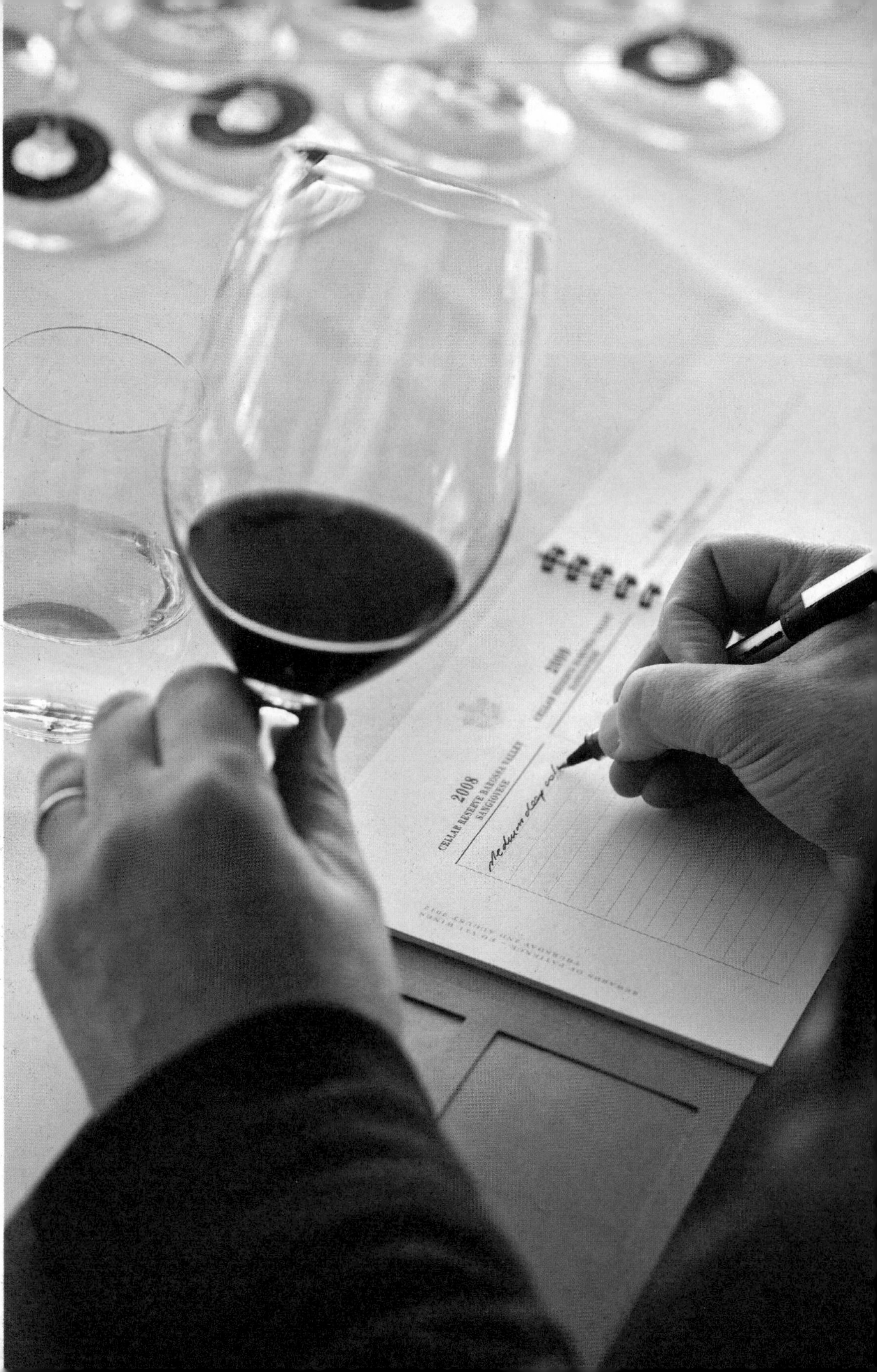
2008
CELLAR RESERVE BAROSSA VALLEY
SANGIOVESE

Penfolds Panel

(Beijing, Berlin, New York and
Penfolds Magill Estate, Adelaide)

Peter Gago (PG)
Penfolds Chief Winemaker

Steve Lienert (SL)
Penfolds Senior Winemaker

Kym Schroeter (KS) (Adelaide only)
Penfolds Senior White Winemaker

Jamie Sach
Penfolds Ambassador

Andrew Caillard, MW (AC)
author and tasting editor

Asia Panel

Beijing, China – November 2011

Chantal Chi (CC)
Shanghai-based wine writer, China

Li Demei (LD)
oenologist and wine writer, China

Yu Sen Lin (YSL)
wine writer and wine educator, Taiwan

Dr Edward Ragg (ER)
academic and wine writer, China

Simon Tam (ST)
auction director and wine educator, Hong Kong

Ch'ng Poh Tiong (CPT)
wine writer, publisher and wine educator, Singapore

Fongyee Walker (FW)
wine writer and wine educator, China

Europe Panel

Berlin, Germany – July 2012

Dr Neil Beckett (NB)
academic and wine editor, UK

Frank Kämmer, MS (FK)
wine critic and educator, Germany

Peter Keller (PK)
wine journalist and columnist, Switzerland

Andreas Larsson (AL)
sommelier and wine writer, Sweden

Neal Martin (NM)
wine writer and author, UK

Peter Moser (PM)
wine journalist and author, Austria

Mario Scheuermann (MS)
wine and food critic, Germany

Americas Panel

New York City, USA – July 2012

Marcelo Copello (MC)
wine journalist and television presenter, Brazil

Joe Czerwinski (JC)
wine journalist and wine educator, USA

Anthony Gismondi (TG)
broadcaster and wine journalist, Canada

Martin Gillam (MG)
*wine journalist and documentary filmmaker,
USA/Australia*

Joshua Greene (JG)
wine editor and publisher, USA

Ray Isle (RI)
wine editor and critic, USA

Dave McIntyre (DM)
wine correspondent and blogger, USA

Linda Murphy (LM)
wine writer and correspondent, USA

Josh Raynolds (JR)
wine editor and critic, USA

Australasian Panel

Adelaide, South Australia – September 2012

Jo Burzynska (JB)
wine writer and educator, New Zealand

Jane Faulkner (JF)
wine, food and lifestyle journalist, Australia

James Halliday, AM (JH)
wine writer and author, Australia

Huon Hooke (HH)
wine critic and author, Australia

Ralph Kyte-Powell (RKP)
wine writer and broadcaster, Australia

Tony Love (TL)
wine journalist and commentator, Australia

Tyson Stelzer (TS)
wine writer, author and publisher, Australia

★★★★★	Classic Penfolds vintage
★★★★	A very fine vintage
★★★	A moderately good vintage
★★	A difficult or ordinary year
★ / ★	Reds / Whites
◈	Special wines

1951 ★★★

A hot and dry growing season • Max Schubert makes the first experimental Grange at Magill • Cellar Foreman Bill Vincent celebrates thirty-two vintages at Penfolds; he joined the firm in 1919 after serving with the 48th Battalion in World War I • Spy scandal as British secret agents Burgess and Maclean defect to the Soviet Union • Drive-in movies become a new form of entertainment in the US • Juan Manuel Fangio is Formula One World Drivers' Champion.

Bin 1 Grange Hermitage   NOW ••• PAST

Not tasted at the Rewards of Patience Tasting (New York), but several bottles opened during the 2012 Penfolds Re-corking Clinics. The first experimental Grange and extremely rare. A valuable collector's item because of its historical significance. Rare hand-blown bottles. The wine itself is past its peak although some bottles still have fruit sweetness and flavour length. Largely the wine has a dull tawny colour and skeletal palate structure with little flesh and fading tannins. 100% shiraz.

50% Magill Estate (Adelaide), 50% Morphett Vale (Adelaide). 100 cases/3 hogsheads made. Released as Bin 1.

1952 ★★★★★

Average to normal rainfall and weather conditions • 1952 Bin 4 Grange Hermitage is made; the first commercial release of this line • Alfred Vesey, at aged eighty-nine, who worked as Mary Penfold's assistant, retires after sixty-nine vintages • Anne Frank's *Diary* is published • Death of Eva Peron (Evita) • The world's first hydrogen bomb test takes place in the Pacific.

Bin 4 Grange Hermitage ◆   NOW

First 'commercial' vintage. Medium-deep amber to brick red. Intense fragrant panforte, dark chocolate orange peel, apricot, graphite aromas with hint of herb garden. Mature complex palate with panforte, praline, apricot, orange peel flavours and fine slinky, lacy tannins. Finishes silky sweet and long. Incredible freshness, buoyancy and weight. Great bottles still holding up well. Very rare. 100% shiraz.

Magill Estate (Adelaide), Morphett Vale (Adelaide). Around 100–150 cases made at less than $1 a bottle at release. Some half-bottle 'pints' were also produced. Released as Bin 4 and Bin 4A.

'Medium amber. Smoky, Christmas spice, pot pourri, cherry liqueur, apricot aromas. Has the sweetness of very old Burgundy. Ageless and seamless!' (JR) 'Glorious fresh spice, savoury floral aromas and silky tannins. Still has time left.' (LM) 'It is in a great moment. A very expressive, soft, rich and delicious wine showing evolution of colour, notes of leather, herbs, coriander.' (MC)

1953 ★★★★★

Dry, mild to warm weather followed a cool, but even, growing season; regarded as a great Grange vintage • John Davoren's first experimental St Henri Claret • Ian Fleming publishes his first James Bond novel, *Casino Royale* • Hillary and Tenzing reach the summit of Mount Everest • Coronation of Queen Elizabeth II.

Bin 2 Grange Hermitage ◆   NOW

Medium-deep amber to brick red. Fragrant red fruits, mocha, praline, orange peel, apricot, spicy aromas with cedar, garrigue, sage, dried lavender notes. Finely textured wine with sweet dried fruit, mocha, roasted chestnut, mushroom, herb flavours and fine loose-knit lacy tannins. Finishes savoury, silky textured and long. Fully mature wine but still showing freshness, elegance, energy and balance. Only the best-cellared bottles will hold. Drink now. Very rare. 87% shiraz, 13% cabernet sauvignon.

Magill Estate (Adelaide), Morphett Vale (Adelaide), Kalimna Vineyard (Barossa). 260 cases made. Some half bottles (375 ml) were released. First vintage – and then uninterrupted – use of Kalimna fruit; hence the term 'mother vineyard'. Released as Bin 2 (also Bins 10, 86C and 145). A Jancis Robinson, MW, 20/20 wine.

'Orange peel, bergamot, mint characters and soft tannins.' (LM) 'A freak wine, simply sublime. Eclipses 1953 Château Margaux as my favourite birth-year wine!' (TG) 'Lightly gamey, dried cherry, umami characters with a touch more austerity than 1952.' (RI) 'Very sound and intense, wonderfully deep and vibrant with power and persistence. It finishes right out. A tremendous wine of big firm tannin structure and great complexity. A great wine indeed.' (HH) 'As the air comes to play, the wine reveals lovely old shiraz notes, a discreet dusting of tannins and vibrant acidity. A beautiful wine.' (TL)

St Henri

The first experimental Penfolds St Henri Claret was made in 1953. Rare bottles surfaced at a Penfolds Red Wine Re-corking Clinic in Adelaide. Unlike the experimental postage stamp labels of Grange, these bottles are monickered St Henri, the design harking back to the original nineteenth-century label. These authentic bottles suggest that John Davoren experimented and released small batches of St Henri mostly to friends and acquaintances. The wines were often half-bottles, further suggesting non-commercial releases.

Bin 9 Grange Cabernet Sauvignon (Block 42) ◈

NOW

Medium-deep brick red. Well past its best now with oxidised, amontillado, leather aromas, sweet nutty, leather flavours and fine chalky texture. It finishes long and savoury. This bottle is just holding on by a thread, but other examples have looked superb, typically showing fully developed polished leather, demi-glace, tobacco aromas, plenty of sweet fruit complexity and lacy dry tannins.

'Nutty oxidative sherry-like aromas. The palate shows hints of former richness, but it's now oxidised with rancio notes, slightly sour acid and drying tannins.' (JB) 'Sadly, oxidised and vinegary, yet with a core of sweet fruit and fine, velvety tannins.' (TS)

Bin 86A SA/HR Dry Red

NOW ••• PAST

Medium-deep colour. Tobacco, wet bitumen, roasted aromas and flavours. Still has sweet fruit and chalky textures. But drink now.

‖ 1954 ★★★

Cool to mild growing season followed by a warm vintage; the Grange style was lightened with only nine months in oak, to meet internal criticism • 'Rock Around the Clock' is recorded by Bill Haley and His Comets • Roger Bannister becomes the first person to run a mile in under four minutes • Vietnam is partitioned into North and South after the end of colonial rule.

Bin 12 Grange Hermitage

NOW ••• PAST

Light medium amber to brick red. The wine has faded now with overdeveloped nutty, briny, earthy, dark chocolate, liquorice aromas, gentle sweet amontillado, toffee flavours and fine chalky dry textures. Finishes lacy firm, long and sweet. Just holding together. A curio now. Very rare. 98% shiraz, 2% cabernet sauvignon.

Magill Estate (Adelaide), Kalimna Vineyard (Barossa Valley). Internal criticism of Grange led Max Schubert to lighten the style slightly. Only nine months in oak. Released as Bin 11 and Bin 12.

'Like an old tawny port. It has its charm yet lacks density, palate weight and complexity.' (LM) 'Nutty, rancio, almost sherry-like. Has now faded.' (TG)

‖ 1955 ★★★★★

Mild to warm growing season with intermittent, but well above average rainfall; a warm, dry vintage • 1955 Grange wins twelve Trophies and fifty-two Gold medals on the Australian show circuit; in 2000, the *Wine Spectator* hails it as 'one of the top twelve wines of the century' • Walt Disney opens his first theme park, Disneyland • Salk polio vaccine is approved by the US Food and Drug Administration • British speed-ace Donald Campbell sets his first world water-speed record in *Bluebird*.

Bin 95 Grange Hermitage ◈

NOW

Medium-deep brick red. Fresh roasted coffee, panforte, meaty aromas with gun flint, chinotto notes. Very complex and beautifully balanced palate with fresh chinotto, roasted coffee, panforte, sweet fruit, leafy flavours, lovely mid-palate richness and persistent fine-grained tannins. Finishes chalky firm but long and sweet. A famously great vintage. Still showing plenty of freshness, fruit sweetness and volume. Now quite rare. 90% shiraz, 10% cabernet sauvignon.

Magill Estate (Adelaide), Morphett Vale (Adelaide), Kalimna Vineyard (Barossa Valley), McLaren Vale. The most decorated Grange – winner of twelve Trophies and fifty-two Gold medals during its relatively short career on the Australian wine show circuit. Spent only nine months in oak. A favourite of Max Schubert's, partly because it won a Gold medal in the Open Claret Class at the 1962 Sydney Wine Show – some members of the judging panel had previously been vocally critical of the style. Chosen by the US magazine, *Wine Spectator*, as one of the twelve 'Wines of the Millennium'. Most common but later release (after show success) is Bin 95 (also Bins 13, 14, 53, 54 and 148A).

'*Opens up like a young wine with sweet dried red fruits and spicy, mocha, caramel notes. Pliant and long with weight and breadth.*' (JR) '*Perfectly proportioned, elegant and complex with finesse, structure and balance to live for many years.*' (MC) '*Still vibrant but lacks silkiness and polish.*' (JC)

1956 ★★

A moderate vintage with a cooler than average growing season • Elvis Presley releases his first hit single, 'Heartbreak Hotel' • Actress Grace Kelly marries Prince Rainier III of Monaco • Olympic Games held in Melbourne, Australia, the first to be staged in the Southern Hemisphere.

Bin 14 Grange Hermitage NOW ••• PAST

Medium amber to brick red. Very perfumed roasted coffee, herb garden, liquorice aromas with touches of tobacco, cedar. Medium concentrated palate with roasted coffee, treacle, bitumen, oyster shell, herb flavours and supple, fine savoury tannins. Finishes firm but long and sweet. Looked very fresh and balanced in this tasting, but many bottles are now past. Very rare. 96% shiraz, 4% cabernet sauvignon.

Magill Estate (Adelaide), Morphett Vale (Adelaide), Kalimna Vineyard (Barossa Valley). Only nine months in oak.

'*A Lazarus wine; the air brings this wine to life. Red fruits, tobacco, cherry stone, mocha, earthy, orange peel.*' (JR) '*Pleasantly earthy, very spicy and savoury with velvety tannins. A seamless/no elbows wine. Hard to believe its age.*' (LM) '*Scorched citrus peel, meaty, mushroom characters with sturdy tannins.*' (JC)

St Henri NOW ••• PAST

The first experimental release.

Bin 136 Magill Burgundy ◈ NOW ••• PAST

Medium-deep brick red. Intense and complex tar, amontillado, roasted chestnut, leather polish aromas with herb garden, praline notes. Fully mature, beautifully concentrated wine with amontillado, chestnut, leather, spice, orange rind notes, smoky nuances and loose-knit, lacy tannins. Finishes chalky firm with lovely density of fruit and flavour length. Almost perfect. Also released as Bin S56. A great bottle.

'*Smells of an antique writing desk. Soft supple palate with inky, old spice, sweet fruit, cinnamon notes, a lift of cinnamon and graceful long finish. A very beautiful wine; glorious in its age. But it can't be forever.*' (JF) '*Old leather armchairs, gentle raspberry fruit and old spice. A silky smooth wine with melted fruit, tannins and acid. This gentle elegant wine is living history, with an ethereal fragrance that lasts beautifully in the mouth and the glass.*' (RKP) '*100 points.*' (TS)

1957 ★★

A mild, dry growing season with very low rainfall; a dry, mild to warm vintage • Winemakers Murray Marchant and Gordon Colquist assist Max Schubert in keeping his dream alive after Penfolds management had ordered production of Grange to cease; the first of the 'hidden Granges' • Danish architect Joern Utzon's design is chosen for Sydney's new opera house at Bennelong • The Treaty of Rome creates the European Economic Community • The Soviets reach space with the world's first artificial satellite, *Sputnik 1*.

Bin 50 Grange Hermitage

Medium-deep brick red. Fresh mature toffee, charcoal, liquorice, wood varnish aromas. The palate is showing advanced age with amontillado, nutty, panforte, smoky flavours and fine chalky, cedary tannins. The wine has now faded. Very rare. 88% shiraz, 12% cabernet sauvignon.

Magill Estate (Adelaide), Morphett Vale (Adelaide), McLaren Vale. The first of the three 'hidden Granges'; the wine was made without the knowledge of Penfolds management, who had ordered Max Schubert to cease production. Matured in the previous year's Grange barrels.

'Black fruits and spice with a rustic edge.' (LM) 'Grange unplugged with age and grace.' (TG) 'Not as complex as the 1956 but still quite different and delicious.' (RI)

St Henri

The first commercial Penfolds St Henri vintage. Fragrant, mint, smoky, old leather aromas. Medium concentrated, soupy and slightly oxidised with dusty dry tannins and only some fragments of fruit remaining. The wine has faded but it's still fascinating to experience.

'A sedate, slightly fatigued menthol, Chinese tea, rosewater and liquorice nose. Medium-bodied, crisp on entry and fleshy in texture with orange peel, strawberry, cedar and leather notes. Tapers towards the finish. Although lacking a little energy, it is beautifully balanced. Remarkable!' (NM) 'Round, soft, warm and velvety textured. A faded glory now.' (NB) 'Meaty, dried fig, smoky "speck" aromas. Still has sweet fruit but fading tannins and fine acidity. Enjoying its last days.' (FK)

Bin 14 Minchinbury Dry Red

Medium-deep colour. Very old and complex wine. It's well past its prime now but it's still an interesting curio. Bottled and matured before release at Penfolds Tempe, NSW.

Described by Max Lake in 1966 as 'developing now into a superb true claret; complex perfume, austere flavour and austere grip on the finish'.

1958

An even growing season followed by a mild to warm vintage • The second 'hidden Grange' • The legendary Magill cellar supervisor Jim Warner, who worked for Mary Penfold, retires after sixty-five vintages • Lake Eucumbene, NSW, Australia's biggest reservoir, is completed • The hovercraft is invented • Anthropologists Louis and Mary Leakey find skeletal evidence of the world's earliest known human in the Olduvai Gorge, Tanzania.

Bin 46 Grange Hermitage

Medium-deep brick red. Intense earthy, sandalwood, roasted chestnut, demi-glace, amontillado, pipe tobacco aromas and flavours. Well-concentrated sweet fruit, roasted chestnut, orange chocolate, amontillado flavours, fine bitter-sweet tannins. Finishes gritty firm with minerally notes. Past its prime now. Very rare. 94% shiraz, 6% cabernet sauvignon.

Magill Estate (Adelaide), Morphett Vale (Adelaide), Kalimna Vineyard (Barossa Valley), Barossa Valley, McLaren Vale. The second 'hidden Grange'.

'Fading but silky and harmonious.' (JR) 'Expressive, wide and soft, with balsamic, medicinal, animal notes.' (MC) 'Elegant, feminine style with lifted stewed fruit, peppery aromas and round, supple, meaty, mushroom/umami flavours.' (JC)

St Henri

Medium brick red. Fresh and complex herb garden, panforte, dark chocolate, apricot, smoky aromas with demi-glace notes. Medium concentrated wine with savoury, cedar, apricot, leathery flavours and fine silky/chalky tannins. Acidity at the finish gives line and length. Still has vinosity and length, but beginning to fragment. A lovely old wine; best to drink now though. John Davoren's favourite vintage.

'Fresh and rustic with earthy, subtle terracotta notes and dried herbs, spice box, stalky, Marmite elements. After a few minutes it springs a startling orange blossom scent. Lively, fresh and quite tart with supple tannins and faded red cherry fruit. It is harmonious and nicely focused.' (NM) 'Gentle but pungent wine with sea breeze, smoky bacon aromas, intriguing elegance, remarkable freshness and finesse. Acidity is starting to poke through. More Burgundy than Rhône in character.' (NB) 'Reminiscent of an old northern Rhône with notes of incense, smoke, spice, tobacco and leather. The palate is starting to fade, with a hint of oxidation and quite high acidity. Fragrant but meagre.' (AL)

Bin 393 Special Bottling Graves   NOW ••• PAST

Medium-deep rusty golden colour. Fresh off-dry baked apricot, prune, marzipan flavours, with touches of brine. Sweet apricot, marzipan, toffee flavours, plenty of silky richness, and mineral acidity. Brassy notes are beginning to appear. Drink up.

1959   ★★★★

Cool to mild growing season with intermittent rains in February and March; a warm vintage • The third 'hidden Grange' • Bin 28 is first made • 'High fidelity' or 'hi-fi' becomes the new stereophonic recording technique • British Motor Corporation releases its economy car, the Mini • Charles de Gaulle inaugurated as President of France's Fifth Republic.

Bin 46 Grange Hermitage   NOW

Medium-deep brick red. Complex fragrant praline, nougat, redcurrant aromas with hints of mint, herb and graphite. Generous praline, panforte, demi-glace flavours, plenty of mid-palate richness and fine sinewy tannins. Finishes chalky firm and long. Sinuous and seductive. Rare. 90% shiraz, 10% cabernet sauvignon.

Magill Estate (Adelaide), Morphett Vale (Adelaide), Kalimna Vineyard (Barossa Valley). The third and final 'hidden Grange'. Released as Bin 46 (also Bins 49 and 95).

'Mouth-filling and generous with earthy rhubarb, plum, cocoa, anise notes and soft tannins. An exceptionally long finish.' (LM) 'A mature red with tomato leaf, orange peel, dried berry, meaty aromas, tangy/resinous notes and a long finish.' (RI) 'Smoky, gamey, bacon aromas reminiscent of northern Rhône. Elegant and refined.' (DM)

St Henri   NOW

Medium crimson brown. A very complex old wine with fresh herb, sage, incense, charcuterie, apricot, barnyard aromas and briny, cedar, graphite notes. Delicate and sinuous with developed cedar, sage, meaty, praline flavours, plenty of volume and richness on the mid-palate and slinky, loose-knit tannins. Finishes chalky firm with lovely flavour length. Still in good condition and should hold for a while.

'A profound, almost Châteauneuf-du-Pape inspired bouquet and a palate that dared to show a touch of exoticism, albeit abraded by the passing decades. Opulent and ravishing!' (NM) 'Saline, savoury and very claret-like. Ample for its age with glossy silkiness and a fresh appetising finish. Astonishingly vigorous without being jagged or ragged.' (NB) 'Very intense characterful nose with herbal, leather, gamey notes and dried fruit. The palate is dissipating with a hint of bitterness but it still delivers a spicy complexity.' (AL)

Kalimna Bin 28 Shiraz   PAST

Brick red. Very evolved wine which fades quickly in the glass. It first showed delicate plum, tangerine, mocha, mint aromas and flavours and lacy fine tannins but amontillado characters developed in the glass. Thinning and tired but interesting curio all the same (*Penfolds The Rewards of Patience*, 2008).

1960 ★★★★

A hot and moderately dry vintage; drought conditions in South Australia led to low yields • Grange production officially recommences • Penfolds acquires its first Coonawarra vineyard, Sharam's Block • Bin 389 is first made • One of Penfolds longest serving winemakers, John Bird, starts work as a laboratory technician • The birth control pill (the Pill) becomes available in the US • Alfred Hitchcock's *Psycho* is released • D.H. Lawrence's *Lady Chatterley's Lover* trial returns a verdict of 'not guilty' under the new Obscene Publications Act 1959, making the novel available to the public.

Bin 95 Grange Hermitage ⬦ NOW

Light-medium brick red. Intense panforte, dark chocolate, molasses aromas with touches of rose petal, orange peel, cloves and spices. Well-concentrated panforte, spicy flavours, satin-textured tannins and underlying savoury notes. Finishes chalky firm but fruit richness persists. Becoming quite fragile. Rare. 92% shiraz, 8% cabernet sauvignon.

   Magill Estate (Adelaide), Morphett Vale (Adelaide), Kalimna Vineyard (Barossa Valley). Released as Bin 95 (also as Bin 45).

   'Soft and supple with melt-in-your-mouth tannins and sweet finish.' (TG) 'Well proportioned and evolved with more elegance than impact.' (MC) 'Leathery, dried tomato, beef broth aromas with some coffee, dried fruit notes. It has a supple texture that makes these older Granges so appealing.' (RI)

St Henri PAST

Not tasted.

1961 ★★★

A hot, dry vintage • The first Grange to be made that included grapes sourced from Coonawarra • Russian cosmonaut Yuri Gagarin becomes the first man in space, aboard *Vostok 1* • The Berlin Wall is constructed to prevent East Berliners from defecting to West Berlin • Miss Joan Barry, twenty-six, pleads not guilty to offensive behaviour for wearing a bikini at Bondi Beach, Sydney.

Bin 395 Grange Hermitage NOW

Deep amber to brick red. Lovely fresh sandalwood, dark chocolate, cedar, herb, mint aromas with underlying dried leaf notes. Ample dark chocolate, panforte, cedar, herb flavours, fine graphite tannins and some leafy notes. Builds up quite muscular and firm at the finish, but lovely freshness, generosity and persistency of fruit. At peak of maturity, but good bottles will hold. Rare. 88% shiraz, 12% cabernet sauvignon.

   Magill Estate (Adelaide), Morphett Vale (Adelaide), Modbury Vineyard (Adelaide), Kalimna Vineyard (Barossa Valley), Coonawarra. Generally released as Bin 95 (also as Bin 395).

   'Smoky, pot pourri, minty, tobacco aromas and long cedary texture, reminiscent of Bordeaux!' (JR) 'Twice as dark as 1960. Comes alive on the palate with focused cherry, plum, herb flavours, gentle tannins and acidity. A quietly beautiful wine.' (LM) 'Wonderful polished wine with sweet dried fruits and a distinct menthol, eucalypt note.' (RI)

St Henri NOW ••• PAST

Medium-deep brick red. Fresh dark chocolate, roasted walnut, dried fig, prune, orange rind aromas with hints of incense. Lovely concentrated, buoyant and supple wine with complex praline, walnut, mint, tobacco flavours and fine muscular tannins. Finishes firm and long. Drink up.

'Evolved leather, mint, weihrauch (incense) aromas. Complex and rustic with a strong assertive grip at the finish.' (FK) 'Denser and richer than the 1959. The palate is soft and fleshy with impressive overall harmony. A surprisingly robust finish.' (NB)

Bin 389 Cabernet Shiraz 　　　　　　　　　　　　　　　　　　　　　NOW ••• PAST

Medium-deep brick red. Flinty, demi-glace, leather, spice aromas with a hint of herb garden and dark chocolate. A supple sweet-fruited palate with dark chocolate, panforte, demi-glace flavours, lovely mid-palate volume and richness, fine loose-knit chalky tannins and mineral acidity. Finishes lacy, firm and dry with some leather, spice notes. Has now reached the end of its drinking window. Some bottles are now past. Nonetheless this is a remarkable old wine. Drink now.

'Fully secondary, with game, leather and a note of bitter hazelnuts. Almost fully receded tannins although admirable fruit length. At the close of its life, though by no means falling apart.' (TS) 'The fruit is incredibly sweet and supple with fully resolved tannins. A wholly remarkable 50+ year-old wine.' (JH) 'Still hanging in with black caramel, toffee aromas and sweet old dark fruits. Lovely flow and richness with melted tannins and a whisper of coconut. Dries out towards the finish.' (RKP)

1962　　　★★★★★

A warm, moderately dry and even growing season followed by a dry, low-yielding vintage; regarded as a great Grange vintage • The Kalimna Vineyard – purchased by Penfolds in 1945 – is now the mother vineyard of Grange • 1962 Bin 60A Coonawarra Cabernet Sauvignon Kalimna Shiraz, regarded by many as Australia's greatest wine of the twentieth century, is made • The first standard-gauge freight train arrives in Melbourne from Sydney, ending eighty years of changing gauges at Albury on the VIC/NSW border • John Glenn becomes the first American to orbit the Earth, aboard *Friendship 7* • Andy Warhol exhibits his 'Campbell's Soup Cans'.

Bin 95 Grange Hermitage ▽　　　　　　　　　　　　　　　　　　　　　　　NOW ••• 2020

Deep red brick. Fragrant and complex dark chocolate, espresso, blackberry, mocha aromas with some herb/sage notes. Fresh supple dark chocolate, dark berry, roasted coffee flavours, plentiful grainy, granular tannins and underlying mocha, vanilla notes. Finishes chalky firm, long and sweet. Still holding very well. Impressively mature classic Grange with 'claret-like structure'. Rare. 87% shiraz, 13% cabernet sauvignon.

Magill Estate (Adelaide), Kalimna Vineyard (Barossa Valley), Adelaide Hills. Barossa Valley fruit becomes ascendant component. Released as Bin 95 (also as Bins 59, 59A and 456).

'An expressive wine with exotic floral, sweet red fruit, dried herbs, musky, pot pourri character, sweet middle-palate but drying finish.' (JR) 'A chocolatey orange concoction with bourbon, cola notes. A substantial beefy style.' (TG) 'Lush, smooth and concentrated with soy, herbs and spices. Shows a perfect integration of aromas and flavours.' (MC)

St Henri ▽　　　　　　　　　　　　　　　　　　　　　　　　　　　　　　NOW

Medium-deep brick red. Fresh and classically mature with prune, mocha, praline, orange peel, clove, spicy aromas and touches of mint. Supple, rich and concentrated with fresh prune, panforte, mocha flavours, some leather, sandalwood notes and fine bitter/silky tannins. Finishes chalky and long with savoury, minerally, tobacco notes. A great Penfolds vintage year. Will probably hold but best to drink soon.

'Conservative, Bordeaux-like in style with exquisite balance, although it does not quite reach the ethereal heights of either the Grange or Bin 60A that year.' (NM) 'The best of the early vintages with red fruit, truffle aromas. It has beautiful sweetness and fruit complexity, roundness and power. Like an old Pomerol or even an old Burgundy!' (MS) 'The nose is intense and mature with a presence of fruit, smoke, fine spices, dried fruit, fig, dates and prunes. The palate is concentrated, vigorous and complex with a beautiful spiciness, silky mouth-feel and utterly long finish. Very complete and beautiful.' (AL) 'Half bottles of 1962 St Henri remain noticeably better than standard bottles. Why?' (PG)

Bin 60A Coonawarra Cabernet Kalimna Shiraz ⬙ NOW

Medium-deep brick red. Fully mature wine with infinite complexity and richness. Ethereal praline, mocha, apricot, espresso, herb garden, polished leather aromas. Supple and fine-grained with seductive espresso, herb garden, dried fruits, apricot flavours, fine, loose-knit tannins and lovely clear gentle acidity. Finishes minerally and long with plenty of *sous bois*, earthy, leather notes. The most famous Australian wine of the 1960s, with an astonishing wine show record and subsequently a contemporary legend. Still holding up well, with enough substance and balance to last another twenty years.

'Marvellous old-wine bouquet with aspects of roasted meats, raisins, prunes, toffee and liquorice. Ethereal, soft and mellow, but intense with more power and drive than 1956 Magill Burgundy. Still has a core of fruit sweetness. Great balance and endless aftertaste.' (HH) 'Nose is mature with notes of spice and bitumen. Mellow, rounded and silky textured with lovely mid-palate sweetness and savoury, spice, liquorice, tobacco notes. Very fragile and harmonious.' (JB)

'An utterly superb wine, a glorious freak of nature and Man; ethereal and beguiling, yet the palate is virtually endless, with a peacock's tail stolen from the greatest of Burgundies; the fruit sweetness perfectly offset by acidity rather than VA. The 100-point dry red? Why not! This is possibly the greatest red wine tasted in our times in Australia.' James Halliday's classic tasting note in Penfolds The Rewards of Patience, 2008.

Bin 434 Coonawarra Claret PAST

Medium brick red. Intense leather polish, tarry, smoky chestnut aromas with a touch of nail varnish, praline and apricot. Very complex, supple and developed with leather, earthy, roasted chestnut, orange peel flavours, powdery tannins and mineral acidity. Walnut, amontillado notes pervade across the palate to the finish. A cabernet shiraz blend. Bottled at Penfolds Auldana Cellars. It's well past its best now.

'Fractured but fascinating with remnants of opulent fruit clutching to its foundations. It's still alive and structurally upright.' (TL) 'A curio with Madeira-like caramel notes and good mouth-feel, but it has faded now.' (RKP) 'Milk chocolate, savoury spice, mushroom and leather, even a touch of caramel on the bouquet. A core of pan juices, gamey flavours and smoky logs crackling on the fireplace. Dry, touch astringent finish.' (TS)

Bin 60 Kalimna Shiraz Coonawarra Cabernet Sauvignon NOW ••• PAST

Medium-deep crimson. Fresh, dark chocolate, prune, roasted chestnut, sweet leather, apricot aromas. Sweet plum, roasted chestnut, liquorice, fruit flavours with lovely chocolatey, lacy tannins and gentle acidity. Finishes long and minerally. A supple and elegant wine. A surprise as many examples are now past. The less famous 'off-blend' of Bin 60A. From the personal cellar of Peter Gago.

'Very mature earthy, polish, meat-stock, slight raisin characters with some cabernet herb notes. Savoury, smooth and soft with powdery tannins. Great impact and length.' (HH) 'Still robust black fruits from blackcurrant to blackberry with earthy, mocha, dark chocolate, spice notes and fully functioning tannins. Very Penfolds in style.' (JH)

Kalimna Bin 28 Shiraz, Bin 128 Shiraz PAST

Most bottles are past.

Bin 414 Sauternes NOW

Golden colour. Fragrant barley water, briny, tonic water, brassy aromas. Sweet, tangy and medium concentrated palate with barley water, lemon tonic flavours, lovely nutty notes and fresh acidity. Superb viscosity and length.

1963 ★★★★★

A warm and dry growing season and vintage • Penfolds and Wine Research Institute carry out malolactic trials • Russian cosmonaut Valentina Vladimirovna Tereshkova orbits the Earth forty-eight times aboard *Vostok 6* and becomes the first woman in space • Martin Luther King, Jr, delivers his fight for civil rights speech, 'I have a dream', in Washington • President John F. Kennedy is assassinated as he rides in a motorcade through Dallas, Texas.

Bin 95 Grange Hermitage ▽
NOW ••• 2025

Deep crimson to brick red. A magnificent wine with intense earthy, dried roses, dark chocolate, cedar aromas. The palate is richly powerful with deep-set dark berry, dark chocolate flavours, fine grainy, al dente tannins and underlying mocha, spicy nuances. Finishes velvety and long with some minerally graphite notes. One of the loveliest early Grange vintages. Still beautifully complex, richly flavoured and balanced. A very great Australian wine. Will continue to hold. Rare. 100% shiraz.

Kalimna Vineyard (Barossa Valley), Barossa Valley, Magill Estate (Adelaide), Morphett Vale (Adelaide), Modbury Vineyard (Adelaide). Released as Bin 95 (also as Bin 65).

'Powerful and full-bodied with perfectly ripened fruit, gentle oak and sturdy tannins. Has another decade at least.' (LM) 'Stewed raspberry, rhubarb, vanilla, cedar aromas. Mouth-filling, lush, creamy texture with notable supple tannins and great length.' (JC) 'Fragrant and appealing with blackberry, blueberry, espresso notes and robust tannins.' (RI) '100% Shiraz, 100% Grange, 100% delivered.' (PG)

St Henri ▽
NOW

Medium-deep brick amber. Fresh roasted earth, fired pottery, panforte aromas with herb, sage notes. Well-concentrated but medium-bodied wine with generous sweet earthy, panforte, redcurrant flavours and fine chalky dry tannins. Finishes chalky dry and minerally. Still has good drive and vinosity, but starting to fade. Drink now.

'Matured molasses, caramel, chocolate, toffee, mocha aromas and flavours. Simple and overdeveloped with a decaying sweetness of fruit.' (FK) 'Leather, mushroom aromas with overripe notes, dry bitter tannins and aggressive acidity.' (PK) 'An oxidative nose with notes of walnut and dried fruit. The palate is medium-bodied with high acidity, dried fruit character, good length and a hint of complexity.' (AL)

Bin 64 Cabernet Sauvignon
NOW

Medium-deep brick red. Fresh geranium, dried roses, cassis, vanilla, spice aromas with fresh bitumen notes. Well-concentrated, sweet dried roses, cassis, chocolatey flavours and plentiful chalky tannins. Firm, muscular and long with bitter-sweet notes at the finish. Definitely more evolved now, but still well balanced. Best to drink now. 1964 Jimmy Watson Memorial Trophy winner. The precursor wine to Bin 707. From the personal cellar of Peter Gago. 100% Block 42 cabernet sauvignon.

'Green stalky notes dominate the nose. The palate is juicy with plenty of savoury, tobacco flavours, but persistent bitterness and varnish/metallic notes diminish the wine.' (JB) 'The bouquet retains primary blueberry, blackberry, pepper and cassis with notes of game and earth, cedar and wood smoke. The palate is equally primary with concentration, drive and internal harmony, complex secondary fruit and fine, velvety tannins. A profound 1963 with decades left in it yet.' (TS)

Kalimna Bin 28 Shiraz
NOW ••• PAST

Pale-medium brick red. Fragrant leather, chinotto, spicy aromas with apricot notes. A fully mature wine with salted liquorice, chinotto, earthy, toffee flavours and fine slinky, lacy tannins. Finishes minerally and dry. Still holding well, but the fruit is beginning to fade now.

'Intense fresh raisin, muscat aromas with some honey notes. The palate is still alive although the fruit is now overdeveloped with a rich, custardy texture.' (TL) 'Lightening off considerably now. Toasty, roasting pan scrapings with cigar box, tobacco and raisin notes. Lean but soft and easy-going. Still a nice drink.' (HH)

Bin 128 Shiraz
NOW ••• PAST

Medium-deep crimson to brick red. Barnyard, leather, amontillado aromas. Supple, smooth, leather, amontillado, nutty flavours and fine loose-knit, chalky tannins. Finishes flavourful and long with some mineral notes. Surprisingly fresh and concentrated. Most bottles are not as good as this. Generally past.

'A gorgeous wine with terrific complexity of bouquet; roasting pan juices, raisin, liquorice and espresso coffee too! The palate is lively and firm but the fruit has mellowed substantially. Still has marvellous fruit sweetness.' (HH)

1964 ★★★

A wet growing season followed by a dry but cool vintage; not a bad vintage • Alfred Scholz, who built the underground drives at Magill and created 'Grandfather Port', retires • 1963 Penfolds Bin 64 Kalimna Cabernet Sauvignon, using Block 42 material, wins the prestigious Jimmy Watson Memorial Trophy at the Melbourne Wine Show • The Beatles' Australian tour following their spurt to fame in the US only months earlier • Nelson Mandela found guilty of sabotage and treason and sentenced to life imprisonment • Constantine II becomes King of Greece.

Bin 95 Grange Hermitage — NOW ••• PAST

Medium-deep brick red. Rustic wine with intense mushroom/*sous bois*, graphite, bitumen, mint aromas. The palate is well concentrated with mushroom, panforte, sweet fruit flavours and pronounced leafy dry tannins. Some pine needle notes. Tenuous quality now with overdeveloped fruit and drying textures. Rare. 90% shiraz, 10% cabernet sauvignon.

Magill Estate (Adelaide), Kalimna Vineyard (Barossa Valley), Barossa Valley. Released as Bin 95 (also as Bins 395, 66, 67 and 68).

'*Slight volatility with exotic orange, furniture polish, characters.*' (TG) '*Camphor, iodine, savoury aromas and flavours.*' (DM)

Bin 707 Cabernet Sauvignon ◈ — NOW

Medium-deep tawny brick red. Beautiful panforte, sage, mint, praline, sandalwood aromas with some primary black fruits, cassis notes. Well-concentrated, mature sweet fruit, panforte, sage, superb mature dark chocolate, blackberry fruit, briny, liquorice flavours with lovely mid-palate richness and underlying hints of malt and vanilla. Fine loose-knit, chalky, touch bitter tannins and a lovely tail of sweet fruit. Very complex and evolved with superb volume and persistency. A remarkable old wine that should hold for some years. 100% Block 42 cabernet sauvignon. (Tasted in Beijing and Adelaide.)

'*Youthful nose dried herb, Chinese medicine store, faded mulberry, cedar notes. Elegant, linear and fine-boned with lots of dried Chinese herbs/haw* (shanzha)*/spice, silky texture and lingering length.*' (FW) '*Vivacious sandalwood, liquorice, red cherries, prunes and mocha aromas. Amazing delicate sweetness of fruit. Evolved, mature but with solid legs to carry it into the future. Stubborn, obstinate freshness. Deserves all six stars out of my normal maximum five-star rating!*' (CPT) '*Has aged very gracefully with complex herbal, liquorice, cedar notes and beautiful concentration, but the tannins are beginning to fade.*' (YSL) '*Showing wonderful depth and complexity with a meaty, charcuterie, stock-like bouquet. Rich, full-bodied and luscious with a twinge of volatile acidity. A staggeringly powerful and intense wine with monumental depth of flavour. Still impressively robust and flavourful at this great age.*' (HH) '*A lovely mature nose of syrupy black fruit and dusty aged complexity. Subtle, long and restrained with seamless mouth-feel, plump succulent fruit and melted tannins. Finishes firm but long and fine.*' (RKP)

St Henri — NOW ••• PAST

Medium-deep brick red. Fragrant, shoe polish, leather, redcurrant, apricot, salted liquorice aromas. Fresh supple and fully mature with redcurrant, leather, praline, apricot, sandalwood flavours and fine long loose-knit savoury dry tannins. Finishes chalky firm and minerally with apricot notes. At the end of its life but still drinking well.

'*A hint of Turkish Delight on the nose and a feminine, sensual palate; more Henrietta than Henri!*' (NM) '*Very Rhônish in style with fine spicy, smoky, black olive, roasted currant aromas. Loose-knit and slimly built on the palate with noticeable acidity at the finish.*' (FK) '*Salty, sweet liquorice aromas with hints of smoke, spices and dried fruits. A nicely smooth and balanced palate with velvety texture, complex spiciness and walnut, leather notes. Nice length and grippy finish.*' (AL)

Bin 389 Cabernet Shiraz

NOW ••• PAST

Medium-deep brick red. Fragrant and complex toffee, leather, graphite, dark chocolate aromas with some herb notes. Well-concentrated and developed leather, sweet dark chocolate, leafy flavours with fine lacy dry tannins. Finishes quite chalky and lean. Starting to fragment. Some bottles are now on the downward slide. Drink up.

'It's soft and graceful, with mushroom, truffles and game flowing delightfully through a long palate, tannins gentle and calm.' (TS) 'Dried fruit, dark berry, plum, spiced fruitcake aromas. Rich and slightly porty but works well as a generous, smooth mature red wine.' (RKP)

Kalimna Bin 28 Shiraz

NOW ••• PAST

Medium brick red. Intense black fruits, roasted coffee, orange peel aromas with a touch of panforte. Roasted coffee, meaty, apricot flavours, chalky dry tannins and some leafy notes. Finishes firm with fresh acidity. It still has some richness of fruit, but the wine is losing its lustre now. Drink up.

'A rich savoury wine with intense roasted spice, dark chocolate, meaty notes. It falls away a little towards the finish but it's drinking surprisingly well; a great older wine.' (JB) 'Deeper coloured than 1963, but a little musty, old and tired.' (RKP)

Bin 128 Shiraz

PAST

Most bottles are past.

1965

★★★

A warm, dry growing season and vintage • Penfolds introduces the very first 'bag in the box' packaging • Death of Sir Winston Churchill at the age of ninety • The Rolling Stones' mega-hit song, '(I Can't Get No) Satisfaction' • Prime Minister Sir Robert Menzies commits to sending Australian troops to join forces with the US in Vietnam.

Bin 95 Grange Hermitage

NOW ••• 2020

Medium-deep brick red. Fresh leather, tobacco, mint aromas with some panforte notes. A savoury but voluminous wine with cedar, tobacco, leather, orange peel, panforte flavours and fine-grained firm tannins. Finishes leafy firm. Plenty of vinosity and substance. Rare. Jimmy Watson Memorial Trophy Winner. 95% shiraz, 5% cabernet sauvignon.

Kalimna Vineyard (Barossa Valley), Barossa Valley, Magill Estate (Adelaide), Morphett Vale (Adelaide), McLaren Vale. Released as Bin 95 (also as Bins 69, 70 and 71).

'Juicy and exuberant with dark cherry, plum aromas, supple tannins and crisp acidity. Great length, composure and sweetness of fruit. A bigger Grange than most!' (LM) 'Lacks definition on the nose but tastes soft and delicious with notes of underbrush, cedar, chocolate and herbs.' (MC) 'Spicy, orange, Curaçao aromas. Nice texture with gritty, grippy tannins and lovely fruit complexity.' (DM).

Bin 707 Cabernet Sauvignon

NOW ••• PAST

Medium-deep brick red. Fresh sea breezy, mint, cherry, cedar aromas. Some mocha notes with touch of sweet fruit. Fresh-ground coffee, cedar, dark chocolate, red cherry flavours with lacy, bitter, dry tannins and marked acidity. Finishes long and savoury with minerally notes. Starting to fade but still enjoyable.

'Pronounced herbal, cedar, dusty, tea leaf (pu-er), light soya sauce/Chinese medicine notes (jiangyou). Palate is smooth, dusty and slightly drying with medicinal, deep spice, cedar, dried fruit (guopu) notes and well-balanced tannins. A core of fruit complexity emerges at the finish.' (FW) 'Leather, "animal", smoked ham, dried plum aromas with myriad spices (clove, cinnamon, cumin). Weighty ripe chewy/tannic structure, refreshing acidity and very good length. Drinking well now.' (ER) 'Nose is still sweet with dried plum aromas, a little volatile acidity and tough tannins.' (ST)

SPECIAL VINTAGE
ST. HENRI
CLARET
Vintage: 1964
Comments: This is a superb early St. Henri showing the benefits of extended bottle ageing.
Bottled: In April, 1966 at Penfolds Auldana cellars. The wine is a full, dark and very rich style noted for its excellent bouquet and flavour development.
A special Limited Release.
PENFOLDS WINES PTY. LTD.
Sydney and Adelaide
PRODUCE OF AUSTRALIA
730ml
VINTAGE 1965
SPECIAL VINTAGE
ST. HENRI
CLARET
PENFOLDS
Auldana
CELLARS
Bottled by
PENFOLDS WINES PTY. LTD.
MAGILL
1 PINT 6 FL OZ
VINTAGE 1966
SPECIAL VINTAGE
ST. HENRI
CLARET
PENFOLDS
Auldana
CELLARS
Bottled by
PENFOLDS WINES PTY. LTD.
MAGILL
1 PINT 6 FL OZ

St Henri

Medium-deep brick red. Complex and fragrant panforte, orange peel, mocha, praline, apricot and prune aromas with *sous bois*/floral/herb notes. Beautifully concentrated and buoyant with rich panforte, mushroom notes and ripe chocolatey tannins. Finishes long and sweet. Fully mature. Drink up.

'Medium-bodied with a pleasant, supple entry. It's very well balanced and defined but quite youthful compared to the 1964. This is a dainty Shiraz with a Burgundian texture. An elegant finish with touches of allspice, balsamic, cumin and white pepper. If the aromatics were more complex, this would be a very fine St Henri. Drink now.' (NM) 'Fresh ripe chocolate, herbal aromas, medium concentration and strong tannins.' (PK)

Bin 28 Shiraz, Bin 128 Shiraz

Most bottles are past.

1966 ★★★★

A dryish vintage with intermittent rain during the growing season; vintage was warm and dry; considered by many as a great Grange vintage • The 1965 Grange wins the Jimmy Watson Memorial Trophy • 1966 Bin 620 Coonawarra Cabernet Shiraz is made; it blitzes the wine shows and becomes a legend • Decimal currency introduced to Australia by Prime Minister Sir Robert Menzies • England claims its first FIFA World Cup win, beating Germany 4–2 in the final at Wembley Stadium • Chairman Mao Tse-tung launches the Cultural Revolution with his Red Guard army.

Bin 95 Grange Hermitage ♈

Deep brick red. Very classic dark chocolate, dark berry aromas with fresh espresso, cedar, mushroom notes. Gorgeously concentrated with praline, dark berry, apricot, orange peel flavours and plentiful sweet chocolatey tannins. A powerful, dense and richly textured wine with superb extract and length. At peak of maturity; good bottles will continue to hold. Rare. 88% shiraz, 12% cabernet sauvignon.

Kalimna Vineyard (Barossa Valley), Barossa Valley, Magill Estate (Adelaide), Morphett Vale (Adelaide). Released as Bin 95 (also as Bins 71 and 72).

'Fabulously floral with pot pourri, bacon fat, blackberry, liquorice aromas and leafy, herb complexity. Lush and juicy with sweet fruit and big tannins. A hedonistic wine, especially for its age.' (LM) 'Substantial weight and volume, lovely balance and power. The star of the 1960s.' (TG) 'Full-bodied, tannic and unyielding. Terrific concentration but does it have the balance of great Grange?' (JC)

Bin 707 Cabernet Sauvignon ♈

Deep crimson to brick red. Classic cassis, roasted chestnut, leafy, green pea, asparagus aromas. Fresh and intense but quite soupy. Delicious, roasted chestnut, praline flavours and fine, loose-knit, chocolatey tannins. Finishes long and sweet with bitter tea leaf, sage notes. Lovely mid-palate buoyancy and richness. Drinking superbly now.

'Lovely, graceful integration of evolved sweet oak and sweetness of blue/black fruits. Mature and supremely balanced with obvious further potential to age. Tenacious, inveterate freshness on the finish. Along with 1964, 1966 is one of the greatest expressions of pure Cabernet Sauvignon anywhere on this planet.' (CPT) 'Sweet sandalwood, Chinese bark medicine, incense aromas. Supple, textured and rich with slightly bitter sweet notes and mocha oak.' (ST) 'Interesting wine, with smoky bacon, toffee, soy aromas. It has plenty of aged characters with date, tobacco leaf, sweet liquorice notes. It's like an old movie with a fluid plot and full of nostalgia.' (CC)

St Henri ♈

Medium-deep brick red. Complex panforte, prune, marmalade aromas with touches of toffee, smoky nuances. Supple and generous with panforte, mocha, redcurrant flavours, fine grainy/al dente tannins and lovely sweet fruit notes. Finishes long and sweet with a touch of bay leaf. Fully mature but classically proportioned and beautiful to drink. Holding well but drink soon.

*'Very deep granite red. Dark spices, toffee and exotic notes. Seamless and integrated. Just a beauty.' (MS)
'Ripe and fresh with complex ripe cedar, chocolate, tobacco, leather notes, supple tannins, good balance and finish.' (PK) 'Shows energy and poise with sweet fruit, confiture, sweet spices and chocolate praline notes. The palate is nicely structured, full and juicy with a firm tannic backbone. Great grip and persistence on the finish.' (AL)*

Bin 620 Coonawarra Cabernet Shiraz NOW ••• PAST

Medium-deep brick red. Fresh redcurrant, leather, crème brûlée, graphite aromas. Surprisingly fresh and vibrant with blackcurrant, mocha, crème brûlée, sweet fruit flavours, fine graphite tannins and soft but balanced acidity. Finishes lacy firm with lovely espresso, roasted coffee, dark berry notes. An exceptional bottle showing remarkable freshness, complexity and harmony. From the personal cellar of Peter Gago.

'A lovely contemplative wine with expressive leather, tobacco notes and smooth, rounded texture. Like an old chesterfield sofa.' (TL) 'A fragrant bouquet is followed by a supple, fine-textured palate. The Coonawarra earth dovetails with the fruit. It finishes exceptionally long, silky and harmonious.' (JH) 'Layers of anise, sweet liquorice, pan juices and wood smoke. The palate is supple and complex with a core of sweet fruit, drying out just a little on the finish. The line and persistence are unrelenting.' (TS)

Bin 389 Cabernet Shiraz ♆ NOW ••• PAST

Medium-deep brick red. Salted liquorice, seaweed aromas with some red fruit notes. A touch oxidised with salted liquorice, dried fruit, earthy flavours and sinewy dry, al dente tannins. Leafy firm at the finish. The fruit has dropped out now. A great old vintage diminished by time. Drink up.

*'Not convinced by this bottle. Slight wet-cork notes and oxidation over a rich underlay of flavours.' (JH)
'Richly concentrated and savoury wine with roasted meats, spice, tobacco notes, mellow but present tannins and fresh acidity. Underlying cork, wood flavours.' (JB)*

Bin 128 Shiraz PAST

Medium-deep crimson brick red. Fresh panforte, leather, walnut aromas. Beautiful concentration, fresh panforte, dark chocolate, sweet fruit flavours, some walnut notes and slinky, dry tannins. Finishes grippy, long and savoury with sweet fruit notes. Generally past!

'Pleasant old-wine bouquet with slight raisin, meat-stock notes. Good depth and substantial grip. Still tasty but doesn't have quite the finesse of the 1963.' (HH)

Bin 2 Shiraz Mourvèdre NOW ••• PAST

Crimson brick red. Fragrant, redcurrants, chestnut, earthy aromas with a touch of espresso/leaf. Beginning to dry out with leather, shoe polish, dark chocolate flavours, some amontillado/maderised notes and fine chalky, plentiful tannins. Finishes dry with medium length. A fading old wine yet still pleasant to drink.

'Very earthy, dusty, leathery old-wine bouquet with tinges of liquorice, anise, crushed ants and meat stock. Lean and light in the mouth but it no longer has much depth or impact. Nonetheless a lovely drink.' (HH)

1967 ★★★★

A relatively dry growing season followed by a warm vintage • The Clare Valley is now a source of grapes for Grange • Penfolds works with Australian Wine Research Institute on the role of yeasts and oxygen in wine-making • 1967 Bin 7 Coonawarra Cabernet Sauvignon Kalimna Shiraz, another legendary bottling, is made; it wins more critical acclaim for Penfolds. • Israeli forces take the old city of Jerusalem during the Six-Day War • Iconic communist hero Che Guevara is killed in Bolivia • South African surgeon Christiaan Barnard performs the first successful human heart transplant • Australian Prime Minister Harold Holt disappears while swimming near Portsea, Victoria, presumed drowned.

Bin 95 Grange Hermitage

NOW ••• 2020

Deep brick red. Intense dried plum, dark chocolate, espresso, liquorice, herb aromas and flavours. The palate is supple and fruit-sweet with velvety – slightly leafy – tannins and savoury, mocha, cedar notes. It finishes chalky firm with a long tannin plume. Starting to fragment and losing its finish. Increasingly rare. 94% shiraz, 6% cabernet sauvignon.

Kalimna Vineyard (Barossa Valley), Barossa Valley, Clare Valley, Magill Estate (Adelaide). Won the Jimmy Watson Memorial Trophy at the Melbourne Show, the second time in three years. Released as Bin 95 (also Bin 74).

'Reminiscent of old Grand Cru Claret. Fresh and supple with meaty, blueberry fruit, savoury notes and old cork flavours.' (TG) 'Coffee, dark fruit, soy, milk chocolate aromas. Rich, round and solid with seamless tannins. One of the better wines of the 1960s.' (RI) 'Full-bodied, richly textured with meaty, savoury, earth flavours and cedar vanilla notes. More resolved and friendly than 1966.' (DM)

Bin 707 Cabernet Sauvignon

NOW ••• PAST

Medium brick red. Fresh, intense, minerally, earthy, spicy aromas with some dark chocolate notes. Well-concentrated and elegantly structured wine with praline, spice, mocha, vanilla, earthy flavours, fine cedary tannins and underlying savoury notes. Finishes bitter-sweet. The fruit is beginning to fade.

'Developed animal, "sweaty" leather, cheesy notes with mellowed red and black plum fruit. More fruit on the palate with a meaty dimension, a big structure of coating, drying tannins and very good length.' (ER) 'Some plum notes and sweet residual oak but astringent and drying out.' (ST) 'It's exuberant and fresh at first with great fruit character and tannin presence, and looks ten years younger. But it fades quickly and just looks like an old wine.' (CC)

St Henri

NOW ••• PAST

Medium-deep brick red. Developed praline, roasted chestnut, dried fruit aromas with rustic, herb garden notes. Surprisingly rich and flavourful with roasted chestnut, praline, fruit-sweet flavours and chalky, slightly sappy tannins. Finishes vigorous and chocolatey firm with an acid twist. Has reached its vertex. Other examples are fading. Drink soon.

'Composed, settled and savoury. Plush, rich and fruit-sweet with powerful grippy tannins. Refreshing acidity runs like a silver thread through the wine.' (NB) 'Roasted coffee, black truffle aromas. Lean, rustic wine with strong tannic structure. No further development expected.' (FK)

Bin 7 Coonawarra Cabernet Sauvignon Kalimna Shiraz ▽

NOW

Medium-deep brick red. Fresh, complex and fully mature panforte, *sous bois*, praline, violet, earthy aromas. Seductive and sinuous with sweet panforte, praline, earthy flavours and loose-knit chalky tannins. A beautiful al dente structure with lovely complex sweet fruit notes and great flavour length. A classic Penfolds wine in remarkable condition.

'The fragrance is so dramatic and expressive. The high-toned cabernet and sweet velvety shiraz are locked perfectly hand-in-hand. A glorious wine with timeless beauty.' (TL) 'Luscious and opulent wine. It builds momentum and the elements coalesce as it passes the mid-palate. Very much in the Bin 60A finishing school. It lost the plot during the 1970s, then came back and hasn't faltered since.' (JH)

Bin 389 Cabernet Shiraz

NOW ••• PAST

Medium brick red. Fresh toasty, marmalade, roasted chestnut, redcurrant, sage aromas. Very complex and concentrated with toasty, roasted chestnut, leather, redcurrant flavours and fine chalky, slightly chewy tannins. Still tangy and vibrant with some graphite, smoky notes at the finish. A lovely old wine but now near its end. Drink up.

'Expressive and pure with a savoury, earthy, chewy, woodiness. It's solid in nature with pretty makeup!' (TL) 'Wonderful fragrant spice, herb and mint notes. A fragile but still lively palate with attractive thyme/mint nuances, savoury undertones, lovely mid-palate richness and sweetness. A beautiful, silky textured and understated wine.' (JB)

Kalimna Bin 28 Shiraz
NOW ••• PAST

Medium-deep brick red. Fresh prune, dried roses, graphite, smoked meaty aromas. Gorgeously seductive and fully mature wine with prune, dried roses, panforte, praline, smoky flavours and loose-knit but fine graphite tannins. A silky textured wine nearing the end of its life, but holding surprisingly well. Drink up.

'Slightly earthy, briary bouquet but the palate has much to offer with gently sweet shiraz fruit, good length and mouth-feel.' (JH) 'An animal and gamey bouquet, with a hint of coal dust and compost. The palate has all of the same, with a contracted finish.' (TS)

Bin 128 Shiraz
PAST

Most bottles are past.

Bin 868 Sauternes
NOW

Medium light mahogany green colour. Fresh barley water, orange peel, apricot, marzipan, rancio, nutty aromas. Intense baked apricot, orange peel, lemon glacé flavours, lovely sweet barley water notes and fine long refreshing acidity. A beautiful wine. Like Sauternes but a touch of sherry! (Tasted at Penfolds Red Wine Re-corking Clinic, Perth 2012.)

1968 ★★

A hot, dry and prolonged vintage, but with an alcohol level of only 12.1% for Grange • 1967 Grange Hermitage wins the Jimmy Watson Memorial Trophy • Civil rights leader Martin Luther King, Jr, is assassinated in Memphis, Tennessee • The discovery of the 'Mungo Man' skeletal remains in the Willandra Lakes region, NSW, proves that Aboriginal people lived in Australia at least 40 000 years ago • Rock band Led Zeppelin plays its first show, at the University of Surrey, England.

Bin 95 Grange Hermitage
NOW

Medium brick red. Panforte, *sous bois*, mushroom aromas with some spicy, orange peel, savoury notes. Seductively smooth rich panforte, praline, soy flavours with al dente, leafy textures and roasted earthy notes. Finishes brambly dry with a slick of tannins. A fresh example, but generally variable in quality. Drink up. 94% shiraz, 6% cabernet sauvignon.

Kalimna Vineyard (Barossa Valley), Barossa Valley, Magill Estate (Adelaide), Clare Valley, Adelaide Hills, Coonawarra. Released as Bin 95 (also as Bin 826).

'Complete, seamless, effortless wine with red fruits, peppery, spice aromas, silky textures and smoky notes.' (JR) 'The surprise of the 1960s, with its amazing supple sweet fruit, soy, orange peel characters and grainy/gripping texture.' (TG) 'Fresh flora, herb, smoked bacon, leather notes. A bit rustic but complex, richly aromatic and flavourful.' (MC)

Bin 707 Cabernet Sauvignon
NOW ••• PAST

Medium-deep brick red/brown. Beautiful dark cherry, tarry, cedar, tobacco aromas with sweet fruit notes. Sinewy dry, long tannins and dark cherry, tarry, tobacco flavours. Still has plenty of chalky substance. Herb-like, chalky finish. Still minerally and fresh. Surprisingly fresh and balanced. Many bottles have faded. Drink up.

'A mature savoury nose with cedar, cigar box, tobacco, pine resin/herb notes with mellowing cassis and black plum fruit. Well balanced with coating angular tannins, ripe sweet fruits and very good length. More savoury in style but impressive nonetheless.' (ER)

St Henri
NOW ••• PAST

Medium-deep brick red. Intense praline, leather, apricot, bitter orange aromas with sage, leafy nuances. Well-concentrated redcurrant, praline, leather flavours with earthy, sweet fruit notes and fine, plentiful, slinky tannins. Finishes firm but long and sweet. Not a classic, but holding up well.

'Potently stemmy but clearly not green or vegetal. A lovely savoury quality. Full-bodied and structured with slightly aggressive underlying tannins and nervy acidity. Still far from falling apart.' (NB) 'Mono-dimensional but finely crafted wine with appealing minerality.' (PM) 'Absolutely enjoyable to drink now.' (MS)

Bin 28 Shiraz, Bin 128 Shiraz PAST

Most bottles are past.

1969 ★

A very wet growing season followed by a mild, wet vintage • Grange did well on the wine show circuit, but a poor vintage. • 1969 Bin 707 Cabernet Sauvignon is the last of the line until 1976 • Maiden flight of Concorde, the world's first supersonic aeroplane, takes off from Toulouse, France • Neil Armstrong becomes the first man to set foot on the moon, watched by an audience of 450 million people • Over 450 000 people attend the most publicised counterculture rock festival, Woodstock.

Bin 95 Grange Hermitage NOW ••• PAST

Medium brick red. Intense and lifted dark chocolate, amontillado, roasted capsicum aromas. The palate is well concentrated and mature with dark chocolate, amontillado flavours and savoury, bitter tannins. Finishes chalky firm and a touch leafy. On the downward slide. Drink up. 95% shiraz, 5% cabernet sauvignon.

Kalimna Vineyard (Barossa Valley), Barossa Valley, Magill Estate (Adelaide), Clare Valley, Morphett Vale (Adelaide), Coonawarra. Released as Bin 95 (also as Bin 826).

'Medium-bodied wine with lifted cocoa, dark fruit aromas and flavours. Tannins are sharp and the volatile acidity is not blowing off.' (LM) 'Deliciously decadent, dry, complex and profound.' (MC) 'Vaguely beef stroganoff, cocoa powder aromas with slick, grippy tannins.' (RI)

Bin 707 Cabernet Sauvignon PAST

Brick red. The wine is leathery, dry and skeletal. Most bottles presented at the Wine Clinics are past.

St Henri PAST

Medium-deep brick red. Very floral, leafy, earthy, strawberry, sappy aromas with a touch of liquorice. Quite firm but still fresh with leafy, redcurrant flavours, dark chocolate, candied orange notes and fine lacy, firm tannins. Finishes long and chewy. Quite a surprise as most examples are now past.

'Seductive Burgundy-like bouquet that offered immense purity and super-fine tannins that underpinned the savoury-tinged palate. A surprise package.' (NM) 'A very elegant Burgundian style with its perfumed aromas, mature silky texture, beautiful composure and liquorice, mint notes.' (PM)

Kalimna Bin 28 Shiraz NOW ••• PAST

Medium-deep brick red. Light roasted hazelnut, chocolate aromas. Lightly concentrated with tobacco, hazelnut, chocolate flavours and loose-knit chalky dry tannins. Finishes firm and minerally. Drink now. (Tasted at Penfolds Red Wine Re-corking Clinic, Perth 2012.)

Bin 128 Coonawarra Shiraz PAST

Most bottles are past.

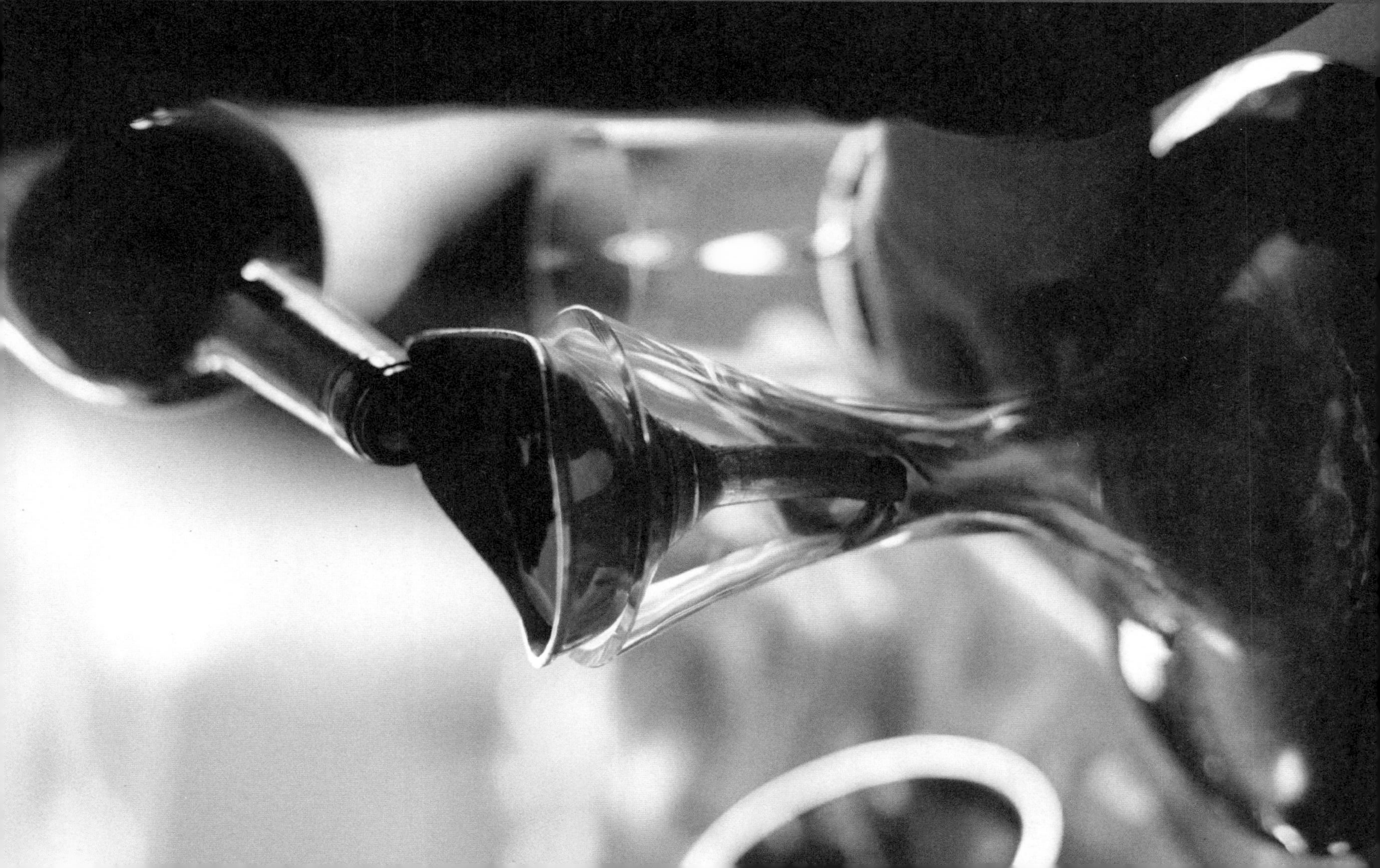

A dry, mild growing season and vintage • Grange bottlings are standardised as Bin 95 • Poseidon 'fever' grips the London Stock Exchange as the Australian nickel share soars by year's end to hit a record as the biggest performer in memory • 'Houston, we have a problem' – *Apollo 13* is forced to abort its lunar mission after an onboard explosion • Australia celebrates the bicentenary of Captain James Cook's landing at Botany Bay • Death of electric blues guitarist Jimi Hendrix in Great Britain.

Bin 95 Grange Hermitage   NOW

Medium-deep brick red. Intensely fragrant and complex walnut, fresh dark truffle, panforte aromas with hints of herb garden. Well-concentrated and supple with walnut/roasted coffee, praline flavours, fine loose-knit and slightly chalky dry tannins. Finishes leafy firm at the finish. Drink soon. 90% shiraz, 10% cabernet sauvignon.

Kalimna Vineyard (Barossa Valley), Barossa Valley, Magill Estate (Adelaide).

'Smoky, herbs, dried cherry, cassis, tobacco aromas. Round and plush with sweet fruit, smoky, vanilla flavours and velvety textures. Becomes spicier and more vibrant with aeration.' (JR) 'Lovely delicate mature wine with meaty balsamic notes. Primary fruit has faded.' (TG) 'Sweet tawny aromas with cedar, wood, spice notes. Still fresh and lively with silky textures.' (DM)

St Henri   PAST

Medium-deep crimson to brick red. Rusty, leather, sandalwood, leafy aromas with some gamey notes. Medium concentrated and mature with earthy, redcurrant, leather, *sous bois* flavours and fine slinky firm tannins. Finishes bitter and leafy dry. The fruit is beginning to fade now but it is still nicely discreet and balanced.

'The nose is rustic and ferrous, with hints of dried rose petal. Medium-bodied with a sour note on the entry. It's a little aggressive and coarse, with savoury, meaty, decayed fruit and a tapered, autumnal finish. This should be drunk soon. Past its best.' (NM) 'Charming and elegant with lovely delicacy, restrained tannins and well-integrated fine acid structure.' (FK)

Bin 389 Cabernet Shiraz   NOW

Medium-deep brick red. Intense camphor, herb garden, redcurrant, leather, mushroom aromas with lovely toasty, marmalade notes. Well-concentrated redcurrant, orange peel, demi-glace, leather flavours with fine loose-knit, muscular tannins. Finishes chalky firm with some leafy notes. Still has concentration and richness. Some bottles are beginning to dry out now.

'A fully secondary game and mushroom profile with tannins still holding out; carrying the fruit long and confidently through to the finish.' (TS) 'Mature and rich, with light mushroom, red berry aromas along with old spice. Still quite lively and appealing but with slightly musty overlay.' (RKP)

Kalimna Bin 28 Shiraz   NOW ••• PAST

Medium-deep brick red. Roasted, earthy, liquorice aromas with coffee notes. Concentrated, roasted earthy, liquorice, panforte flavours with some lacy, graphite notes. Finishes sinewy firm with excellent length. Best to drink soon.

'Looking old and faded with sweet fruit, leather, fruitcake, spice notes. The palate is earthy with ripe tannins and marked acidity.' (JF) 'This has truly developed to a tertiary phase but the palate texture and grip is superb. The flavours still show their vinous origins.' (TL)

Bin 128 Shiraz   NOW ••• PAST

Medium-deep crimson to brick red. Very fresh, mushroom, truffle, panforte aromas with dark chocolate notes. Fresh mushroom, truffle, panforte flavours with a touch of leather and spice, fine-grained, muscular tannins and fresh mineral acidity. Finishes chalky with plenty of flavour length.

'A spicy bouquet with mushroom notes and hints of cork mustiness. A touch disjointed with a firm grip. Good bottles would still be in good drinking condition.' (HH)

1971

★★★★★

A great South Australian vintage; ideal, generally warm conditions throughout the growing season and vintage • The legendary 1971 Grange Hermitage, with only 12.3% alcohol, is made; it becomes an 'alternative investment' indicator during the 1990s • Wine packaging goes metric in Australia • East Pakistan secedes from West Pakistan, creating the new, independent nation of Bangladesh • China launches its first space satellite • IBM introduces the first 'floppy disk' data storage system.

Bin 95 Grange Hermitage ⬙

NOW ••• 2020

Medium-deep brick red. A classic Penfolds year with an unusually low alcohol, around 12.3%. Very beautiful, fresh, developed apricot, panforte, vanilla, cedar, sweet fruit aromas with touches of tar/bitumen. Fully mature wine with complex apricot, panforte, roasted chestnut, smoky, espresso flavours and superb lacy dry tannins. Finishes long and sappy with graphite firmness. Looking quite fragile. 87% shiraz, 13% cabernet sauvignon.

Kalimna Vineyard (Barossa Valley), Barossa Valley, Magill Estate (Adelaide), Clare Valley, Coonawarra. 'If you had to point to a wine which fulfilled the ambitions of Grange it would have to be the 1971,' Max Schubert, 1993. Topped the Gault-Millau Wine Olympiad in Paris in 1979 – beating some of the best Rhône wines and creating a sensation.

'Like an old school Southern Rhône with wild meaty, sweet liqueur aromas and long sappy sweet flavours. A wild gamey earthy expression for the natural wine crowd!' (JR) 'Extraordinary youthful bouquet with luxurious texture, mouth-coating richness and lingering finish.' (LM) 'A classic Grange. Complex and elegant, harmonic and lively.' (MC)

St Henri ⬙

NOW ••• 2018

Medium-deep crimson to brick red. A beautifully mature and aromatic wine with redcurrant, roasted walnut, espresso, *sous bois* aromas and hints of gun flint/smoke. The palate is beautifully balanced with concentrated roasted walnut, espresso, demi-glace, panforte flavours, wonderful mid-palate richness and fine silky tannins. Finishes chalky firm and long. Superb fruit complexity and volume. A classic Penfolds St Henri vintage. Drink soon.

'Like the restrained exoticism of a veiled belly dancer. Exquisite pure sandalwood, meaty aromas. Gloriously dense with unparalleled richness and harmony. Great inner life and latent energy. Supreme textural refinement allied to sheer beauty of flavour. Composed, settled finish. Gently lingering.' (NB) 'Very hedonistic wine showing its wonderful best.' (MS) 'Burgundian-like nose with gently spicy red fruit aromas and floral touches. The palate is far more powerful than the nose suggests with good grip, persistency and intense aftertaste. A discreet powerhouse.' (AL) 'This St Henri always gives '71 Grange a run for its money!' (PG)

Bin 389 Cabernet Shiraz ⬙

NOW

Medium-deep crimson to brick red. Fully mature wine with praline, apricot, dark chocolate, earthy, polished leather aromas. Supple, dense, chocolatey wine with developed dark chocolate, espresso, herb flavours and lacy dry tannins. Finishes flavourful and minerally with some mushroom notes. At the brink of its life, but still impressively complex and balanced.

'A triumphant old Bin 389 with dark chocolate, beef jus and mixed herb aromas, some secondary berry fruit notes and a wonderful core of sweet intensity, perfectly upheld by the final remnants of tannin.' (TS) 'A truly lovely mature wine, still with a commanding presence. The line and length are admirable, as is the balance of fruit, oak and tannins.' (JH)

Kalimna Bin 28 Shiraz ⬙

NOW •• PAST

Medium-deep brick red. Geranium, dried roses, celery, coffee aromas. Buoyant redcurrant, sweet fruit, sage, liquorice flavours and fine gritty tannins. The fruit is beginning to dry out. It's now past its best.

'Wonderful aromas of anise, sarsaparilla, prunes, blueberries and notes of exotic spice. The palate carries all of the same, with enticing complexity, finishing a touch dry and contracted, yet with lingering liquorice notes.' (TS) 'Developed colour, of course, but it possesses that extra dimension of flavour, typical of the 1971 vintage. The fruit, however, is starting to dry out. Drink as soon as possible.' (JH)

Bin 128 Shiraz
NOW ••◦ PAST

Medium-deep brick red. Intense saddle-leather, mint, dark chocolate aromas with earthy, smoky, meaty, bitumen notes. Fresh, but a touch metallic with developed leather, mint, tarry flavours, chalky, loose-knit tannins and some mineral notes. Finishes chalky, firm and long. Drink soon.

'Complex savoury, earthy meat-stock flavours with pepper, spice, old leather notes and drying tannins. Starting to taste very mature, but still a very good drink.' (HH)

1972
★★★

A dry growing season followed by a mild vintage • Jeffrey Penfold Hyland is awarded an OBE for services to the wine industry • Beginning of the Watergate scandal that would ultimately bring President Nixon down • Storming of the Munich Olympic Games by Palestinian 'Black September' terrorists leaves nine hostages massacred • Australian wool prices hit a twenty-year high with a bale fetching up to $7.20.

Bin 95 Grange Hermitage
NOW ••◦ 2022

Medium-deep brick red. Quite developed with earthy, *sous bois*, sweet fruit, roasted chestnut, sandalwood aromas. Beautifully textured wine with cedar, sweet fruit, apricot, earthy flavours and fine supple tannins. Finishes al dente firm but long and sweet. This has always been a controversial wine, yet always performs well at Rewards of Patience tastings! Drink now. 90% shiraz, 10% cabernet sauvignon.

Kalimna Vineyard (Barossa Valley), Barossa Valley, Magill Estate (Adelaide), Modbury Vineyard (Adelaide), Coonawarra. A very good Grange vintage. Reportedly a batch was unintentionally oxidised during bottling, resulting in significant bottle variation.

'Wild, heady, expressive, suave and well-knit wine with lovely cherry, vanilla, cassis, mocha aromas and velvety sweet palate. Amazing clarity of fruit, long and seamless.' (JR) 'Roasted fruits with savoury, soy notes. Complex sweet fruit notes and grippy tannins.' (RI)

St Henri
NOW

Medium brick red. Slightly corky at first but it blows off to reveal fresh truffle, rusty, chesterfield leather, polish, butterscotch toffee aromas with hints of dried roses and tar. Fully mature wine with praline, prune, leather flavours with spicy, cinnamon notes and fine loose-knit chalky but supple tannins. Finishes long and sweet. Still holding up very well.

Bin 389, Bin 28, Bin 128
PAST

Most bottles are past their optimum drinking window.

1973
★★

A dry growing season followed by a cool vintage • Last Grange Hermitage to be vinified at Magill in Adelaide although maturation of most Grange and Bin wines still takes place at Magill until 1996 • Refrigeration breaks down halfway through vintage, but wines prevail at future wine shows • Penfolds acquires the Koonunga Hill property in the Barossa Valley • Ray Beckwith retires • Ceasefire in Vietnam as an agreement is reached between the US and North Vietnam • End of the UK–Australia Trade Pact, a preferential tariff agreement set up in 1932 • Australian swimmer Shane Gould becomes the first woman in history to swim 1500 metres in less than 17 minutes • The voting age in Australian federal elections is lowered from twenty-one to eighteen • The Arab States increase oil prices by 70 per cent and cut back production in protest at US and allied support for Israel in the Yom Kippur war • Queen Elizabeth II opens the Sydney Opera House, declaring it one of the wonders of the world.

Bin 95 Grange Hermitage NOW

Medium-deep brick red. Smoky, dark chocolate, roasted chestnut, eucalypt/mint aromas. The palate is well concentrated with rich voluminous praline, panforte flavours, and lacy dry/al dente tannins. Finishes cedary but long. Still showing freshness. Drink up, but best bottles will hold. 98% shiraz, 2% cabernet sauvignon.

Kalimna Vineyard (Barossa Valley), Barossa Valley, Magill Estate (Adelaide), Modbury Vineyard (Adelaide). Last vintage entirely vinified in open wax-lined concrete fermenters (before barrel fermentation) at Magill Estate.

'Expansive, plush, sweet silky wine, almost pinot-like, with powerful cherry, spicy, pipe tobacco, underbrush aromas and graceful pliant textures.' (JR) 'Tightly tannic with just enough fruit to support.' (LM) 'Delicate, elegant and understated with meaty, mushroom notes and a long, silky textured palate.' (JC)

St Henri NOW ••• PAST

Medium-deep crimson to brick red. Complex roasted earth aromas with leathery, orange peel, camphor notes. Developed and well concentrated with earthy, roasted chestnut flavours and lacy dry, slightly coarse tannins. Finishes firm, minerally and dry with a fleeting hint of fruit sweetness.

'Warm rich nose of dried fruits, black olives, mahogany, minty aromas. Good fruit presence on the palate with grippy tannin structure and lively length.' (FK) 'A very complex wine with spicy, herb garden, dried fruit (Amaro) aromas and bitter-sweetness on the palate.' (MS) 'Flowery with aromas of sweet raspberry and fine spices. A vigorous full-bodied palate with rich fruit and hints of liquorice and mint. Very good grip and length. Truly good.' (AL)

Bin 170 Kalimna Shiraz NOW

Medium-deep crimson to brick red. Lovely dark chocolate, dark berry, roasted chestnut aromas with silage notes. A sweet-fruited wine, almost glossy, with roasted chestnut, dark berry, mocha, panforte, spice flavours and lovely velvety tannins. Finishes firm but long and sweet. Impressively balanced and in remarkably good shape. Originally destined for blending with 1973 Bin 169 Coonawarra Cabernet Sauvignon to create a Bin 60A homage wine.

'Overt spice and smoked meat aromas. Plenty of fruit sweetness on the palate with milk chocolate, spice, smoked meat notes, juicy acidity and balanced tannins.' (JB) 'Roasted nuts, black olives, toasted corn, roast meats and chocolate notes with an explosion of Vegemite and crushed ants. Robust tannins and a dry finish.' (TS)

Bin 169 Coonawarra Cabernet Sauvignon NOW

Medium-deep crimson to brick red. Dark chocolate, chinotto, graphite, herb garden, violet aromas. Supple, sweet-fruited wine with abundant musky, dark chocolate, cassis, chinotto, fennel flavours, superb mid-palate density and plentiful fine loose-knit graphite tannins. It finishes muscular firm with a long chocolatey plume. Also bottled as Bin CC (Coonawarra Claret). The original Bin 169.

'Flooded with sweet fruit, sarsaparilla and aniseed. A lovely focused wine with extraordinary depth and complexity, young tannins and lively acidity. It still has years to go.' (JF) 'A mature nose with lush cabernet aromas, tarry, chocolate, earthy dimensions and lovely evolved fruit sweetness and complexity. Velvet-smooth and plush with a grippy firm structure. A powerful masculine style.' (RKP)

Bin 389, Bin 28, Bin 128 PAST

Most bottles are past their optimum drinking window.

1974 ★

Difficult wet vintage and growing season; downy mildew outbreak just before vintage created new challenges; the grapes are rigorously triaged, ensuring a good, but low-volume Grange (only 2300 dozen) • John Duval joins Penfolds at Nuriootpa as an assistant winemaker • The Great Train Robber, Ronald Biggs, is arrested in Brazil, which refuses to extradite him to the UK because he has a Brazilian child • Australian opera singer Joan Sutherland is made a Dame of the British Empire • Cyclone Tracy devastates most of Darwin during the early hours of Christmas morning, leaving sixty-five people dead.

Bin 95 Grange Hermitage
NOW ••• PAST

Medium-deep brick red. Fresh cedar, panforte, dark chocolate, plum, tobacco aromas with herb garden notes. Supple sweet fruit, panforte, dried citrus peel, cedar wood flavours with fine plentiful chocolatey tannins. Finishes chalky firm and long. The wine is fading and some bottles are past. 93% shiraz, 7% cabernet sauvignon.

Kalimna Vineyard (Barossa Valley), Barossa Valley, Magill Estate (Adelaide). Winemaking transferred to Nuriootpa. Vinification in small stainless steel fermenters – completed in barrel.

'Mature, musky, tobacco, orange peel, peat aromas. Shy and unforthcoming with spicy notes and a gentle silky texture.' (JR) 'A murky wine, one-dimensional and ungenerous.' (LM) 'Old mature wine with dry grippy mid-palate and wood polish/balsamic notes. Dinner with old friends is the best way to go with this.' (TG)

St Henri
NOW ••• PAST

Medium-deep brick red. Perfumed redcurrant, apricot, chocolate aromas with spicy, herb garden, tomato leaf notes. A fleshy mouth-filling wine with dense redcurrant, praline, dried orange peel, liquorice flavours and al dente tannins. Finishes muscular firm, long and sweet. Quite solid and impactful but without the complexity, harmony and richness of great vintages. Penfolds 'wine of the vintage'.

'Still, it does unfurl with aeration and should be considered a very respectable St Henri considering this was the infamous "downy mildew vintage".' (NM) 'Deep rich roasted coffee, mocha, cherry compote aromas with spicy note. Intense and juicy with grippy tannins and harsh muscular finish.' (FK) 'Very Burgundian in style with a confiture of red berries, really beautiful structure and length.' (AL).

Bin 389, Bin 28, Bin 128
PAST

Most bottles are past.

1975 ★★★★

Good winter rains were followed by a warm summer with a few hot spells; cooler weather and intermittent rainfall prior to a dry harvest resulted in a high-quality crop • Max Schubert retires • Don Ditter is appointed National Production Manager • Microsoft founded by Bill Gates and Paul Allen • Steven Spielberg's blockbuster film *Jaws* opens in the US • China reveals the archaeological discovery of the life-sized 'Terracotta Warriors' near Xian • Prime Minister Gough Whitlam is dismissed by the Australian Governor-General Sir John Kerr in a move that shocked the nation • Papua New Guinea declares independence from Australia.

Bin 95 Grange Hermitage
NOW ••• 2025

Deep brick red. Intense panforte, cedar, sage, wood varnish aromas. Well-concentrated complex wine with developed panforte, cedar, praline flavours and muscular/grippy dry tannins. Finishes firm and dry. Superb drive and vinosity. A 'dark horse' vintage. 90% shiraz, 10% cabernet sauvignon.

Kalimna Vineyard (Barossa Valley), Barossa Valley, Coonawarra.

'Gorgeous, plump succulent fruit with good grip and a fresh brisk close. Perfect drinking now. So much pleasure.' (LM) 'Barnyard, sweet roasted fruit, orange peel aromas and big soft multidimensional tannins.' (RI) 'Still impressively dark with lifted meaty, mushroom, undergrowth aromas and richly textured palate.' (JC)

St Henri NOW ••• PAST

Medium-deep brick red. Minty, herbal, redcurrant aromas with praline, leafy notes. Rich, fleshy wine with dark chocolate, mocha flavours and dominating muscular dry tannins. Nonetheless, it has great concentration and flavour length. 75% shiraz, 25% cabernet sauvignon. Drink soon.

'Impressive clarity with slight farmyard notes. Lovely light syrup textures, good density and intensity of fruit. Pronounced dry tannins and very good length. Harmonious with integrity, energy and elegance.' (NB) 'Ripe chocolate, tobacco notes with ripe tannins, fresh mature fruit and plenty of length.' (PK)

Bin 389 Cabernet Shiraz NOW

Medium-deep brick red. Lovely dark chocolate, meaty, dark berry aromas with some apricot notes. Well-concentrated and fully mature wine with ample praline, meaty, earthy flavours and soupy, dry, muscular tannins. Finishes minerally and long. A robust style with plenty of sweet fruit notes. Probably best to drink soon.

'A shy bouquet and tightly restrained with roast capsicum and boot polish notes and astringent/robust tannins.' (TS) 'Vibrant blackcurrant, spice, demi-glace, crushed herb aromas. Complex, profound and rich with grippy drying tannins.' (RKP)

Kalimna Bin 28 Shiraz NOW

Medium-deep brick red. Fresh and complex redcurrant, leather, vellum, smoked meat, graphite aromas. Well-concentrated wine with fleshy, redcurrant, apricot, leather, smoked meaty flavours and loose-knit dusty tannins. Finishing chalky firm with lively acidity. The fruit is powering off now. Drink up.

'Aged and complex with a bouquet of roasted nuts and roasted meats. Deep and rich with stern grippy tannins and some ironstone characters. It still has density, flesh and guts but the fruit has receded.' (HH) 'Light and developed tawny red with minty leafy nuances. It has been a long and dusty road.' (JH)

Bin 128 Shiraz NOW

Medium-deep crimson to brick red. Fresh grilled nuts, amontillado, earthy aromas with underlying red fruit characters. Concentrated redcurrant, bitumen, graphite flavours with saddle-leather notes, fine slinky, dry tannins and underlying silage nuances. Finishes grippy firm with good flavour length.

'Earthy, dusty aromas with peppery, earthy, medicinal notes and drying tannins. Still a decent drink.' (HH)

1976 ★★★★★

A remarkable year; perfect growing conditions during spring were followed by a warm/hot summer; then top-up rains in February and a warm dry vintage resulted in wines of immense power and richness • The Penfold Hyland family loses control of Penfolds, ending four generations of family ownership • Don Ditter is given the new title, Chief Winemaker • Bin 707 Cabernet Sauvignon is produced for the first time since 1969 • First vintage of Koonunga Hill • Apple Computer Company is founded by Steve Wozniak and Steve Jobs • America's *Viking* spacecraft lands on Mars and sends back the first close-up pictures of the planet's surface • Chairman Mao Tse-tung, the leader of communist China, dies.

Bin 95 Grange Hermitage ⬦ NOW ••• 2040

Deep brick red. Fresh dark chocolate, toffee, panforte, dark berry aromas with liquorice notes. Concentrated and plush with blackberry, dark plum, praline flavours and rich, plentiful chocolatey tannins. Finishes long and fruit-sweet with a strong plume of tannins. Superb richness, density, harmony and energy. Will continue to develop more bottle-age complexity. 'Ethereal and buoyant.' Regarded as a great Grange vintage, but bottles are increasingly variable. The best bottles have levels at base or into the neck or have passed successfully through Penfolds Re-corking Clinics. 89% shiraz, 11% cabernet sauvignon.

Kalimna Vineyard (Barossa Valley), Barossa Valley, Magill Estate (Adelaide), Modbury Vineyard (Adelaide). The twenty-fifth anniversary of Grange. Max Schubert considered it 'More in the old style: a good vintage'. The first Australian wine to cross the $20 barrier at release. Robert Parker scores it 100 points.

'Medium red. Seductive and heady floral, spice, cinnamon, mocha bouquet. Expansive, plush and youthful with silky round tannins.' (JR) 'Lush and concentrated with balsamic notes, fresh herbs, South American tonka bean and sweet spices.' (MC) 'Wonderfully round and seamless in texture.' (JC)

Bin 707 Cabernet Sauvignon ⬨   NOW ••• 2025

Deep brick red. Intense dark cherry, cassis, vanilla, malt aromas with herb, liquorice notes. Lovely rich dark cherry, blackcurrant, coffee, mocha, liquorice fruit with fine velvety structure and underlying malty notes. Finishes fine-grained and fruit-sweet with a long tannin plume. Mature and complex with plenty of power, substance and generosity. Should continue to develop.

'Powerfully fresh mulberry aromas with strong notes of sweet cedar, Chinese medicine. Lots of red fruit and oaky notes with drying tannins. Silky textured with mouth-watering acidity and surprisingly gentle finish.' (FW) 'Liquorice, Szechuan peppercorn, dry spices, cinnamon, nutmeg aromas; evocative of a Shanghai grocery store. Textured, full, rich and rounded with balanced tannins.' (ST) 'The Year of the Dragon – strong!' (CPT)

St Henri ⬨   NOW ••• 2018

Deep brick crimson red. Intense panforte, dark chocolate, leafy aromas with dried raisin, liquorice notes. Well-concentrated panforte, chocolatey, roasted walnut, spicy flavours, plenty of sweet fruit notes and fine, slightly drying tannins. Finishes chalky firm, long and savoury. A great Penfolds vintage although some bottles are more advanced. Best to drink soon. 74% shiraz, 26% cabernet sauvignon.

'Complex leather, earthy, dusty aromas. Tightly knit wine with lovely texture, density of fruit and finely integrated acidity. A touch of staleness.' (FK) 'Flirts with Burgundy with its sweetness, fruit compote and mild spicy aromas. Rich and full palate with youthful dark plummy fruit, liquorice, spicy flavours, strong tannin backbone and a truly lingering aftertaste.' (AL)

Bin 389 Cabernet Shiraz ⬨   NOW

Medium-deep crimson to brick red. Fragrant dark berry, salted liquorice aromas with some ironstone, rusty notes. Reticent at first with dark berry, liquorice and praline flavours, fine chalky dry tannins and some herb garden notes. It builds up richness and length at the finish. It has lost its magic now, but it will hold for a while.

'On the bridge between youth and maturity. Plenty of fruit but the structure is gnarly and almost ungainly.' (TL) 'Powerful wine with black fruits, earth, mocha aromas and plenty of tannins. However, this won't improve from this point on.' (JH)

Kalimna Bin 28 Shiraz   NOW ••• PAST

Medium-deep crimson to brick red. Lifted panforte, wood varnish, dried fruit aromas. Generous redcurrant, panforte, espresso flavours and fine, sinewy dry tannins. Finishes leafy firm with earthy, rusty notes. Solidly built, but nearing the end. Bottles are extremely varied, so best to drink up.

'Fragrant sandalwood, liquorice aromas. Dried fruit, candied peel flavours are balanced with attractive savoury nutty undertones and sandalwood spice notes. Still drinking nicely.' (JB) 'Still vibrant, ripe and fresh, with mixed spiced fruit aromas with hint of lemon. Plump, round and long with grippy tannins forming a solid foundation.' (RKP)

Bin 128 Shiraz   NOW

Medium-deep crimson to brick red. Fresh, walnut, meaty aromas with a whiff of volatile acidity. Walnut, praline, slightly salty flavours with some redcurrant, prune notes. Fine chalky, almost graphite, rusty tannins. Finishes firm and long with savoury, amontillado-like notes. At the end of its days now.

'Very developed chocolate, vanilla aromas. Lots of body weight and grip, some sweet raisin flavours. A big and chunky wine without much finesse or complexity.' (HH)

Koonunga Hill Shiraz Cabernet ⬙

Medium-deep brick to tawny red. Intense espresso, boot polish, cigar-box aromas with some leafy, soupy nuances. Well-concentrated, expansive, smooth, supple wine, almost glossy, deep-set espresso, dark chocolate, boot polish, panforte, tobacco flavours. Finishes chalky firm, long and sweet. Acid and alcohol starting to poke out.

'Delicious wine that seems to be only in middle age.' (DM) 'Flawless and enjoyable.' (MC) 'Reaching its end, but still soft, round and harmonious. Inspirational for the modest collector.' (TG) 'A Penfolds classic; collectors will pay up to $300 a bottle for this!' (PG)

1977

★★★

A very dry growing season with hot spells during summer; milder weather in March was followed by a cool, dry but late vintage • The last Grange to be entered into the Australian Wine Show system • The Centre Pompidou for modern arts opens in Paris • The first instalment of George Lucas's *Star Wars* is released in US theatres • Smallpox becomes the first disease to be officially wiped out across the world • Elvis is found dead at his Graceland mansion in Memphis.

Bin 95 Grange Hermitage

NOW

Deep brick red. Fresh grilled meats, cedar, chesterfield leather, truffle, mint aromas and flavours with abundant sweet fruit, plentiful chalky dry tannins and underlying savoury notes. Fruit tapers off leaving a firm leaf finish. Doubtful it will improve. Drink soon. 91% shiraz, 9% cabernet sauvignon.

Kalimna Vineyard (Barossa Valley), Barossa Valley, Magill Estate (Adelaide), Clare Valley.

'Peak drinking with nicely mature fruit, mouth-coating texture and crackling acidity. Starting to dry out at the finish.' (LM) 'Earthy, funky, eucalypt, roasted berry aromas. Very polished wine with slick shellac notes. Compellingly elegant.' (RI) 'A lovely seamless wine with fresh cocoa, mint notes.' (DM)

Bin 707 Cabernet Sauvignon

NOW

Medium red brick. Espresso, mint, cassis aromas with dark chocolate, brambly notes. Well-concentrated and buoyant dark chocolate, espresso, mineral flavours with some minty notes and chalky dense tannins. Builds up quite muscular, bitter-sweet at the finish. At its peak of development, but should hold.

'Lovely cassis, black cherry aromas with tobacco, vanilla, oak, cinnamon and other sweet spices. A gorgeous core of sweet ripe cassis, plum, black cherry fruit with plentiful juicy ripe tannins. A wine of great balance, poise and remarkable length.' (ER) 'Perfumed with mint, thyme notes and pure firm tasty tannins.' (ST) 'Some rustic notes.' (YSL)

St Henri

NOW

Medium-deep brick red. Intense liquorice, cedar, orange peel, dark chocolate, red fruit aromas. Richly concentrated and velvety with dense walnut, espresso flavours and strong bitter grainy tannins. Finishes firm and chocolatey dry. A rustic element now emerging. Will continue to hold but drink soon. 51% shiraz, 49% cabernet sauvignon.

'Brilliant clarity and limpidity. Nuanced and gamey with spicy/stemmy notes. Simple well-crafted palate with fruit richness and iron-clad structure. Still potent with plenty of vigour and a long flourishing red liquorice finish. Spectacular length.' (NB) 'Leather, roasted coffee aromas, medium concentration and dry, slightly bitter tannins.' (PK) 'Fine sweet coffee, mocha, herb notes. Round and smooth with lovely fruit density.' (MS)

Bin 389 Cabernet Shiraz

NOW

Medium-deep crimson. Complex expressive orange peel, apricot, praline aromas with some leather, sandalwood, spice notes. Rich supple panforte, apricot, salted liquorice flavours, loose-knit chalky tannins, plenty of mid-palate fruit sweetness and mineral notes at the finish. A fully mature wine with lovely flowing texture and vinosity. Drink now.

'A vivacious wine that smells of roasted fruits, spices and dusty old cupboards. The palate is complex, harmonious, spicy and long.' (RKP) 'Expressive deep dark fruits with a real bond with tannin and acidity. Although does not show the great higher notes, this is a well-balanced wine.' (TL)

Kalimna Bin 28 Shiraz   NOW

Medium-deep crimson to brick red. Fragrant forest floor, redcurrant, espresso, bitumen aromas with some roasted hazelnut notes. Well-concentrated, smooth roasted chestnut, redcurrant, coffee flavours, mid-palate richness and plentiful lacy dry tannins. Finishes earthy and long. Still showing plenty of sweet fruit characters. Still holding.

'Plenty of appeal for an aged wine with coffee/mocha, funky notes dominating the fruit. The palate is solid but lively with supple tannins and crunchy acidity.' (TL) 'This has always been a sleeper and continues to be. Supple and smooth, mint and sweet berry fruit make an unusually pleasant taste blend.' (JH)

Koonunga Hill Shiraz Cabernet   NOW ••• PAST

Deep brick red. Intense burnt coffee, dark chocolate aromas with herb, graphite notes. Burnt coffee, toffee, panforte, briny flavours with fine chalky muscular tannins. Finishes long and sweet but elements starting to unravel. Still drinking well but now past its peak.

'Less expressive than 1976, but amazingly alive and well balanced.' (MC) 'Fresh, well-ripened fruit with sweet fruit flavours and velvety tannins. A slight herbal, eucalyptus edge that I enjoy.' (LM)

1978   ★★★★

A warm, dry season followed by a mild to warm vintage • Experimental trials of machine pruning in the Barossa Valley • Steve Lienert joins Penfolds as a cellar hand • British anthropologist Mary Leakey announces the discovery of 4 million-year-old human-like footprints in Laetoli, Tanzania • The first 'test-tube baby' is born, in England • The late Chairman Mao's *Little Red Book*, first published in 1966, is denounced by the official Chinese Communist Party.

Bin 95 Grange Hermitage ♢   NOW ••• 2020

Deep brick red. Fresh cedar, panforte, dried fruit, herb, violet aromas. Lovely mature and supple palate with dense cedar, panforte, prune, sweet fruit, vanilla flavours and fine chocolatey, al dente tannins. Finishes leafy firm. Lovely vinosity and flavour length. Beautifully balanced wine poised between secondary and tertiary development. 90% shiraz, 10% cabernet sauvignon.

Kalimna Vineyard (Barossa Valley), Barossa Valley, Magill Estate (Adelaide), McLaren Vale, Clare Valley, Coonawarra. Cork is uniquely vintage-dated and printed with 'Grange Hermitage'.

'Expressive dark cherry, Christmas spice aromas. Plush and seamless with superb texture and harmony. An explosive sweet finish with a touch of vanilla. Endless vivacity!' (JR) 'Glossy smooth structure.' (TG) 'This is a terrific Grange with perfect balance. Lovely developed fruit notes with polished but emphatic tannins.' (RI)

Bin 707 Cabernet Sauvignon   NOW

Medium-deep brick red. Lovely dark berry, roasted nut, praline, herb garden aromas. Gorgeously seductive, beautifully balanced wine with cassis, blackcurrant, plummy, roasted chestnut flavours, fine chocolatey sweet tannins and underlying malty oak. Finishes chalky dry, long and minerally. A delicious wine.

'Cigar boxes, sandalwood, red fruit, haw aromas. Intensely fruity with lovely elegant tannins. Good power and length. Vastly enjoyable.' (FW) 'A vivacious wine with evolved blue fruits, cassis aromas supported by wood spice and leathery notes. The tannins are also mature but finishing fresh. Balanced and elegant.' (CPT)

St Henri

Medium-deep brick red. Intense baked red fruits, praline, spice, walnut aromas with a touch of toffee. Supple, almost velvety smooth wine with chocolatey, dark fruit flavours and metallic edges, and pronounced acid. Finishes chalky firm but long sweet and savoury, walnutty. The wine is still drinking well, but the elements are fragmenting. Best to drink soon. 56% shiraz, 44% cabernet sauvignon.

'There is no disguising the brettanomyces *with scents of hung game, sweaty saddle and Provençal herbs. Fortunately it does not subjugate the fruit. The palate is fleshy and rounded on the entry. It is soft and generous with plenty of attractive savoury fruit and a rather four-square finish with a dash of Moroccan spice.' (NM) 'Delicious berry fruit with cabernet herbaceousness and minty freshness. Very balanced with lovely tannins, fruit density and transparency of flavour. Intense grip at the finish.' (FK) 'The wine is definitely complex in aromas and flavours, but my notes do not reveal* brettanomyces.*' (PG)*

Bin 389 Cabernet Shiraz

Medium-deep crimson. Fresh herb garden, graphite, sweet fruit, dark berry, spicy aromas with a touch of mint. Well-concentrated and sturdy with redcurrant, leafy, graphite flavours, tertiary roasted walnut notes, a thread of fruit sweetness and loose-knit slightly dry tannins. Elegant and muscular but still holding. Drink soon.

'A terrific wine with prosciutto, earthy notes, lovely palate weight, ripe silky tannins and plenty of acidity. A slight grip on the finish.' (JF) 'Sweet and mellow with savoury notes of game and herb. Developed prune, meats, dried herbs and spices with a sturdy structure of acid and tannin. There's plenty of mid-palate sweetness and freshness.' (JB)

Kalimna Bin 28 Shiraz

Medium-deep brick red. Intense liquorice, dried fruit, wood varnish, leafy aromas. Complex panforte, dark berry fruit, herb, roasted nut flavours and fine plentiful loose-knit chalky tannins. Finishes al dente firm. Powerful and generous with spicy, savoury notes. Drink soon.

'From a cool growing season with leafy, herb, mocha, sweet fruit notes. A medium-bodied palate with leafy freshness and clean bright acidity. Finishes grippy firm. Still hanging in there!' (JF) 'A powerfully structured wine with complex savoury, herbal, bouillon, graphite, gamey aromas, rich potent flavours, big tannins and dusty spicy notes. Some leather, herb nuances at the finish. There's a lot going on here with the structure holding it all together.' (JB)

Bin 128 Shiraz

Medium-deep crimson to brick red. Intense caramel, toffee, crème brûlée, espresso, redcurrant aromas. Redcurrant, roasted coffee, chestnut flavours with some leather notes. Fine, supple tannins and underlying mineral acidity. A touch underpowered.

'Caramelised sweet bouquet of vanilla and chocolate. Palate is dry and hollow. Starting to fall apart now.' (HH)

Koonunga Hill Shiraz Cabernet

Medium brick red. Developed earthy, dried plum, tobacco aromas and flavours. Palate is beginning to dry out with gritty/chalky tannins and fading fruit. Some good bottles around, but best to drink soon.

'Advanced wine showing oxidation but palate is better, with fruit sweet, caramel notes pushing through.' (TG) 'Very Médoc-like, with subtle floral, raspberry, cassis, herb aromas and fine-grained tannins.' (JR)

1979

★★★

An unusually wet and hot, almost sub-tropical, growing season followed by a mild, wet vintage; nonetheless, a decent vintage • China institutes the one-child-per-family rule to help control its exploding population • Lord Mountbatten assassinated by IRA bomb while on holiday in Ireland • Judy Davis and Sam Neill star in *My Brilliant Career*, an Australian film production.

Bin 95 Grange Hermitage

NOW ••• 2025

Deep brick red. A classically proportioned Grange with intense liquorice, inky, praline, cigar-box aromas with molasses, spicy, bitumen, leather notes. The palate is velvety smooth with dense black fruits, dark chocolate, tar, cedar flavours and fine-grained, loose-knit tannins. Finishes long with a chocolatey tannin plume. A beautiful textural wine. Retasted in Australia. 87% shiraz, 13% cabernet sauvignon.

Kalimna Vineyard (Barossa Valley), Barossa Valley, Clare Valley, Magill Estate (Adelaide), McLaren Vale. Magnums first released. Last vintage using bottles with off-white foil capsules.

Bin 707 Cabernet Sauvignon

NOW ••• PAST

Medium-deep brick red. Intense blackcurrant, malty, cedar aromas with sage, mint, leafy notes. The palate is well balanced with cassis, dried fruit, sage flavours, underlying flinty, smoky notes and chalky dry tannins. Builds up al dente, leafy at the finish, but long and tangy.

'Attractive cassis and black plum aromas with leather, tobacco, cinnamon, vanilla notes. Complex, developed sweet fruit, savoury spice notes and lovely juicy coating tannins. Not as "showy" as the gorgeous 1977 but maturing very well.' (ER) 'Smoky, salted fish, sweet fruit, old farm notes, silky tannins and mouth-watering acidity.' (ST)

St Henri

NOW

Medium-deep brick red. Complex dark plum, leather, sandalwood, minty aromas with liquorice notes. Supple and smooth with generous walnut, prune, dark chocolate flavours, lovely mid-palate richness and plentiful fine grainy tannins. Finishes chocolatey and long. At its peak of development. Drink soon. 67% shiraz, 33% cabernet sauvignon.

'Earthy, slightly vegetal aromas with leather, liquorice notes. Ripe and well composed with good concentration, fresh acidity and a long finish. Still has potential.' (PK) 'Intense and inviting perfume with liquorice, dark plum, mint and cassis notes. Generous palate with spicy fruit, mint, liquorice flavours and chewy tannins. Thick and viscous texture with great length. A certain complexity, warmth, generosity and spiciness.' (AL) 'A beautiful wine with opulent fruit and superb freshness. A great bottle.' (MS)

Bin 389 Cabernet Shiraz

NOW

Medium-deep brick red. Intense roasted coffee, dark berry, dried fruit, liquorice aromas with mocha, panforte, *sous bois* notes. Generous liquorice, dark chocolate, espresso, herb flavours, sweet mocha notes and muscular dry, touch grippy tannins. Finishes firm and tight. A solid wine with plenty of richness and length. Now entering its dotage. Drink soon.

'Mellow fruit, roasted meats, spicy, dried leaves. A touch hollow and dried out.' (RKP) 'Light ground-coffee, herb aromas. The palate is relatively light but fresh and lifted with depth and mid-palate sweetness.' (JB)

Kalimna Bin 28 Shiraz

NOW

Medium-deep brick red. Fresh leather, mint, smoky, graphite, leather, horse blanket aromas. Well-concentrated leather, blackberry, leafy flavours with fine chalky dry tannins, marked acidity and underlying chestnut, toffee notes. Beginning to fracture. Drink up.

'Smells of crushed sea shells. It lacks fruit definition on the mid-palate and it's hollow. A very drying finish. Past.' (HH) 'An intricate complex bouquet of sweet leather, earth, game and graphite. The palate is fully secondary, savoury and clinging to its final remnants of fruit. A touch dried out, yet maintaining good length and character.' (TS)

Koonunga Hill Shiraz Cabernet

NOW ••• PAST

Brick red. Fresh complex roasted coffee, praline, leather, earthy, orange rind aromas. Lovely subtle, elegantly structured wine with roasted coffee, dried apricot, mushroom flavours and plentiful fine lacy tannins. Finishes al dente firm but long and minerally. Surprisingly fresh, but drink now or soon.

'Very soft and delicate; possibly at the end of its life.' (MC) 'The generosity of shiraz is still showing through.' (TG) 'Complex bouillon aromas, caressing tannins, bright acidity and energy. Not very complex but delicious now.' (LM)

1980

★★★

An ideal growing season with good winter rains and mild conditions over spring, a warm to hot summer and a cool, dry vintage • Infant Azaria Chamberlain disappears at Ayers Rock, starting a dingo hunt • John Lennon assassinated by a fan outside his home in New York • The 'Hand of Faith', the biggest gold nugget found in the twentieth century, is unearthed in Victoria, Australia, weighing 27.2 kilograms • Australia's Alan Jones is Formula One World Drivers' Champion.

Bin 95 Grange Hermitage
NOW ••• 2020

Deep brick red. Fragrant dark chocolate, dark berry, cranberry, panforte, mint aromas. Well-concentrated vigorous dark chocolate, dark plum flavours with savoury mocha, menthol notes, and fine grainy, almost powdery tannins. A lengthy, sinuous finish. Drink now or keep for a while. 96% shiraz, 4% cabernet sauvignon.

Kalimna Vineyard (Barossa Valley), Barossa Valley, Clare Valley, Magill Estate (Adelaide), McLaren Vale, Coonawarra.

'Cranberry, dried cherry nose, but somewhat austere and tight with an astringent finish.' (LM) 'Densely concentrated with mint, earthy, cedar aromas and fine tannins.' (MC)

Bin 707 Cabernet Sauvignon
NOW

Deep brick red. Spicy cassis, dark chocolate, sage, leather aromas with mocha, malt, oak nuances. A tough unyielding wine with cassis, espresso, leather, earthy, tarry flavours and firm chewy tannins. Finishes minerally and lean with fruit fading away. A sweet and savoury style with a strong muscular structure. Doubtful it will improve, but who is to know?

'Cassis, mocha aromas with minty notes, silky tannins and balanced acidity. So fresh, young and delicious.' (ST) 'Very attractive mellowing wine with cassis, tobacco, leather, spicy aromas, rounded supple fruit and strong chewy tannins.' (ER)

St Henri
NOW ••• PAST

Medium-deep brick red. Fragrant cedar, tobacco, smoky, leafy aromas with redcurrant notes. Medium-concentrated wine with redcurrant, tobacco, prune flavours and fine loose-knit chalky tannins. A slightly leafy finish. Very elegant wine with an emphasis on structure rather than fruit. Drink now. Many examples are now past. 77% shiraz, 23% cabernet sauvignon.

'Mature claret-style wine with fresh juicy flavours, fine tannins and balanced acidity.' (FK) 'Refined spicy, incense, pepper, liquorice and dried fruit aromas. The palate is polished with a soft texture, yet with a good bite and very pleasant fruit. Long and complex on the finish. Very fine.' (AL)

Bin 80A Coonawarra Cabernet Barossa Shiraz
NOW ••• 2020

Medium-deep crimson. Intense redcurrant, dark chocolate aromas with some espresso and herb notes. Expressive palate with intense blackberry, blackcurrant, praline, tarry flavours and plentiful chocolatey tannins. Finishes sinewy, almost gritty, but there's plenty of fruit volume and persistency. It is showing lovely vinosity and energy, but really needs a few more years to get through this phase of life. Drink now or keep.

'Fragrant sandalwood, spicy, savoury notes with hints of herb and mint. A juicy palate with plenty of fruit sweetness, savoury undertone and slightly rough combination of tannins and acidity. A smart wine nonetheless.' (JB) 'Profound bouquet of crunchy red- and blackcurrants, layers of mixed spice and a hint of cold smoke. The palate shows remarkable fruit poise and distinction with wood smoke, antique leather sofa, cedar and a touch of black olive notes. The tannins are confident, assured, firm and drying, yet promise plenty of life ahead.' (TS)

Bin 389 Cabernet Shiraz
NOW

Medium-deep crimson to brick red. Intense, leafy, cassis, praline, panforte aromas with polished leather, spicy notes. The palate is richly concentrated with deep-set praline, dark fruit, panforte flavours, underlying savoury *sous bois* notes and gritty, dry, leafy tannins. Finishes granular and minerally. It still has energy and persistency but the structure dominates the fruit. Drink soon.

'A real jump here with more freshness and vibrancy than earlier vintages. A lovely combination of primary fruit aromas, wood spices and maturity.' (TL) 'Stewed plums, spice, vanilla and caramel aromas. Smooth and mature with an attractive thread of bright fruit and slight cork wood flavour. Nice mouth-feel, flow and finish.' (RKP)

Kalimna Bin 28 Shiraz NOW

Pale-medium brick red. Intense coffee, dried raisins, milk chocolate aromas. Well-concentrated, smooth chocolatey wine with roasted coffee, chestnut, redcurrant flavours and fine slinky tannins. Finishes sweet and sour with burnt toffee notes. At the brink of life now.

'Has no shortage of flavour in a distinctly jammy spectrum on the mid- to back-palate, but also some savoury anchors that stop the boat from tipping over.' (JH) 'Raspberry jam, leather and dusty floors with a touch of caramel. Tightly structured, but supple with generous fruit and fine dry grippy tannins.' (RKP)

Bin 128 Shiraz NOW

Medium-deep crimson to brick red. Fresh espresso, mocha, earthy, sweet fruit aromas. Earthy, sweet fruit flavours with lingering espresso, nutty, aniseed notes and savoury loose-knit, slinky tannins. Finishes chalky firm with acidity starting to poke out. On the downward slide now. Drink up.

'A peppery, cool year. It's bony and lacking freshness and softness. A lean and ungenerous wine.' (HH)

Koonunga Hill Shiraz Cabernet NOW ••• PAST

Brick red. Red cherry, leather, mushroom, cigar box, dried herb aromas. The palate is still quite buoyant and fresh, with red cherry, roasted coffee, mushroom flavours and fine lacy dry tannins. Finishes savoury firm with pronounced acidity. The fruit is fading now.

'Harmonious wine with solid acid/fruit balance and grippy tannins.' (TG) 'Sumptuous nose with juicy red fruit, orange peel notes and solid tannin structure. Pure pleasure.' (LM)

1981 ★★★

A warm to hot drought-affected growing season followed by ideal vintage conditions; a very big tannic vintage • End of the Cultural Revolution in China following the trial of the 'Gang of Four' • Royal Wedding of Lady Diana Spencer and Prince Charles, watched by 700 million television viewers worldwide • IBM starts manufacturing the personal computer • Scientists identify acquired immune deficiency syndrome (AIDS).

Bin 95 Grange Hermitage NOW ••• 2025

Deep brick red. Fresh complex roasted meats, tobacco, prune, dark chocolate aromas. Lovely concentrated dark chocolate, panforte, tobacco flavours and firm graphite tannins. Finishes minerally with some herb garden notes. Solidly structured wine. Will hold for many years, but will it improve? 89% shiraz, 11% cabernet sauvignon.

Kalimna Vineyard (Barossa Valley), Barossa Valley, Magill Estate (Adelaide), Modbury Vineyard (Adelaide), Clare Valley, Coonawarra.

'Intensely floral, graceful, elegant and precise with superb silky finesse.' (JR) 'Very complete wine with red and black berry, coffee, chocolate notes, powerful tannins and sustained flavours.' (RI) 'Full-bodied, firmly tannic and built for extended cellaring.' (JC)

St Henri NOW ••• PAST

Medium-deep brick red. Lifted mint, sage, wood varnish, forest floor aromas with faint red fruit notes. Surprisingly buoyant sweet-fruited palate with fresh redcurrant, strawberry flavours, some leather, wood

Penfolds

COONAWARRA CLARET
BIN 128

This wine was vintaged in 1963 from Shiraz grapes grown in the Coonawarra District of South Australia. It was matured in oak casks for eighteen months prior to bottling in October, 1964

BOTTLED BY PENFOLDS WINES PTY. LTD.

NET 1 PINT 6 FL. OZS.

Penfolds.

BIN 128

SHIRAZ

2010 · COONAWARRA

The Penfolds Bin collection of wines began in 1959,
part of a modern era of Penfolds innovation that has
resulted in a diverse family of wines with ageless
appeal. Each Bin wine bears the distinctive
Penfolds quality stamp whilst exhibiting
a unique style of its own.

750ML

VINTAGE 1966

Penfolds

COONAWARRA CLARET
BIN 128

This wine was vintaged in 1966 from Shiraz grapes grown in the Coonawarra District of South Australia. It was matured in oak casks for eighteen months and bottled at Tempe Cellars, N.S.W.

Bottled by PENFOLDS WINES PTY. LTD.

NET 1 PINT 6 FL OZ

varnish notes and plentiful but slightly bitter al dente tannins. Finishes firm and long with a chalky tannin plume. Drink up. 74% shiraz, 26% cabernet sauvignon.

'Autumnal, chestnut nose reminiscent of Brunello. Better density and vivacity than the 1980 with better freshness and surface tension. Harmonious with resolved tannins and a faint stab of acidity at the finish. Unlikely to improve.' (NB) 'A touch volatile and lean. The palate is dissipated, bitter and short.' (AL)

Bin 389 Cabernet Shiraz   NOW ••• PAST

Medium-deep brick red. Roasted walnut, leather, smoky, sandalwood, toffee, black fruit aromas with some minty notes. A complex, dense chocolatey wine with plenty of panforte, roasted chestnut, black fruit flavours, abundant granular, touch leafy, tannins and underlying savoury, roasted walnut, malty notes. A firm style with plenty of spicy fruit sweetness and richness.

'Fresh and alive with developed beef stock, meaty, umami aromas and fresh cherry plum notes. Prominent tannins and saturated cork adding mushroom, forest floor edges.' (TL) 'A wine with stacks of personality and plenty in the tank. The tannins are firm but the ample fruit provides balance.' (JH)

Kalimna Bin 28 Shiraz   NOW

Medium brick red. Fresh coffee, leather, spice, praline aromas. Well-developed espresso, leather, chinotto, spice flavours with a touch of pepper, fine supple loose-knit tannins and savoury/amontillado notes. A minerally, bitter, dry, chalky finish. Some underlying meaty, barnyard characters. Starting to fade. Drink soon.

'A gentle wine with attractive spicy, sandalwood, dark chocolate notes and meaty nuances. The palate is juicy with balanced tannins and fresh acidity. A complex wine still holding up well.' (JB) 'Boot polish and mushroom aromas. A good front palate presence but the sweet fruit notes diminish towards a drying finish.' (TS)

Bin 128 Shiraz   PAST

Most bottles are past.

Koonunga Hill Shiraz Cabernet   NOW ••• PAST

Brick red. Fresh polished leather, earthy, herb garden, tobacco, panforte, liquorice aromas. Substantial palate with dark berry, panforte flavours and fine chocolatey dry tannins. Finishes savoury firm with panforte, orange peel, mint notes. Tannins are beginning to dominate.

'Expressive earthy, with vegetal notes, chocolate spices, slightly bitter in the aftertaste with good tannins and acidity. Very lively.' (MC) 'Smooth and juicy but tannins are dry at the finish.' (LM)

1982   ★★★

A mild growing season followed by a hot vintage – an atypical vintage for Penfolds and Grange • Argentina invades the Falkland Islands • Steven Spielberg's blockbuster film *E.T. the Extra-Terrestrial* is released, to become the highest grossing film at the time • Seattle dentist Barney Clark is the world's first human recipient of a permanent artificial heart • The green movement blockades the Franklin River in southwest Tasmania to fight the Government's hydro-electric dam project.

Bin 95 Grange Hermitage   NOW ••• 2025

Deep brick red. Perfumed vanilla, mocha, seaweed, wild strawberry, redcurrant aromas. The palate is supple with plenty of redcurrant, red plum, cranberry flavours, sweet roasted hazelnut, mocha nuances and fine bitter chalky tannins. Finishes firm and tight. An atypical yet generously proportioned Grange with developed sweet fruit complexity and sinuous texture. Will continue to mature for some years. 94% shiraz, 6% cabernet sauvignon.

Kalimna Vineyard (Barossa Valley), Barossa Valley, Magill Estate (Adelaide), Modbury Vineyard (Adelaide), Clare Valley.

'Sweet and savoury with opulent fruit and great persistency.' (TG) 'Round, rich, sweet and generous.' (RI) 'After 30 years this wine still delivers!' (PG)

Bin 707 Cabernet Sauvignon ◈ NOW

Medium-deep brick red. Highly perfumed, cassis, boot polish, leather/spice aromas with some plummy notes. Luscious plum, cassis, red cherry, roasted chestnut flavours, fine plentiful lacy, al dente tannins and underlying new vanilla oak. A really well-balanced wine with superb fruit complexity, richness, length and volume. A highlight of the tasting.

'Mulberry and boysenberry aromas matched with sweet cedary oak. Powerfully fruity wine with layers of plum and mulberry, very nice integration of vanilla oak and lush tannins. Quite delicious and plump.' (FW) 'Textbook Cabernet Sauvignon brimming, bursting, overflowing with varietal DNA. Ripe, sweet fruit and ripe, succulent tannins. Incredibly youthful, fine, elegant and supremely balanced. A piece of velvet. Marathon finish.' (CPT)

St Henri NOW

Medium-deep brick red. Intense praline, roasted hazelnut, red cherry, forest floor aromas. Sweet-fruited palate with praline, strawberry, cassis, herb garden notes and plentiful fine-grained, slightly chalky tannins. Lovely concentration, fruit complexity and flavour length. Has developed really well. Drink now. 61% shiraz, 39% cabernet sauvignon.

'A Northern Rhône-like bouquet with scents of mulberry, sloe, sandalwood, cooked meats and Provençal herbs. The palate is sharp and precise although the tannins are more obtrusive than the previous two vintages, rendering the finish rather abrupt and severe. This is a conservative, correct St Henri, but without the charm or personality of the 1980 or 1981.' (NM) 'Medium-intense with leather, spicy, floral notes, fully mature fruit, fragmented tannins and slightly aggressive acidity.' (PK)

Bin 820 Coonawarra Cabernet Shiraz NOW

Medium-deep crimson. Fresh cassis, red cherry, mulberry aromas with some herb, sappy notes. Sweet supple wine with plenty of squashy cassis/red berry fruits, some herb notes and loose-knit, grainy tannins. Finishes chalky, slightly al dente with a touch of blackcurrant cordial. Looking its best in years. A pleasantly surprising performance in the tasting. Drink now nonetheless.

'A finely structured wine with a flood of black fruits, including black cherry and mulberry. Where did all that shaded fruit character of previous Rewards of Patience tastings go?' (JH) 'Incredibly concentrated summer berries on the nose. Fresh and succulent with juicy crushed berries, dark chocolate flavours, a plump mid-palate and long, fine but dry grippy tannins. Totally out of left field but it has plenty of charm.' (RKP) 'A true love-or-hate wine oscillating across the decades.' (PG)

Bin 389 Cabernet Shiraz NOW

Medium-deep brick red. Fragrant red cherry, redcurrant, sage aromas with leafy, herb garden, dark chocolate notes. Smooth, silky palate with redcurrant pastille, chocolate, herb, tea leaf flavours, lovely mid-palate fruit sweetness and supple, fine-grained tannins. Builds up chalky dry but the flavours lengthen out at the finish. The wine is evolving in the most surprising and delightful way. Drink now or possibly wait for a while.

'A gloriously seductive wine with sweet fruit, cassis, spice and peppery nuances, silky tannins and great length and line.' (JF) 'Supple and aromatic with spicy herb notes. A really attractive silky mid-weight palate with spicy, savoury flavours, hints of thyme and ripe tannins. A fresh sweet-fruit finish.' (JB)

Kalimna Bin 28 Shiraz NOW ••• PAST

Medium brick red. Intense dried roses, contrived red cherry aromas with some malty notes. Richly concentrated wine with generous red cherry, dark chocolate, pot pourri, tobacco flavours and plentiful chalky lacy tannins. Finishes a touch leafy, but long and fruit-sweet. Smooth and supple, but an atypical Bin 28. Has reached its optimum drinking plateau.

'A ripe, almost porty wine. The big rich sweet fruit harmonises with the powerful structure and aged notes. The Bin 28 style seems to reach a new level of concentration around this point.' (TL) 'Showing more maturity than 1980 and 1981. Attractive dark berry jam, sweet spice aromas. Balanced, plump and round with drying tannins seamlessly integrated.' (RKP)

Bin 128 Shiraz

NOW ••• PAST

Pale-medium brick red. Fragrant redcurrant, cherry, floral aromas with a touch of mint, sage and chocolate. A supple, sweet-fruited wine with red cherry, blackberry pastille flavours, roasted walnut, herb garden, earthy notes and loose-knit lacy tannins. Finishes chalky but long and fruit-sweet. Not as dense as earlier vintages but it still has vinosity and freshness. Drink up.

'Fully developed walnut, roasted chestnut bouquet with a touch of caramel/toffee. Palate is a touch bony and lacks flesh and density. Just ordinary but still drinking well.' (HH)

Koonunga Hill Shiraz Cabernet

NOW ••• PAST

Brick red. Fully mature walnut, leather, sweet fruit, herb garden aromas. Very fresh and buoyant with red cherry, redcurrant, walnut flavours, herb garden nuances and fine lacy/silky tannins. Still has richness and volume. Finishes with an alcoholic kick. Surprisingly in good shape but the fruit is drying out. Drink up.

'Simply delicious with lively secondary fruit, vibrant acidity, wood flavours and a lingering finish.' (DM) 'Impressive Grange-like nose with silky dry textures.' (TG) 'So juicy and fresh, pure sweet cherry and plum. Redwood bark notes! Long finish. Beautiful wine.' (LM)

1983

★★★★

Dry northerly winds over a tinder-dry landscape led to the devastating Ash Wednesday bushfires • Bizarrely, these were followed by February rains and March flooding; nonetheless, a monumental Grange and other impressive wines were made • A group of Australian wine collectors attempts to corner the 1983 Grange on release • Penfolds Magill Estate Shiraz is first made • American physicist Sally Ride becomes the first US woman in space, aboard the space shuttle *Challenger* • Australia wins the America's Cup with *Australia II* and its controversial winged keel, breaking a 132-year domination by the US • A military coup takes place in Grenada, followed by invasion by US military forces • Bob Hawke becomes Prime Minister of Australia after the Labor Party wins the Federal election • The Australian dollar is floated by Federal Treasurer Paul Keating.

Bin 95 Grange Hermitage ▽

NOW ••• 2035

Deep crimson. Intense blackcurrant, panforte, liquorice aromas with hints of mocha, sage. Powerful and richly concentrated with deep-set dark berry, panforte, mint flavours and fine plentiful graphite, muscular tannins. Finishes firm and tight with a tannin plume. A massively rich and expansive wine with years of cellaring potential. 94% shiraz, 6% cabernet sauvignon.

Kalimna Vineyard (Barossa Valley), Barossa Valley, Magill Estate (Adelaide), Modbury Vineyard (Adelaide).

'Opulent fruit, supple texture and strong barrel influence. Engaging now and I expect for another 20 years!' (LM) 'Very powerful and concentrated with herbal, earthy, dark chocolate, coffee notes and fine dry tannins. The best of the 1980s.' (MC) 'Black fruits and herbs. Still rugged and young with dominating tannins.' (DM)

Bin 707 Cabernet Sauvignon

NOW

Medium-deep brick red. Intense fragrant redcurrant, dark chocolate, eucalypt, sweet fruit aromas. Concentrated, almost soupy thick, redcurrant, cedar, dark chocolate, dark fruit flavours with plentiful graphite/muscular strong tannins and underlying savoury nuances. Finishes cedary and long with a tannin plume. 11.2% alcohol – picked early to avoid extreme heat. A tough, astringent wine.

'Expressive red plum, red and black cherry fruit with sous bois, slight barnyardy notes. A very tannic structure with ripe sweet fruit notes and mellowed development.' (ER) 'Lush fruit with Chinese medicine notes and astringent tannins.' (ST)

Magill Estate Shiraz

NOW ••• PAST

Pale-medium crimson to brick red. Fragrant redcurrant, roasted chestnut, cedar aromas with some herb notes. Well-sustained, fully mature roasted chestnut, redcurrant, tobacco flavours with loose-knit grainy tannins, underlying savoury notes and minerally acidity. Dries off at the finish. Beginning to fragment. Drink up.

'A glorious wine evoking both sadness and joy. It still has remnants of sweet fruit, and some leather/cigar-box nuances. Although it has faded there's still a vitality to it.' (JF) 'An attractive, mature wine with sweet leather, lovely savoury spice and Vegemite notes. Still composed but just beginning to become a little tired on the finish.' (TS)

St Henri NOW

Medium-deep brick red. Fragrant and complex minty, sage, salted liquorice, cassis, mocha, spicy aromas. Mature, full-bodied and supple with salted liquorice, cassis, mocha flavours, layers of sweet fruit and fine, plentiful chalky tannins. A bitter, al dente, minerally finish with lovely savoury fruit complexity and length. Perfect drinking now but will keep for a few years. 81% shiraz, 19% cabernet sauvignon.

'Textbook St Henri with lovely deep minerally and complex aromas, superb tannins quality and fine acidic structure. Still has great potential.' (FK) 'A pleasurable wine with complex ripe sweet plums, chocolate aromas and flavours and fine tannins. Wonderful to drink now.' (PK)

Bin 389 Cabernet Shiraz NOW

Medium-deep brick red. Intense panforte, roasted earth, liquorice aromas with blackberry, hazelnut, toffee notes. Well-concentrated blackberry, panforte, earthy, graphite flavours, plentiful savoury dry tannins and underlying crème brûlée, orange peel notes. Finishes chalky firm but long and sweet. It has reached its full potential now but will hold for a few more years.

'Plenty of flavour but the oak and fruit combine to present a sweet veneer and leave the texture stranded. Nice wine nonetheless.' (JH) 'Very attractive but forward wine with dark fruit, floral caramel aromas. Textured, ripe and concentrated with balanced, dry tannins.' (RKP)

Kalimna Bin 28 Shiraz NOW

Medium brick red. Intense espresso, leather, herb garden, sweet fruit aromas. Developed choco-berry, leather, espresso, demi-glace flavours, strong chocolatey dry tannins and underlying roasted hazelnut notes. Finishes grippy firm but sweet fruit flavours drive through. Robust and tannic. This is unlikely ever to soften out. Drink soon.

'Earthy, cold roast meat aromas. The fruit has ebbed away leaving the tannins exposed. Quite a big, rich burnished wine with a fair whack of tannin. Still drinkable.' (HH) 'Rough as bags!' (JH)

Bin 128 Shiraz NOW

Pale-medium brick red. Intense graphite, praline, apricot, leafy aromas with smoky coal dust, barnyard notes. Earthy, dark chocolate flavours with chocolatey dry tannins and underlying savoury notes. Finishes firm but long and flavourful.

'Earthy, leathery characters with some peppery, herbal notes. Savoury and a little lean but it does have some palate weight.' (HH)

Koonunga Hill Shiraz Cabernet NOW

Brick red. Fresh red cherry, panforte, herb garden aromas with touch of forest floor, chesterfield leather. Well-concentrated palate with panforte, roasted chestnut flavours and loose-knit silky tannins. Finishes long, smooth and minerally. Fully mature wine. Holding well but best to drink now.

'Impressive with lots of spices, bitter chocolate.' (MC) 'Tannic and astringent.' (LM) 'Suave and juicy with silky tannins and very good energy.' (JR)

1984 ★★

A cool summer followed by a cool, dry, late vintage • Max Schubert is awarded the Order of Australia (AM) for his contribution to the wine industry • Medicare, a universal health-care system, is established in

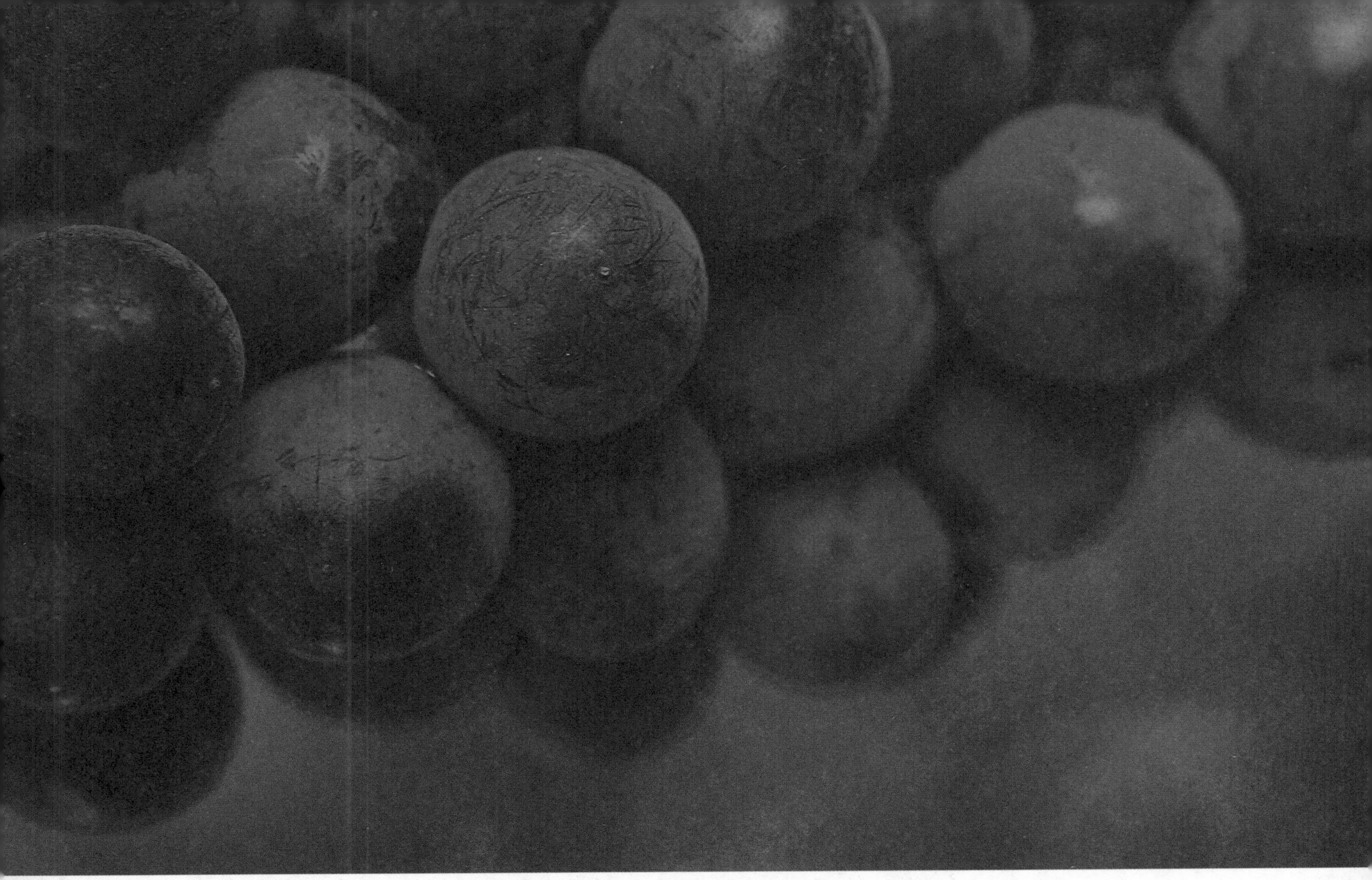

Australia • In Sarajevo, Jayne Torvill and Christopher Dean win Olympic gold for ice dancing to the rhythm of Ravel's *Bolero*, being awarded maximum points by all nine judges • 'Advance Australia Fair' proclaimed as official Australian national anthem, and green and gold as the national colours.

Bin 95 Grange Hermitage
NOW ••• 2020

Deep crimson. Fragrant dark plum, praline aromas with vanilla, mocha nuances. Expressive dark berry, dark chocolate, roasted coffee, spicy flavours with superb tannin richness. Finishes chalky firm with plenty of sweet fruit flavour length. An underlying muscular structure. Not a classic Grange but still drinks very well in its fourth decade. 95% shiraz, 5% cabernet sauvignon.

Kalimna Vineyard (Barossa Valley), Barossa Valley, Magill Estate (Adelaide), McLaren Vale, Clare Valley, Coonawarra.

'Old Australian shiraz characters with leather, spice, cumin notes and sweet mid-palate.' (TG) 'Dark ripe fruits and underlying spicy, peppery notes. Very good long finish.' (RI)

Bin 707 Cabernet Sauvignon
NOW

Medium-deep brick red. Ripe blackberry, dried plum, herb aromas with tobacco notes. Well-balanced and velvety smooth with sustained dark berry, plum, vanilla, malt, smoky flavours, loose-knit chalky tannins and plenty of savoury oak. Finishes minerally, spicy and long. Drinking really well.

'Toasty, smoky aromas with crushed blackberry/mulberry, dark chocolate notes. Powerful, plush and gorgeous on the palate, with lovely fruit richness and plush tannins.' (FW) 'Mellow black fruit cassis, black pu-er tea, leather and tobacco aromas and flavours with high-coating juicy tannins.' (ER)

Magill Estate Shiraz
PAST

Most bottles are past.

St Henri
NOW

Medium-deep brick red. Complex, earthy, baked fruit, prunes, leafy aromas with some rusty notes. Sweet prune, dark chocolate flavours, and fine chocolatey dry tannins. Finishes savoury, firm and slightly metallic. Not getting any better. Drink up. 77% shiraz, 23% cabernet sauvignon.

'Rather faded, vapid and slightly varnishy. Underpowered wine with moderate fruit density, silky textures and a drying finish. Has lost its vigour.' (NB) 'Earthy, mushroom nose and dry bitter tannins. Slightly metallic.' (PK)

Bin 389 Cabernet Shiraz
NOW

Medium-deep brick crimson. Intense dark chocolate, dark plum, vanilla aromas with some liquorice, herb garden notes. Rich soupy wine with deep-set dark choco-berry, panforte, vanilla flavours, underlying savoury notes and loose-knit, grainy, touch rusty tannins. Finishes firm and flavourful with confectionary, glacé fruit nuances. Drinking well now; should hold for a while.

'A gentle wine with secondary complexity and soft sweet fruit notes. Finishes short and simple.' (TS) 'Beautiful marriage between fruit and oak with dark fruits, spices and black tea notes. Quite heavy in body strength with muscular tannins. At its best now.' (TL)

Kalimna Bin 28 Shiraz
NOW ••• PAST

Pale-medium brick red. Fully mature coffee, salted liquorice, brambly, praline aromas with a touch of herb/ pepper. Moderately concentrated espresso, praline, brambly flavours, underlying cherry, boiled sweet notes and fine chocolatey dry tannins. Finishes chalky firm, very dry and savoury. Fading away now.

'Woody, savoury, bilgy and tannic. Past.' (JH) 'Some blackberry and pepper notes linger, although the finish is fading and beginning to dry out.' (TS)

Bin 128 Shiraz
PAST

Most bottles are past.

Koonunga Hill Shiraz Cabernet NOW

Brick red. Developed dark chocolate, prune aromas with some truffle, leather, spice notes. The palate is supple, almost soapy, with dried fruits, panforte flavours and almost melted tannins. Finishes parch dry and chalky. At its cusp of age. Drink now.

'A seamless old wine; round and harmonious.' (TG) 'Gorgeously plump wine with juicy red fruit, spicy notes, wonderful depth and smooth lingering tannins.' (LM) 'Pliant, sweet and round with expansive fleshy fruit.' (JR)

1985 ★★★

A cool to mild growing season and vintage, although intermittent rains delayed picking • Andrew Baldwin joins Penfolds as a cellar hand • The sinking of Greenpeace's *Rainbow Warrior* by the French government in Auckland, New Zealand • Live Aid Concert at Wembley Stadium, London, a globally televised charity event organised by Bob Geldof for the famine in Ethiopia • British scientist Alec Jeffreys announces a way of positively identifying individuals based on their DNA.

Bin 95 Grange Hermitage NOW ••• 2018

Deep crimson. Intense redcurrant, mocha, tobacco aromas with some wet bitumen, iodine notes. Well-concentrated redcurrant, tobacco, graphite flavours with chewy fine tannins and underlying mocha notes. 99% shiraz, 1% cabernet sauvignon.

Kalimna Vineyard (Barossa Valley), Barossa Valley, Clare Valley, Modbury Vineyard (Adelaide).

'Smoky cherry, dried rose, dusty floral, herb aromas and fine-grained but pronounced tannins. A youthful structure but evolved character.' (JR) 'Full-bodied, round and mouth-filling yet not overbearing. Fades at the finish.' (JC)

Bin 707 Cabernet Sauvignon NOW ••• 2018

Medium-deep garnet to brick red. Intense violet, cherry, cedar, praline aromas with 'Chinese medicine shop', *sous bois* notes. Lovely buoyant wine with bitter chocolate, dark berry, sage, dried leaf flavours, fine loose-knit, chalky tannins and underlying savoury oak. Finishes cedary and earthy.

'Ethereal Chinese medicine shop bouquet with roasted coffee, blackberry notes. Grippy fine tannins wrap a core of blackberry/mulberry fruit and Chinese herbs and spices.' (FW) 'Complex cocoa, chocolate, tobacco aromas with mellowing cassis, black cherry notes. Rich, lush cassis and black cherry flavours with a big structure of chewy coating tannins.' (ER)

Magill Estate Shiraz PAST

Most bottles are dried out and too austere in structure to really enjoy.

St Henri NOW ••• 2016

Medium-deep brick red. Aromatic camomile, dark cherry, liquorice, cedar aromas with some leather notes. Beautifully concentrated and generous with dark fruit, liquorice, cedar flavours, fine-grained, slightly gritty tannins and plenty of richness, fleshiness, glossiness on the mid-palate. Finishes savoury and firm with some mineral nuances. Drinking well now but should develop further. 99% shiraz, 1% cabernet sauvignon.

'Fragrant red cherry, plum aromas with floral notes. Very charming and harmonious with alluring rich sweet fruit and fine tannin structure.' (FK) 'Generous and aromatic with floral, sweet dark-berry, cassis and spice box notes. A vigorous palate, with good bite, concentrated fruit, a rigid backbone and very good length. Still feels youthful on the finish.' (AL)

Bin 389 Cabernet Shiraz NOW

Medium-deep crimson to brick red. Fresh blackberry, sage, tar, pot pourri aromas. Concentrated and juicy with vanilla, crème caramel, blackberry, roasted earth, aniseed flavours, underlying spicy notes and chocolatey rich, slightly muscular tannins. Finishes tangy with plenty of sweet fruit. Not a classic vintage but shows lovely fruit complexity, density and flavour length. Drink now or hold for a while.

'Medium-bodied wine with lifted boot polish, spicy, pepper aromas, grippy tannins and pinching acidity. A touch hard on the finish.' (JF) 'Intense cassis, liquorice, spice aromas. A really attractive juicy palate with beguiling dark fruits, sandalwood, rose, roasted spice, star anise flavours with game, leather notes, chalky tannins and long floral length.' (JB)

Kalimna Bin 28 Shiraz   NOW ••• PAST

Medium-deep brick red. Lifted floral, panforte aromas with salty mineral notes. Panforte, dried fruit, espresso, redcurrant flavours and sinewy dry tannins. Finishes walnutty firm and tight. Lacks harmony and generosity.

'Intense raisin, dried apricot, sweet fruitcake notes, but the palate is a touch dry and hard.' (JB) 'Sweet old shiraz aromas with jammy, prune, porty notes. It's still firm, taut, big and solid with a grippy finish. A relatively simple wine beginning to lose balance.' (HH)

Bin 128 Shiraz   NOW

Medium crimson to brick red. Fragrant and mature redcurrant, marzipan, roasted chestnut, coffee aromas. Supple, smooth, almost glossy palate with buoyant, redcurrant, panforte flavours and supple, sweet tannins. Finishes long and fruit-sweet with plenty of vinosity and freshness. Lovely balance and richness of flavour. Best to drink soon.

'Attractive mellow-aged toasty bouquet with roasting pan juices, a touch of raisin, leather and roasted nuts. Complex, savoury, balanced and smooth with nice fruit-sweetness. It's very ripe but delicious.' (HH)

Koonunga Hill Shiraz Cabernet   NOW ••• PAST

Brick red. Dark chocolate, marzipan aromas with some farmyard notes. Medium-concentrated wine with earthy, dark chocolate, roasted nut, tobacco flavours and loose-knit bitter-sweet tannins. Fruit loses momentum at the finish. Drink now but beginning to fade.

'Still a pleasant soft little red.' (JC) 'Elegant, classy with leather/spice aromas and dry tannins.' (MC) 'Composed with a spicy kick at the end!' (LM)

1986   ★★★★★

A warm, dry growing season with perfect ripening conditions: a great Penfolds vintage • Don Ditter retires; John Duval is appointed Chief Winemaker • Space shuttle *Challenger* mission ends in tragedy 70 seconds after lift-off • Halley's comet is visible from Earth as it comes closest to the sun on its latest seventy-five-year orbit • Chernobyl nuclear reactor explodes, causing massive radioactive leakage • Super-light plane *Voyager* circles the globe non-stop in nine days.

Bin 95 Grange Hermitage ▽   NOW ••• 2035

Deep crimson. Fragrant roasted chestnut, dark chocolate, cedar spice aromas. Rich expansive wine with roasted chestnut, praline, dark cherry fruit, fine plentiful chocolatey tannins and underlying mocha oak. Finishes al dente firm with plenty of sweet fruit, panforte notes. The defining vintage of the 1980s. Will continue to mature gracefully for at least another two decades. 87% shiraz, 13% cabernet sauvignon.

Kalimna Vineyard (Barossa Valley), Barossa Valley, Clare Valley, McLaren Vale, Modbury Vineyard (Adelaide). Don Ditter's last Grange.

'Highly aromatic wine with floral, Asian spice, raspberry, dried roses, lavender notes. Juicy and silky with a remarkable interplay of richness and vivacity.' (JR) 'Full-bodied with earthy, meaty, dark fruit aromas, full-bodied velvety texture and enormous concentration. A long berry-inflected finish.' (JC)

Bin 707 Cabernet Sauvignon ◈ NOW ••• 2028

Medium-deep brick red. Ripe blackcurrant, cedar/malty aromas with some dark chocolate notes. Beautifully complex with blackcurrant, malty, vanilla, sage flavours, fine-grained cedar tannins and superbly integrated, meaty, malt oak notes. Finishes long and sweet with a lacy tannin plume. A great Bin 707 with amazing fruit power, complexity and balance.

'Chinese medicine shop, toast, sweet cedar and black fruits. Butch and burly wine with strong generous tannins followed by rich black, crushed mulberry fruits and layered spice notes. Lovely length and balance. Yummy!' (FW) 'Minty, herbaceous, blackcurrants and spiced. Wonderfully vivid and vivacious with ripe, contoured, lifted tannins on the finish. Deliciously fresh and invigorating.' (CPT) 'Very stylish, expressive and powerful wine with layers of flavours and textured tannins.' (CC)

Magill Estate Shiraz NOW ••• PAST

Medium crimson to brick red. Fresh black olive, wet bitumen, cedar aromas with a touch of liquorice. Savoury dry palate with blackcurrant, leafy, *sous bois*, silage flavours, fine loose-knit muscular tannins. Earthy notes at the finish. Well past its prime, but still holding.

'Fully developed, light-bodied and softly textured.' (JH) 'Meaty, savoury, animal, dried herb notes with fine velvety tannins, good persistence and slightly tart acidity.' (TS)

St Henri NOW ••• 2018

Medium-deep crimson. Ethereal and complex wine with blackberry, plum, praline, roasted walnut, tobacco aromas and aniseed, cedar notes. Fresh, voluminous and balanced with plenty of sweet fruit, blackberry, plum, redcurrant, walnut, liquorice flavours and smooth, supple, almost velvety tannins. Finishes chalky firm with lovely buoyant sweet and cedar notes. A stylish wine with superb fruit complexity and definition. 86% shiraz, 14% cabernet sauvignon. Drink now or keep for a while.

'Engaging ripe black fruit, mulberry, dark plum, dried rose petals and a hint of star anise. Medium-bodied with ripe, fleshy, gently gripping tannins. Exquisitely balanced with beautiful poise and finesse. Supremely fresh and well-defined finish of blackberry, tobacco and mint. Drinking perfectly now but will age with style.' (NM) 'Beautifully sweet and perfumed with floral tones and red berries. The palate is really juicy with a tannic presence, plenty of freshness, an impression of fruit sweetness, very good length and a fair amount of complexity.' (AL)

Bin 389 Cabernet Shiraz NOW ••• 2018

Medium-deep crimson. Complex and classic with dark chocolate, praline, espresso, dark berry aromas with sage, demi-glace notes. Ripe, expressive and generous with deep-set praline, dark berry, spicy flavours, underlying sweet vanilla notes, cedary, almost briary tannins and gentle mineral acidity. Finishes al dente firm but richness of fruit gives an expansive roundness. Lovely vinosity and personality. In top form, but could develop further.

'Just about at the peak of its form. Marries the complexity of development and the fruit richness and definition of the vintage. The tannins are firm, balanced and integrated.' (JH) 'Attractive dark fruits with a layer of vanilla, toffee and demi-glace. Smooth, round and rich with finely dovetailed tannins and a long and sustained flow.' (RKP)

Kalimna Bin 28 Shiraz NOW

Medium-deep crimson to brick red. Intense dark berry, dark chocolate, mocha, prune aromas with a touch of sage. Concentrated dark berry, dark chocolate, prune, sweet fruit flavours with fine chalky loose-knit tannins. Finishes gritty firm with an aniseed/peppery kick. Buoyant and fresh with some tertiary polished leather notes. A classic Penfolds vintage but unlikely to improve.

'What was once an exceptional wine is now a warrior, still refusing to surrender its armoury. Fractured but not entirely broken.' (JH) 'Mint, blueberry, sweet black fruits and liquorice notes against an evolving backdrop of savoury, leather, gamey complexity. Very finely structured tannins and excellent persistence.' (TS)

Bin 128 Shiraz ♦ NOW

Medium-deep brick red. Intense cassis, cedar, roasted coffee aromas with some graphite, leafy, green pepper notes. Sweet, roasted coffee, cassis, cedar flavours, herb-garden nuances and fine-grained, slightly leafy tannins. Quite minerally with leafy, graphite-textured finish. Still has a youthful freshness.

'Herbal, pepper, spice aromas with earthy, meaty complexity. It still has fruit-sweetness and mid-palate richness. A really delicious wine of a riper style with lots of charm, sweetness and flavour.' (HH)

Koonunga Hill Shiraz Cabernet NOW

Brick red. Fresh red cherry, amontillado, nutty, *sous bois* aromas and flavours. The palate is soft and fleshy with red cherry, nutty flavours and lacy dry tannins. Finishes firm with some minerally notes. Best to drink now.

'An easy-drinking wine with sweet fruit flavours and grippy tannins at the finish.' (TG) 'Several years of life left (ten years) thanks to the tannin/acid structure. Quite charming and inviting.' (LM) 'A lush, creamy, expressive wine with great freshness and energy.' (JR)

1987 ★★

A cool, even growing season with crop levels moderated by October 1986 hailstorms – the coolest summer since the 1890s • The Kalimna Vineyard is redeveloped with new plantings and restoration of old vineyard blocks • Kym Schroeter joins Penfolds as a laboratory assistant • Australia wins the cricket World Cup for the first time • The stock market plummets in a tidal wave of panic selling • Van Gogh's *Irises* sells at auction for a treble record price of £30 million.

Bin 95 Grange Hermitage NOW ••• 2020

Medium crimson. Intense plum, dark chocolate, herb garden, white pepper, mint aromas. Well-concentrated wine with praline, herb garden, mint flavours and fine plentiful chalky texture. A strong tannic finish with some sappy, leafy notes. 90% shiraz, 10% cabernet sauvignon.

Kalimna Vineyard (Barossa Valley), Barossa Valley, McLaren Vale.

'Tons of black fruits, black olives, sarsaparilla aromas. A big wine with tannins to match.' (LM) 'Very tannic with cool spicy blackberry fruit and leafy notes.' (RI) 'Powerful fruit and supporting tannins. Oak and alcohol are integrated.' (DM)

Bin 707 Cabernet Sauvignon NOW

Medium-deep brick red. Dark chocolate, redcurrant aromas with some minty, eucalypt, earthy wood-floor notes. Very savoury, earthy style with eucalypt, mint, sage flavours and fine-grained/sinewy tannins. Finishes al dente with leafy notes. Evolving quickly but has good depth and concentration. An Adelaide Wine Show trophy-winning wine. Drink up.

'Fresh cassis, herbal notes but lacks texture and length.' (ST) 'Complex cassis, herbal, mint notes. Great balance and flesh with polished tannins.' (LD) 'Stylish like the 1986, but in a very different way. Fresh liquorice, red berry, cassis aromas with herbal notes. It has length rather than power.' (CC) 'An elegant style with polished tannins.' (YSL)

Magill Estate Shiraz PAST

Most bottles are past.

St Henri NOW

Medium-deep brick red. Fresh redcurrant, leafy, sage aromas with touches of praline and forest floor. Elegant, 'cool vintage' style with redcurrant, praline, herbal flavours, lacy dry tannins and plenty of *sous bois*, brambly, polished leather notes. Finishes long and grainy. Drink now. 87% shiraz, 13% cabernet sauvignon.

'Back to the dry tomato notes and antique room. Fresh and supple enough but quite grippy.' (NB) 'Medium-concentrated with vegetal, leather, spicy, animal notes and balanced but obvious tannins.' (PK)

Bin 389 Cabernet Shiraz NOW

Medium-deep crimson. Intense redcurrant, leafy, roasted chestnut aromas with *sous bois* notes. Well-concentrated wine with ample redcurrant, cranberry flavours, underlying mocha notes and plentiful chalky, lacy tannins. Finishes chewy/leafy dry. A difficult vintage. Drink soon.

'Quite a surprise, with an impressive profile of primary red and black berries, crashing into a wall of oak and tannins.' (TS) 'Full-bodied, rich and ripe with dense fruit, grippy unforgiving tannins and plenty of cleansing acidity. There's a touch of volatility which gives a hard veneer finish.' (JF)

Kalimna Bin 28 Shiraz NOW

Medium-deep crimson to brick red. Intense redcurrant, panforte, roasted walnut aromas with salty demi-glace notes. Sweet-fruited wine with redcurrant, sage, herb flavours, pronounced linear acidity and lacy dry tannins. Finishes firm, drying and tight. Now past its best.

'Subtle, fully savoury and secondary, with a short, dried-out finish. Fading.' (TS) 'Lacks richness and definition. The acidity is starting to poke out.' (JF)

Bin 128 Shiraz PAST

Most bottles are past.

Koonunga Hill Shiraz Cabernet NOW

Brick red. Redcurrant, panforte, leather, tobacco, liquorice aromas. Redcurrant, panforte, tobacco flavours and fine lacy tannins. Finishes leafy dry. Fruit is fading. Drink now.

'Persistent but awkward.' (TG)

1988 ★★★

A warm, high-quality 'unhurried' vintage with below-average rainfall • Max Schubert is named *Decanter* magazine's Man of the Year • Veteran winemaker Gordon Colquist retires after fifty vintages • Australia celebrates its bicentenary of the arrival of the First Fleet to Sydney Harbour • Australia's new Parliament House, Canberra, is opened by Queen Elizabeth II • Australian Kay Cottee becomes the first woman to sail solo, single-handed and non-stop around the globe • Olympic athlete and record breaker Ben Johnson is stripped of his gold medal after positive drug testing.

Bin 95 Grange Hermitage NOW ••• 2030

Medium crimson. Praline, prune, dark plum aromas with vanilla, mocha notes. The palate is rich and buoyant with sweet dark chocolate, prune, vanilla flavours and supple, graphite tannins. Builds up chewy firm, long and sweet at the finish. 94% shiraz, 6% cabernet sauvignon.

Kalimna Vineyard (Barossa Valley), Barossa Valley, Padthaway, McLaren Vale.

'Dark berry, root beer, medicinal cherry, liquorice aromas and flavours. Structured and focused with bright acidity. Still youthful and needs further time to age.' (JR) 'Dark chocolate, sweet fruit, mint aromas with generous deep fruit, mocha, spice notes and tangy acidity.' (RI) 'Meaty, savoury with hints of cedar, mint. Full-bodied, open-knit and accessible. Finishes dry and astringent.' (JC)

Bin 707 Cabernet Sauvignon NOW ••• 2025

Medium-deep brick red. Lovely fresh ripe blackcurrant, dark cherry, boysenberry, malty aromas with some cedar notes. Fresh, sinuous and generous with plenty of blackcurrant, dark cherry fruit, underlying vanilla, savoury oak and fine, plentiful, chalky, almost chocolatey, tannins. Lovely weight and balance. Although not regarded as a classic, this has developed really well. Will continue to hold.

'Cassis, spice, leather, tobacco aromas. Rich cassis, black plum fruit on the palate with a big structure of chewy tannins. A little closed.' (ER) 'Well concentrated with lovely sweet fruit, ripe tannins and balanced oak.' (LD)

Magill Estate Shiraz

NOW ••• PAST

Most bottles are past.

St Henri

NOW

Medium-deep crimson. Mature and complex redcurrant, orange peel, eucalypt, mint aromas with touches of forest floor. The palate is generously proportioned with plenty of redcurrant, sweet fruit, orange peel, cloves, spicy, liquorice flavours and fine, lacy dry tannins. It finishes chalky, minerally and long. Into its tertiary stage. Drink soon. 88% shiraz, 12% cabernet sauvignon.

'Round and supple with good clarity and density, but it still lacks real conviction and excitement of a top St Henri!' (NB) 'Mature spicy, sweet fruit aromas with "benchmark" liquorice, mint, peppery notes. Nice bite on the palate with sweet plummy, pepper, liquorice flavours and lovely persistency. Well balanced with complexity and character.' (AL)

Bin 389 Cabernet Shiraz

NOW

Medium-deep crimson. Fragrant cassis, herb garden aromas with spearmint, sage notes. Rich and flavourful with blackcurrant, plum, minty flavours, underlying savoury notes and velvety, touch leafy, tannins. Finishes firm and minerally with some rich panforte notes. A substantial wine with plenty of fruit complexity and volume. Drink now.

'Seems darker and more broody than others, with dense blackberry, exotic fruit compote and spices. The fruit really lingers on the palate.' (TL) 'Very friendly and classically styled Bin 389, with liquorice, chocolate, spice, black caramel aromas. Velvety, rich and complete palate with good persistence.' (RKP)

Kalimna Bin 28 Shiraz

NOW

Medium-deep crimson. Intense redcurrant, roasted walnut, mocha, panforte aromas with spicy/cedar notes. A gentle sinuous palate with buoyant sweet-fruit flavours, lovely mid-palate richness and fine graphite tannins. Finishes firm and long. At the peak of its development.

'A gorgeous wine with wood spice box, mocha aromas. Elegant and round with a glorious combination of spice, ripe tannins and drive. The perfect balance of age and youth.' (JF) 'Very definitely alive. Supple and smooth with juicy red and black fruits and fine tannins.' (JH)

Bin 128 Shiraz

PAST

Most bottles are past.

Koonunga Hill Shiraz Cabernet

NOW

Brick red. Intense, complex panforte, ginger, apricot, orange peel aromas with touch of bitumen. Well-developed and fully mature palate with sweet panforte, demi-glace, apricot flavours and fine chocolatey tannins. Finishes lacy dry but long and sweet. Ready to drink.

'A generous reward of wine with mouth-filling fruit and coating tannins.' (LM)

1989

★★★

An odd growing season; generally quite cool to start with but a burst of very hot weather in February that shrivelled grapes; then heavy rain set in making it a difficult vintage • John Duval is the International Wine Challenge's 'Winemaker of the Year' in London • Penfolds is awarded the inaugural Adelaide Wine Show Trophy for 1987 Bin 707 Cabernet Sauvignon; the best red wine of the show • Peter Gago joins Penfolds as a sparkling winemaker • Salman Rushdie receives *fatwah* death threats from Ayatollah Khomeini for his *Satanic Verses* • The supertanker *Exxon Valdez* runs aground in Prince William Sound, Alaska, spilling 11 million gallons of oil • The Berlin Wall is dismantled as the German people force the nation towards reunification.

Bin 95 Grange Hermitage

NOW ••• 2025

Medium crimson. Fragrant redcurrant, cassis, leather, sandalwood aromas. Luxuriant 'tawny port-like' intensity on the palate with fleshy sweet fruit, liquorice, mocha notes and fine-grained muscular tannins. Finishes firm but long and sweet. A very successful wine for the vintage. 91% shiraz, 9% cabernet sauvignon.

Kalimna Vineyard (Barossa Valley), Barossa Valley, McLaren Vale. The last vintage labelled as 'Grange Hermitage'; future vintages are released as 'Grange'.

'Essence of red berries, fleshy, sweet and full of energy.' (JR) 'Gorgeously floral and accessible with powerful dark cherry, liquorice, mandarin, orange peel fruit and plush ripe texture.' (LM) 'Warm meaty, sweet fruit aromas with some lifted notes. Glossy tannins, sweet mid-palate and warm finish. A showy wine.' (TG)

Bin 707 Cabernet Sauvignon

NOW ••• 2016

Medium-deep brick red. Fragrant herb garden, raisin, mocha, liquorice aromas with sour cherry/tomato leaf notes. Sweet plum, raisin, sour cherry, herb flavours, muscular tannins and underlying savoury, malty oak. Finishes chalky, firm, sweet and sour. A solid wine with good extract and concentration, but best to drink soon.

'Ripe blackberries, dark chocolate, coffee bean aromas. Palate is fresh and lively with grippy tannins, lovely balance of fruit, oak and structure.' (FW) 'A full weighty wine with dense deep porty fruit and very alcoholic.' (ST)

Magill Estate Shiraz

NOW ••• PAST

Most bottles are past.

St Henri ▽

NOW ••• 2015

Medium-deep crimson. Well-developed and complex with fresh, roasted walnut, leafy, tobacco, wood varnish aromas and a lovely undercurrent of cassis and sweet fruit. Well-concentrated redcurrant, cassis, sage and vanilla flavours, fine-grained, slightly gutsy hard tannins and lovely supple cassis, mineral notes. Very refreshing fruit-complex and characterful wine. 89% shiraz, 11% cabernet sauvignon. Drink soon.

'Stewed berry, coffee, truffle aromas. Transparent and complex with lovely texture, tannin backbone and fruit expression. Easy to love and enjoy!' (FK) 'Coup de coeur! Intense, perfumed and complex with black truffle, violet, tobacco and cassis notes. Full and structured with perfectly integrated grippy tannin, layers of fruit and persistent spiciness on the finish. A bloody good glass of wine!' (AL)

Bin 389 Cabernet Shiraz

NOW ••• 2018

Medium-deep crimson. Intense ripe, black cherry, raspberry aromas with vanilla, spice, dark chocolatey notes. Concentrated red cherry, blackberry, praline, herb flavours, plush velvety firm tannins and plenty of vanilla oak notes. Finishes long and sweet. Lovely to drink now.

'An impressive presence of black fruits framed in a rigid structure of firm, finely poised tannins; dark chocolate oak in support and a lingering berry fruit finish.' (TS) 'Developed cherry cola aromas and flavours with the complexity, density, harmony and structure that truly works after 20 years.' (TL)

Kalimna Bin 28 Shiraz

NOW

Light-medium crimson. Fresh scented redcurrant, herb, dried roses, contrived cherry aromas. Smooth redcurrant sweet fruit, slightly jammy flavours and grainy loose-knit tannins. Firms up at the finish. Not overly complex but still fresh and alive. Best to drink soon rather than keep.

'One of the better old wines, albeit in a slightly jammy style with a dark chocolate, sweet confit, porty bouquet and some richness and flesh on the palate.' (HH) 'Lacks presence. The fruit is falling away.' (RKP)

Bin 128 Shiraz

NOW

Medium-deep crimson to brick red. Red cherry, herb garden, aniseed aromas with leafy, roasted, earthy notes. Well-concentrated red cherry, herb garden flavours and loose-knit, chalky drying tannins. A touch underpowered with leanness on the mid-palate. Finishes minerally.

'Smooth, supple and savoury with some fruit-sweetness on the mid-palate, quite firm tannins and some peppery elements. It's not fleshy or rich but has depth and flavour length.' (HH)

Koonunga Hill Shiraz Cabernet

NOW ••• PAST

Brick red. Fresh espresso, smoky, herb aromas with hints of panforte and orange peel. The palate is fresh and supple with plenty of choco-berry, leafy, earthy flavours and fine lacy dry tannins. Finishes minerally and slightly sappy. Drink now.

'A modern Koonunga Hill with tighter lines.' (TG) 'A Bordeaux style of wine in the nose with juicy fruit flavours and electric acidity!' (LM)

1990

★★★★★

A dry and even growing season produced shiraz of incredible opulence; a great Grange vintage • Bin 407 Cabernet Sauvignon's first commercial vintage • Grange heads up the first Langton's Classification of Australia, a position it continues to hold • Jeffrey Penfold Hyland dies, aged seventy-nine • Nelson Mandela released from incarceration, signalling the beginning of the end of apartheid in South Africa • Europe and the US join forces to launch the Hubble space telescope • The Conventional Armed Forces Treaty signed in Paris marks the end of the Cold War.

Bin 95 Grange ⬨

NOW ••• 2045

Deep crimson. Powerfully expressive with beautiful dark berry, cranberry, praline, herb garden, graphite, mocha aromas. Lovely seductive dark berry, praline flavours, fine chalky/graphite tannins and roasted chestnut, mocha nuances. Finishes gravelly/chocolatey firm with some aniseed, liquorice notes. A great Grange vintage with superb fruit definition, generosity, balance and structure. 95% shiraz, 5% cabernet sauvignon.

Kalimna Vineyard (Barossa Valley), Barossa Valley, Clare Valley, Coonawarra. Voted Red Wine of the Year by *Wine Spectator* magazine in December 1995.

'Silky and elegant with great energy and racy acidity.' (LM) 'Kaboom! Density, power and length.' (TG) 'Rich, complex and elegant. Perfectly balanced with a solid structure of very fine tannins.' (MC)

Bin 707 Cabernet Sauvignon ⬨

NOW ••• 2030

Medium-deep crimson. Intense dark berry liquorice, roasted chestnut aromas with a touch of sage and praline. The palate is beautifully balanced, complex and buoyant with generous deep-set blackberry, cassis, praline, liquorice flavours, bitter chocolatey tannins and underlying vanilla, nutty oak. Finishes chalky firm and tight with lovely mineral acidity. Fresh and harmonious wine with power and substance.

'Liquorice, cassis and mint. Rich persistent fruit is supported by opulent but very fine, generous tannins. A long, lively finish. Incredibly harmonious.' (CPT) 'Rich sweet fruits with some eucalypt notes. Smooth and harmonious with melting silky tannins.' (LD) 'Complex and delicious with juicy flavours and silky polished tannins.' (YSL)

Magill Estate Shiraz

NOW ••• 2018

Medium crimson. Developed blackberry, dark chocolate aromas with some savoury, panforte, cedar notes. Complex and dense with dark chocolate, black fruits, cedar flavours, wood varnish notes and plentiful chocolatey dry tannins. Finishes earthy firm. Surprisingly fresh with lovely fruit complexity and vinosity.

'The primary fruit has departed; savoury elements now predominate. It's bigger, richer and denser with lovely line and structure.' (TL) 'Lovely ripe black fruits with cedar/aniseed complexity. A lush palate with rich, ripe round tannins.' (JF)

St Henri ⬨

NOW ••• 2025

Medium-deep crimson. Intense ripe blackberry, dark cherry, dark chocolate, black truffle aromas with marzipan, spicy notes. Smooth, sweet, supple wine with abundant blackberry, mocha, panforte, spicy flavours, plentiful fine loose-knit grainy tannins. Finishes bitter-sweet and long. Superb vinosity, fruit complexity and freshness. A great St Henri with marvellous potential. Drink now or keep. 89% shiraz, 11% cabernet sauvignon.

'A warm, inviting, finely delineated bouquet with attractive scents of raspberry, wild strawberry, candied orange peel and cedar. The palate is soft and supple on the entry. It has a fleshy, corpulent mouth-feel and fine acidity with a caressing, silky smooth finish. Does it have the tannic "chassis" to merit long-term ageing as does the splendid 1986? But for now, this is drinking beautifully.' (NM) 'Gloriously opulent wine with silky textures and lovely transparency of flavours. Flamboyant and spectacular.' (NB) 'Fresh, beautifully mature wine with ripe cherries and spices. More advanced than 1990 Grange, but juicy and fresh with beautiful fruit sweetness and length. Still has a long future.' (MS)

Bin 920 Coonawarra Cabernet Shiraz
NOW ••• 2030

Deep dark crimson. Fragrant panforte, liquorice, crème de cassis aromas with some cedar, vanilla notes. Voluminous sweet panforte, cassis, sage flavours with leafy, mint notes, fine-grained, slightly chewy tannins and underlying savoury, grilled nut oak. Finishes firm, rich and flavourful. A powerfully built wine with the density, balance and personality for long-term ageing.

'Astonishingly fresh with bright ripe cabernet, integrated oak, spice notes, lively dancing acidity and long persistent finish. This has years ahead of it.' (JF) 'Lovely deep spicy, earthy aromas with plenty of black fruits. A big, full-bodied wine with firm, stately, persistent tannins, great mid-palate density and length. Very solid, impressive, powerful and emphatic. Another 30 years in this.' (HH)

Bin 90A Coonawarra Cabernet Barossa Shiraz
NOW ••• 2040

Medium-deep crimson to brick red. Intense panforte, praline, herb aromas. Very classic and beautiful with intensely rich roasted coffee, blackberry, praline, mocha flavours, complex tobacco, earthy notes and velvety thick dry tannins. Finishes chalky/chocolatey firm with herb, pepper nuances. Surprisingly elemental but superbly balanced. It's possible to drink now, but the best is still to come. Keep.

'More developed and rustic than Bin 920 with a spicy savoury nose, mellow flavours of meat, tobacco and spice, structured tannins and fresh acidity.' (JB) 'More extract, more complexity, more adjuncts to the fruit and more tannins. The war games will eventually subside over time and richly repay patience.' (JH)

Bin 389 Cabernet Shiraz ◈
NOW ••• 2025

Medium-deep crimson. Intense dark chocolate, dark berry, panforte, mocha aromas with a hint of mint. Rich, voluminous wine with dark berry, panforte flavours, abundant chocolatey tannins, plenty of mid-palate richness and superb fruit sweetness. Finishes chalky firm and long. A beautifully balanced wine with lovely fruit complexity, density and flow. Should last the distance. Drink now or keep for a long while.

'Another shift in weight and focus. The cabernet expresses incredible clarity and dominates the shiraz. It's positively juicy and way beyond the wines that preceded it. Wonderful but slightly atypical. Drink now if you enjoy richness.' (JH) 'Rich, complex and complete with plenty of dark fruits, a seamless integration of components, lovely texture and flow. An excellent and typical 389 style.' (RKP)

Bin 407 Cabernet Sauvignon ◈
NOW ••• 2018

Medium-deep crimson. Intense cassis, walnut, leafy aromas with a hint of mint. Well-concentrated wine with mature cassis, walnut, tarry flavours and some leafy, minty notes. Complex and fine-grained with earthy, savoury nuances. Finishes velvety and long. Very much at the peak of its secondary development. It could well last a few more years.

'Big dark, pure cabernet of imperial proportions. A perfect union of fruit and tannins. Still in its honeymoon; just showing the first notes of its age lines.' (TL) 'A medium-bodied wine with cassis, cigar-box, spice aromas, lovely cabernet leafiness, velvety tannins and savoury nuances. This has real depth, vitality and freshness.' (JF)

Kalimna Bin 28 Shiraz
NOW ••• 2020

Medium crimson. Intense dark choco-berry aromas with panforte, graphite, smoky notes. Beautifully concentrated, dark chocolate, dark berry, sweet fruit flavours with some roasted chestnut, polished leather notes, plentiful chalky/velvety tannins and underlying mineral acidity. Finishes firm with sustained flavour length. A classic Penfolds vintage with great cellaring potential.

'A wine of style and class. Red and black fruits, fine tannins and balanced oak. At the peak of its life.' (JH)

'A lovely fruit profile of black fruits, liquorice and chocolate with notes of sweet leather. A softly structured wine, with fine tannins and savoury complexity building up to the finish.' (TS)

Bin 128 Shiraz ▽

NOW ••• 2020

Medium-deep crimson. Intense ripe blackberry, earthy, panforte aromas. Well-concentrated and generous with blackberry, panforte, earthy flavours, fine, lacy tannins and good mid-palate richness and volume. Finishes chalky firm, long and fruit-sweet with underlying savoury notes. A classic Bin 128 with superb fruit complexity. At the vertex of its development.

'Firm and sound, deep, concentrated and together with excellent line and length. Very good fruit flavour and substantial power. A lovely wine; taut, deep and promising much for the future.' (HH)

Koonunga Hill Shiraz Cabernet ▽

NOW ••• 2018

Brick red. Fresh dark cherry, herb, panforte, praline aromas with wet bitumen/tar notes. Palate is rich and voluminous with panforte, dark cherry, tarry flavours and muscular tannins. Finishes chalky firm and savoury. Wine is now entering tertiary phase. Still impressive, reflecting a classic Penfolds vintage. Best to drink now but it will hold for some more years.

'Elegant juicy wine with dark fruit, toast aromas and plenty of texture.' (DM) 'Expressive and structured.' (MC) 'Juicy, sweet spicy fruit with smooth seamless textures.' (JR)

1991

★★★★★

A warm, dry growing season punctuated by hot weather in late summer produced an early vintage: a brilliant follow up to 1990, another great Grange year • John Duval is the International Wine Challenge's 'Red Winemaker of the Year' in London • John Davoren dies, aged seventy-six • End of the Gulf War as UN forces liberate Kuwait • British computer scientist Tim Berners-Lee puts the World Wide Web online • The Soviet Union ceases to exist.

Bin 95 Grange ▽

NOW ••• 2040

Deep crimson. Complex dark chocolate, roasted chestnut, vanilla, herb aromas. A rich voluminous palate with saturated dark chocolate, mocha, malty, aniseed flavours and fine-grained supple tannins. Finishes chalky firm and long. A rich buoyant and ethereal Grange with tremendous substance and lasting power. A classic vintage. 95% shiraz, 5% cabernet sauvignon.

Kalimna Vineyard (Barossa Valley), Barossa Valley, McLaren Vale.

'High-toned wine with cherry, blackberry, floral fruit and spicy, sappy, juicy flavours.' (JR) 'Delicious, dense, complex and elegant with fine oak, chocolate, cedar, spicy notes and fine tannins and good acidity.' (MC) 'Huge, plush and dramatic with a nearly endless finish. Awesome now and for decades to come.' (JC) 'Which do I prefer – the 1990 or 1991? These two vintages are both evocative expressions of Grange. It's difficult to pick. This week I favour ...' (PG)

Bin 707 Cabernet Sauvignon ▽

NOW ••• 2025

Medium-deep crimson to purple. Intense dark cherry, bitter chocolate, blackcurrant aromas. Concentrated and stylish with rich complex dark cherry, plum, bitter chocolate, roasted coffee flavours, fine loose-knit, slinky tannins and underlying malt oak. Finishes walnutty dry but long and sweet. A classic Bin 707.

'Cassis, black cherry and black plum aromas with black tea, cedar, mushroom notes. A big structure with lovely concentration and juicy, slightly angular tannins. This will easily go on to 2026.' (ER) 'Expressive and complex with sweet currant, liquorice, black tea notes. Fresh, rich and fleshy with polished tannins.' (LD)

Magill Estate Shiraz ▽

NOW ••• 2015

Medium-deep crimson. Fragrant blackberry, panforte, roasted coffee, spicy, tobacco aromas. Beautifully complex wine with plenty of rich roasted coffee/espresso flavours, forest floor, leather/polish notes and plentiful leafy dry tannins. All the elements are in perfect harmony. Understated and classic.

'Gorgeous spice aromatics, plenty of sweet fruit notes and rich ripe round tannins. A wonderful wine.' (JF) 'A great Magill Estate vintage with smoky, roasted meat nuances, magnificent silky flow and soft caressing tannins.' (TS)

St Henri

NOW ••• 2020

Medium-deep crimson. Intense praline, mocha, tobacco, redcurrant, sage aromas. Well-concentrated, generous and smooth with rich mature praline, redcurrant, sweet fruit flavours, lovely mid-palate buoyancy and fine, slinky, graphite texture. Finishes long, sweet and minerally with a lacy tannin plume. Expansive and expressive. Drink now or keep. 90% shiraz, 10% cabernet sauvignon.

'Considerable charm with earthy, sweet fruit aromas, fine-grained tannins, expansive volume and lovely seductive flavour across the palate.' (NB) 'Sweet and developed nose with confiture, pepper, smoky notes. A nicely developed round and unctuous palate with a plethora of ripe fruits and soft texture. A long and juicy aftertaste with salted liquorice, spicy notes.' (AL)

Bin 389 Cabernet Shiraz ⬦

NOW ••• 2025

Medium-deep crimson. Fragrant, roasted coffee, salted liquorice, beef stock aromas with some eucalypt, toffee notes. Intensely concentrated and vigorous with dense roasted coffee, blackberry, liquorice, roasted walnut characters, underlying savoury, mocha notes and supple graphite tannins. Finishes chocolatey long with plenty of sweet fruitiness. A classic Bin 389. Drink now or keep.

'Incredibly vibrant with deep spicy dark red fruits, liquorice, boot polish notes. Medium-bodied with real tannin drive, beautifully integrated acidity and a long persistent finish.' (JF) 'Lives up to its reputation with almost contemptuous ease. Here all the varietal and regional components are welded together superbly. Were any of the earlier vintages as good as 1990 or 1991 when 20 years old? Maybe the 1971 but thereafter I doubt it.' (JH)

Bin 407 Cabernet Sauvignon ⬦

NOW ••• 2016

Medium-deep crimson. Lovely fragrant violet, praline, blackcurrant aromas with a touch of cedar. Earthy, roasted walnut, cassis flavours with fine, chocolatey dry tannins, underlying cedar notes and mineral acidity. Finishes firm, chalky and long. Ready to drink.

'Earthy, gamey bouquet and bright capsicum secondary redcurrant notes. Acidity and softly integrated tannins are still sustaining the palate. Near the end of its life.' (TS) 'A prime example of the tussle between the Penfolds 1990 and 1991 vintage. Here vinosity surges on the finish; lingering finish and aftertaste.' (JH)

Kalimna Bin 28 Shiraz ⬦

NOW ••• 2020

Medium crimson. Classic panforte, cassis, mocha, praline aromas with a touch of sage. Well-developed dark cherry, blackcurrant, apricot flavours with fine lacy tannins and some demi-glace notes. A chewy firm, sweet-fruited finish. Drinking beautifully now, but should hold for a while.

'A very appealing wine with beautiful generosity of fruit. The foundations are solid yet strikingly alive and pleasing.' (TL) 'Mature dark fruits, dark chocolate aromas. Silky textured, fine and elegant. Similar in style to 1990.' (RKP)

Bin 128 Shiraz

NOW

Medium-deep crimson. Intense, fragrant redcurrant, roasted chestnut aromas with aniseed notes. Classically proportioned wine with redcurrant, roasted chestnut flavours, underlying savoury, spicy notes and fine-grained tannins. Finishes supple and fruit-sweet with strong minerally acid notes.

'Caramelised, toffee, biscuity bouquet with oxidised/rancio notes. Quite intense, penetrating and firm with plenty of acidity and piercing red-fruit flavours.' (HH)

Koonunga Hill Shiraz Cabernet

NOW

Brick red. Intense red fruits, praline, prune, tobacco, leather aromas and flavours. Well-concentrated palate with praline, dried fruit, leather, herb flavours and plentiful savoury/grainy tannins. Finishes firm and long. Drink now.

'Plenty of weight and richness.' (JC) 'A pleasant and classic old Penfolds wine; soft and smooth with plenty of fruit sweetness and some spice.' (TG) 'Elegance and refinement.' (LM)

1992 ★★★

A cool growing season, then a late vintage, with intermittent rains • Penfolds releases its first 100% Old Vine Shiraz Grenache Mourvèdre • Australia's High Court *Mabo* judgment overturns the misconception of *terra nullius* and legally recognises the presence of Aboriginal peoples before European settlement • Artist Brett Whiteley dies alone from a heroin overdose in a motel room.

Bin 95 Grange
NOW ••• 2018

Deep crimson. Redcurrant, cassis, sage, herb garden aromas with some mineral notes. The palate is well concentrated with rich layers of redcurrant, sweet fruit, meaty flavours, fine-grained tannins and underlying mocha oak. Muscular dry finish. An elegantly structured Grange with plenty of fruit power and weight. 90% shiraz, 10% cabernet sauvignon.

   Kalimna Vineyard (Barossa Valley), Barossa Valley, Coonawarra, McLaren Vale.

   'Highly aromatic cherry cola, pot pourri, raspberry aromas. Pliant, supple palate with superb clarity and compelling spiciness.' (JR) 'Expressive and intense with liqueur de cassis, dark fruits, earthy, fresh herb notes. Great balance.' (MC)

Bin 707 Cabernet Sauvignon
NOW ••• 2018

Medium-deep crimson. Developed earthy, cassis, red cherry aromas with mint/herb/walnut notes. The palate is muscular and dense with sappy blackcurrant, walnut, earthy flavours and fine gravelly tannins. Finishes leafy firm. A very tannic and unyielding wine.

   'Expressive and mouth-filling with dry green, slightly astringent tannins.' (ST) 'Very intense with some leafy, fine herb notes. Well-balanced, rich and elegant. Easy and delicious.' (LD)

Magill Estate Shiraz
NOW

Medium crimson. Fresh developed chocolate, roasted coffee, black fruits, tobacco aromas with white pepper, sage notes. Quite Rhônish in style with pure black fruits, white pepper, panforte, leafy, mint flavours, underlying vanilla oak and muscular tannins. A crunchy firm finish. Unlikely to further improve but should hold.

   'Superb opulent fruit with rich enveloping tannins, superb richness and complexity. After 20 years, there's still real vintage character and vinosity.' (TL) 'A cooler vintage has introduced some fresher minty notes. An attractive drink-now style.' (JH)

St Henri
NOW ••• 2015

Medium-deep crimson. Fresh salted liquorice, dark chocolate, dark cherry aromas with leafy, violet notes. Moderately concentrated but restrained palate with cherry, salted liquorice, dark chocolate, dried fruit flavours, lacy al dente, slightly leafy tannins and underlying acidity. Drink now. Finishes firm and crisp. 79% shiraz, 21% cabernet sauvignon.

   'Beautifully balanced and seductive wine with vivid cassis-like fruit. Gentle soft entry. Harmonious, round and supple chalky textures.' (NB) 'A cool growing season–type nose with red fruit, green/herbal notes and mint chocolate. A restrained palate with higher acidity and marked tannin, but nicely preserved cassis, currant and plum fruits. Good length with a tannic twist.' (AL)

Bin 389 Cabernet Shiraz
NOW ••• 2016

Medium-deep crimson. Fragrant redcurrant, red liquorice, vanilla aromas with bouquet garni notes. A dense, richly flavoured wine with redcurrant, dried fruit, spicy, herb garden notes and plentiful chalky, dry tannins. It builds up grippy firm at the finish. A solid muscular wine packed with sweet fruit, tannins and mineral notes. Drink soon.

'Intense dark fruits, chocolate, spice, herb aromas. Almost briary with loose-knit grippy tannins.' (RKP)
'Dark chocolate, roasted spice aromas. A richly fruited wine with exotic spices, liqueur chocolate, rum truffle notes and a muscular skeleton.' (JB)

Bin 407 Cabernet Sauvignon NOW

Medium-deep crimson. Fresh black fruits, graphite, cedar, orange peel aromas. Fully mature black fruits, graphite, mushroom, truffle flavours, some leafy notes and sinewy dry tannins. Finishes grippy with mineral notes. Doubtful this will get any better. Drink soon.

'Subdued herb and spices on the nose. The palate is more restrained with moderate fruit weight, an earthy/ savoury core and juicy acidity.' (JB) 'Very appealing loose-knit style with leafy blackcurrant fruit. Juicy and forward on the palate. Not as refined as 1990 or 1991.' (RKP)

Kalimna Bin 28 Shiraz NOW

Medium-deep red. Intense dark chocolate, herb garden, leafy aromas with roasted hazelnut, mint notes. Medium-weighted and slightly underpowered palate with dark chocolate, herb garden flavours and fine slinky tannins. Finishes chalky firm, long and sweet. Drink up.

'Sound, clean and ripe with attractive developed fruit. A soft fleshy wine with sweet fruit notes, good density and silky textures. Lovely line and length.' (HH) 'South Australian rains around vintage time just took the vinosity out of the mid-palate. Nice wine nonetheless.' (JH)

Bin 128 Shiraz NOW

Medium-deep crimson. Complex redcurrant, ferric, graphite aromas with a touch of toffee. Well-concentrated redcurrant, rusty flavours with spicy notes and a slight alcoholic punch. Lacy, dry, chalky tannins with some savoury, walnut notes. Finishes leafy dry and muscular with only medium length.

'Spicy, nettle, toasty old-wine aromas with hints of roast meats. Full-bodied, deep, sinewy and long with some richness but also that firm, focused and slightly lean/bony Coonawarra structure. Still has a future.' (HH)

Old Vine Shiraz Grenache Mourvèdre NOW ••• PAST

Medium-deep crimson to brick red. Developed walnut, leather, wood varnish aromas. Mature wine with earthy, old leather, boot polish, tarry flavours and muscular dry tannins. Finishes firm, tight and a touch acidic. The wine has all but dried out now. It's still drinkable but drink up. 51% mataro, 25% grenache, 24% shiraz.

'Entering a beautiful secondary life with inky, herbal, meaty aromas, an amazing sweet spot of fruit and lively acidity. I really love this.' (TL) 'A meaty, leathery, earthy bouquet and palate, with notes of boot polish and beeswax on the finish. Drying out and past its prime.' (TS)

Koonunga Hill Shiraz Cabernet NOW

Brick red. Intense black cherry, cedar, sandalwood aromas with some leafy notes. Well-concentrated black cherry, cedar flavours and leafy dry tannins. Finishes grippy firm. Tannins overwhelm fruit. Drink soon.

'A briary, slightly rustic edge.' (JC) 'Really lovely, elegantly balanced wine with long soft finish.' (DM)

‖ 1993 ★★

A very wet growing season was followed by warm, drier conditions resulting in a very late, mixed quality vintage fruit; an Indian summer in Coonawarra delivered fully ripened fruit • Representatives of 130 nations sign a treaty to ban the manufacture or possession of chemical weapons • Bill Clinton succeeds George Bush as the 42nd President of the United States of America • Steven Spielberg's film version of *Jurassic Park* is released.

Bin 95 Grange NOW ••• 2018

Deep crimson. Fresh red capsicum, dark chocolate, roasted chestnut, mushroom, pot pourri aromas. Well-concentrated generous wine with jammy dark berry, dark chocolate, espresso flavours and al dente, bitter

tannins. Finishes firm and tight with underlying cedar complexity. 86% shiraz, 14% cabernet sauvignon.

Kalimna Vineyard (Barossa Valley), Barossa Valley, Coonawarra. The highest percentage of cabernet sauvignon in its history.

'*Powerful liqueur blackberry, floral, musky aromas and lush spicy flavours.*' (JR) '*Medium length and power with red fruits, spicy notes and grippy dry tannins at the finish.*' (TG)

Bin 707 Cabernet Sauvignon NOW ••• 2015

Medium-deep crimson. Dark chocolate, sage, silage, mocha, berry aromas with musty notes. The wine is soupy and herbal with dark chocolate, cedar, mocha flavours and robust, astringent stemmy tannins. It's powerful and concentrated but the elements are not resolved. Will keep but it probably won't improve.

'*Very young looking! Quite herbaceous on the nose with cigar box/cedar notes. Generous crushed berry fruit, strong tannins and plenty of oak.*' (FW) '*Restrained cassis, red plum, herbaceous aromas with well-integrated cedar oak and some leather notes. Well-balanced with a core of cassis, black cherry fruit, very drying, coating tannins and lively acidity.*' (ER)

Magill Estate Shiraz NOW

Medium crimson. Fresh complex earthy, silage, tobacco leaf, black fruits, aniseed aromas. Surprisingly buoyant with sweet fruit, earthy, silage, tobacco leaf flavours and chalky loose-knit tannins. Finishes savoury dry and sinewy with some bitter notes. Just past its peak. Best to drink soon.

'*Dried sage, smoky, roasted game and pan juices flow through a silky structured, elegantly poised and persistent palate. Completely transcends the reputation of the vintage.*' (TS) '*Over-ripe stewy/porty characters prevail.*' (JF)

St Henri NOW

Medium-deep crimson. Fresh cedar, sandalwood, dark chocolate, sage, bay leaf aromas. Buoyant, medium-concentrated and fleshy with sweet-fruited, dark chocolate, dark berry, silage flavours and plentiful chalky but lightly grippy/stalky tannins. Great concentration and length. Fresh and delicious to drink. Drink soon. 87% shiraz, 13% cabernet sauvignon.

'*It has an introspective and distant nose: terse and tertiary. It reluctantly unfurls with attractive black truffle notes, but perhaps it demands more time before it yields anything else. The palate is medium-bodied with very fine tannins and it borrows that fleshy, caressing texture of the 1990.*' (NM) '*Characterful sweet fruit, leather, spices, liquorice, peppery, mint aromas. Medium-weighted with an interesting spiciness and rather youthful fruit. A touch lean and acidic at the finish.*' (AL)

Cellar Reserve Coonawarra Cabernet Clare Padthaway Shiraz NOW

Medium-deep crimson. Intense cassis, sage, silage aromas with spicy, malty notes. Developed sweet fruit, liquorice, cassis, silage flavours and plentiful, chocolatey dry tannins. Finishes leafy firm with a touch of acidity. At its cusp of age. Drink very soon.

'*Contrived blackcurrant jam, canned corn aromas with black olive notes. Lean and bony without much flesh, but evolves in the glass. No point in ageing it further.*' (HH) '*A fully mature wine with layers of game, earth and leather. Lingering secondary berry fruits bring sweetness to its core, with soft tannins still elegantly supportive. It's in a happy place and impressively balanced for a lesser vintage.*' (TS)

Bin 389 Cabernet Shiraz NOW

Medium-deep crimson. Fresh roasted walnut, cassis, black olive aromas with some praline, polished leather, barnyard notes. Concentrated dark chocolatey, walnut, panforte flavours and fine-grained, touch leafy tannins. Finishes sappy with complex sweet fruit, silage notes. A difficult year. Unlikely to improve greatly. Best to drink soon.

'*Dark chocolate, dark plum almost prune-like aromas, explosive fruit intensity on the palate, controlled tannins and tight acidity.*' (TL) '*Slightly subdued spicy, earthy aromas. It's more developed than previous three vintages with a slightly more rustic profile over juicy dark fruit.*' (JB)

Bin 407 Cabernet Sauvignon NOW

Medium-deep crimson. Intense blackcurrant, praline, panforte, roasted nut aromas with touches of sage and mint. Classical cabernet characters but quite soupy, with cassis, leafy, violet flavours and sappy/grippy tannins. It still has concentration and vinosity. Will continue to hold but it's unlikely to improve.

'Ready to drink with plenty of dark berry, cassis aromas, sweet fruit and ripe, grainy but powerful tannins. There's lots of depth, background vanilla notes and balanced acidity.' (JF) 'Savoury, dried herb and capsicum notes; a fraction salty. Good fruit length and somewhat robust tannins. The finish is a touch coarse but persistent.' (TS)

Kalimna Bin 28 Shiraz NOW

Medium-deep red. Developed roasted chestnut, panforte, redcurrant, graphite, sage aromas. A slightly sappy wine with redcurrant, prune, herb, silage notes and grainy dry/leafy tannins. It has richness and volume but the tannins are beginning to take over. A peppery/tobacco kick at the finish. Best to drink up.

'Deep earthy, mocha aromas with smoky, tobacco notes. A big wine with plenty of ripe fruit, round tannins and good balanced acidity.' (JF) 'Earthy and slightly vegetal. The palate lacks concentration and definition; the tannins are the only thing holding it together on the finish.' (TS)

Bin 128 Shiraz NOW ••• 2015

Medium-deep crimson. Intense redcurrant, tobacco, sage aromas with vanilla notes. Surprising wine with sweet redcurrant fruit, tobacco, minty notes and loose-knit, savoury tannins. Finishes chalky firm with good persistency. Fruit is beginning to fade and the tannins are taking over. Drink soon.

'A dusty earthy, slightly leathery, old wine bouquet. Intense, savoury, direct and taut, with a leanness that borders on bony.' (HH)

Old Vine Shiraz Grenache Mourvèdre NOW

Medium-deep crimson to brick red. Sweet plum, liquorice, chestnut, sweet fruit aromas with a touch of apricot. Sweet plum, panforte, dried fruit flavours, fine chalky loose-knit tannins and mineral acidity. Finishes chalky firm. The fruit is beginning to power off. Best to drink now. 46% shiraz, 28% grenache, 26% mataro.

'Attractive mature style with rich savoury, sous bois, truffle notes and some mid-palate sweetness. Fresh and drinking well.' (JB) 'A fresher, more incisive envelope for the confit fruit and spicy centre. Good length and balance.' (JH)

Koonunga Hill Shiraz Cabernet NOW

Brick red. Red cherry, cedar, sage aromas with hint of tobacco leaf. Fresh buoyant palate with red cherry, sage, touch confectionary flavours, fine loose-knit/bitter tannins. Finishes leafy firm with an alcoholic kick! Drink very soon.

'Has a restrained austerity.' (JC) 'Rich and spicy with delicious warm fruit profile.' (DM) 'Still quite tannic and backward.' (TG) 'Explosive raspberry, cherry fruit and round supple tannins. Full of vivacity and energy.' (JR)

1994 ★★★★

A dry, mild, even-ripening vintage in the Barossa; intermittent rains – but mild conditions – in McLaren Vale and a warm dry autumn in Coonawarra delivered a very high quality vintage • Max Schubert, AM, dies aged seventy-nine on 6 March after battling emphysema and heart problems • The Labor Keating government passes *The Native Title Act 1993* which provides for determinations of land rights and interests by Indigenous Australians • South Africa holds its first multiracial elections, resulting in the end of white rule and the start of democracy • The Channel Tunnel opens, linking England to France and Europe.

Bin 95 Grange ◈

Medium-deep crimson. Fresh intense plum, liquorice, sandalwood, earthy aromas. The palate is richly concentrated with saturated plum, blackberry, sweet fruit, mocha, herb flavours and fine plentiful, muscular tannins. Showing plenty of tertiary notes, but excellent drive and vinosity. A classic vintage. 89% shiraz, 11% cabernet sauvignon.

Kalimna Vineyard (Barossa Valley), Barossa Valley, McLaren Vale, Coonawarra. Considered by Penfolds winemakers as 'a sleeper Grange vintage'.

'Hugely aromatic. A black hole of dark fruit. Ridiculously unevolved and expansive.' (JR) 'A massive wine with ripe sweet blackberry, animal notes.' (RI) 'Ripe and round, but does it have the extra volume and density of the best years?' (JC)

Bin 707 Cabernet Sauvignon

Medium-deep crimson. Intensely fresh blackberry, aniseed, spicy, frankincense, mint aromas. Well-balanced, generous and voluminous wine with sweet blackcurrant, graphite, cedar flavours, fine chocolatey dry tannins, underlying espresso notes and chalky al dente finish. Lovely richness and savoury persistence. Drinking well now, although will further improve with age.

'Nose is deeply integrated with seamless transition of oak and fruit. Really delicious palate with grippy fine tannins, rich ripe fresh fruit, harmonious oak and velvety strong finish.' (FW) 'A very serious wine with rich black fruits, chocolate and fine herbs. Dense, tannic and lush.' (LD) 'Dark chocolate characters pervade across the wine with some floral sweet fruit notes. The tannins are tight and tough.' (CC) 'A sleeper vintage awakens!' (PG)

Magill Estate Shiraz ◈

Medium-deep crimson. Fragrant redcurrant, tobacco, roasted coffee, prune aromas. Roasted, redcurrant, tobacco flavours, underlying savoury, malty oak and fine-grained, al dente tannins. Finishes firm, a touch brambly, and long. Good concentration, vinosity and fruit sweetness. A classic vintage. Vintage started the day after Max Schubert died.

'Complex, earthy and poised with a core of lovely bright sweet fruit complexity and abundant ripe tannins. A medium-bodied wine but smartly balanced, fresh and lively.' (JF) 'Very stylish and complete with superb fruit definition and beautiful oak, wood spice notes. It's lively, generous and grippy with an amazing persistent finish.' (TL)

St Henri

Medium-deep crimson. Intense cedar, mocha, black cherry, walnut, herb garden, violet aromas. A beautifully balanced wine with deep-set dark cherry, blackberry, praline flavours, plentiful chocolatey tannins and underlying spicy, roasted walnut notes. Finishes very firm with superb richness and length. Plush and voluminous with a strong tannin structure. Labelled Shiraz–Cabernet. Drink now or keep. 77% shiraz, 23% cabernet sauvignon.

'A fresh, charming wine with fresh herbaceous nuances, generous fruit and elegant tannins. Still not fully evolved!' (FK) 'Beautifully perfumed with violet, raspberry, cassis, mild spicy aromas. Very dense and concentrated with incredibly silky tannins, creamy luscious fruit, refined spiciness and magnificent length. Remarkable with huge potential.' (AL)

Bin 389 Cabernet Shiraz

Medium-deep crimson. Intense and complex blackberry, dark chocolate, seaweed, liquorice, minty, bouquet garni aromas. Rich and powerful with deep-set dark chocolate mint flavours, earthy, meaty, black olive nuances and sinewy/leafy tannins. This is an atypical vintage with plenty of stuffing to last the distance, but will it ever really soften out? Drink now or keep.

'A powerfully tannic wine, with cabernet-led fruit cowering in the wake of its structure.' (TS) 'Interesting wine with cool-fruit flavours throughout but not in the least bit green. Bitter dark chocolate, spice, pepper notes with savoury charcuterie characters. Another wine that sits outside the norm.' (JH)

Bin 407 Cabernet Sauvignon ♢

NOW ••• 2018

Medium-deep crimson. Beautifully defined praline, leafy, cassis, herb garden, violet aromas. Generous sweet cassis, blackcurrant, cedar flavours, lovely earthy, praline complexity and classic fine-grained tannins. Finishes firm, savoury and minerally long. Should further develop, although drinking well now.

'Subtle spice and dark fruit nose with bitter chocolate, tobacco notes. Relatively restrained with harsh acidity at the finish.' (JB) 'A full-flavoured robust style with its strong point being the overall balance and mouth-feel. Juicy notes prevail and counter the mid-palate oak.' (JH)

Kalimna Bin 28 Shiraz

NOW ••• 2016

Medium-deep crimson red. Intense, violet, camomile, dark chocolate, walnutty aromas with a hint of mint. Concentrated dark chocolate, prune, walnut flavours, plenty of fruit sweetness on the mid-palate and plentiful velvet tannins. Finishes firm and chocolatey. Still holding very well, but drink soon.

'Ripe, rich and fleshy with plush dense fruit and weighty but not overpowering tannins. There's plenty of life and freshness in this wine.' (TL) 'A fruity nose with plum and rosehip notes. An aromatic, dense and youthful palate, with roasted spice, rosehip, floral notes and big tannins.' (JB)

Bin 128 Shiraz ♢

NOW ••• 2016

Medium-deep crimson brick red. Intense red cherry, blueberry, savoury aromas with hints of mint and vanilla. Sweet-fruited wine with fresh red cherry, blueberry, white pepper flavours and loose-knit, chalky tannins. Finishes tangy with good overall length. Lovely intensity and weight. Drink now or keep for a while.

'Very sound and youthful with charcuterie, toasty, savoury, faintly meaty/spicy aromas, excellent fruit depth, ample tannins, very good generosity and length. This has an extra degree of fleshiness and richness above most vintages of Bin 128. Many years ahead of it!' (HH)

Old Vine Shiraz Grenache Mourvèdre

NOW

Medium-deep crimson brick red. Intense red plum, dark chocolate, sweet fruit aromas with liquorice notes. Ripe red plum, praline, musky flavours and plentiful slinky, graphite tannins. Finishes minerally, slightly grippy and long. Elegant and buoyant, but drink soon. 55% grenache, 28% shiraz, 17% mataro.

'Very fresh yet complex deep plum, cherry compote notes with savoury, spice, liquorice nuances, round ripe tannins and balanced acidity. A lovely medium-bodied wine with some grip and tension.' (JF) 'A deep well of aromatics and flavours with a firm solid body. The fruit is still in balance, but the wine is at the fulcrum point in the ageing cycle.' (TL)

Koonunga Hill Shiraz Cabernet ♢

NOW

Brick red. Perfumed violet, dark cherry, praline aromas with some leafy/sage notes. Fresh dark berry, dark chocolate flavours with some tobacco/walnut notes and fine chalky tannins. Finishes chalky firm and fruit sweet. Still has buoyancy and power. At its peak. Drink soon.

'Pleasing but no future.' (TG) 'Understated. Can't find much complexity.' (LM)

1995

★★★ / ★★★★

A period of drought and September frosts reduced potential yields; warm dry conditions prevailed until late March–early April when a cooler weather pattern marked by drizzle set in • Grange is described as 'a leading candidate for the richest, most concentrated dry table wine on planet Earth' by Robert Parker in the *Wine Advocate* • The inaugural 1995 Bin 144 Yattarna is made; it wins the Tucker–Seabrook Perpetual Trophy at the 1997 Royal Sydney Wine Show. Chairman of Judges, Len Evans, describes it as 'a revelation' and 'a step forward for Australian Chardonnay' • Mother Mary MacKillop of Penola, Coonawarra, is beatified by the Pope • One of London's oldest merchant banks, Barings, is brought down by a single rogue trader • Irish champion racehorse Red Rum, the unmatched treble winner of the UK's Grand National steeplechase, dies.

Bin 95 Grange

NOW ••• 2020

Deep crimson. Intense and exuberant liquorice, roasted chestnut, panforte, dried fruit aromas with some tarry notes. Medium concentrated with redcurrant, roasted chestnut, espresso flavours, fine grainy, al dente tannins and some mocha, cedar notes. Finishes chalky/leafy firm but long and flavourful. Lovely energy and balance. 94% shiraz, 6% cabernet sauvignon.

Kalimna Vineyard (Barossa Valley), Barossa Valley, Magill Estate (Adelaide).

'Highly perfumed cherry, liquorice aromas with seamless, smoky, peppery spice notes.' (JR) 'Powerfully built wine with richness and depth.' (RI) 'Bright fruit, terrific brawny structure and finish.' (DM)

Bin 144 Yattarna Chardonnay

PAST

The first release of Yattarna. Dubbed 'White Grange', it won a Trophy at the Sydney Wine Show and its release created the biggest story in Australian wine media history. It's a curio now. Adelaide Hills, McLaren Vale. The formative 1996–1999 vintages are mostly past.

Magill Estate Shiraz

NOW ••• 2018

Medium-deep crimson. Fragrant wood varnish/polish, black fruits, liquorice aromas with dark chocolate notes. Smooth, quite developed palate with black fruits, aniseed, panforte flavours and chocolatey sweet tannins. Plenty of malty oak fills out the middle palate and lengthens the finish. Drink now but should still hold.

'Light medium-bodied wine with crisp juicy ripe fruits and good length.' (JH) 'A vintage of elegant subtlety, with ripe fruits, vibrant acidity and lingering savoury complexity.' (TS)

St Henri

NOW ••• 2015

Medium-deep crimson. Fresh black cherry, plum, mocha, liquorice aromas with herbal mint notes. Complex and well-concentrated with dense plummy, cassis, mocha, tobacco flavours and fine chalky tannins. Finishes sappy firm but long, sweet and savoury. Drink now but will keep for a while. 85% shiraz, 15% cabernet sauvignon.

'Tertiary, black truffle–tinged aromas that are broody and serious. The palate is medium-bodied with sappy black fruit laced with tobacco, tar and sous bois. This is a St Henri that would not be seen dead without his tie.' (NM) 'Not very complex with some red fruit, mint notes, medium concentration, ripe red berry flavours and dry tannins.' (PK)

Bin 389 Cabernet Shiraz

NOW ••• 2016

Medium-deep red. Intense espresso, dark berry, mulberry aromas with some graphite, sweet fruit notes. Sweet, dark berry, mulberry, ground coffee flavours, chocolatey, al dente textured tannins, underlying mocha notes and a leafy dry finish. A firm, muscular style with plenty of buoyancy, fruit density, richness and length. Drink now or keep for a while.

'A fine elegant nose with light jammy, herb aromas and some mocha notes. A flowing palate with jammy, floral, herb, mocha-rich flavours and big tannin structure. It finishes long, firm and slightly drying.' (RKP) 'Super concentrated, rich and ripe, with cedar oak notes poking through. It's loaded with spice notes and dense, silky tannins. A big powerful wine with years ahead of it.' (JF)

Bin 407 Cabernet Sauvignon

NOW

Medium-deep crimson. Fresh black olive, cassis, graphite, roasted coffee aromas with some sappy notes. Concentrated cassis, black olive, roasted coffee flavours, herb, brambly notes and muscular dry tannins. Finishes very firm and tight. Quite lean and savoury with strong acidity. This is not going to improve.

'I love the lifted cassis, dark chocolate, minty aromatics but the palate is tough and lean. It just doesn't fulfil its promise.' (TL) 'Attractive straightforward wine with concentrated dark fruits and savoury notes. A long tight structure.' (RKP)

Kalimna Bin 28 Shiraz

NOW

Medium-deep crimson. Fragrant herb garden, leafy, mint, redcurrant aromas with a touch of leather/spice. Developed redcurrant, herb garden, leafy flavours and loose-knit chalky dry tannins. Finishes firm and tight with liquorice notes. A touch disjointed. This is not going to improve with age. Drink up.

'Slightly monochromatic wine with peppery nettle aromas, dense concentration, plenty of flesh and chewy tannins.' (HH) 'Elegantly structured and true to style with syrupy berries, spicy touches and vanilla notes. Smooth and not super-ripe.' (RKP)

Bin 128 Shiraz

NOW

Medium-deep crimson to brick red. Fragrant red cherry, cinnamon, cassis, sweet walnut aromas. Medium-weighted wine with cassis, red cherry, cinnamon flavours and loose-knit, chalky tannins. Quite simple but it has good richness and volume.

'Peppery, dry and broad, roughly textured and a touch disjointed. Not a great vintage.' (HH)

Old Vine Shiraz Grenache Mourvèdre

NOW ••• PAST

Medium-deep crimson to brick red. Complex salted liquorice, old leather, cranberry, earthy aromas. The palate is savoury and dried out with overdeveloped fruit and fine-grained chewy tannins. Finishes muscular firm. Well past its prime. 54% shiraz, 25% mataro, 21% grenache.

'Fresh lightly savoury, spicy nose. A juicy palate that straddles primary berry/cherry fruit and mature savoury, spicy characters.' (JB) 'Animal/gamey notes with a core of earthy tertiary development and some persistence, though drying out and past its prime.' (TS)

Koonunga Hill Shiraz Cabernet

NOW ••• PAST

Medium crimson. Developed tomato leaf/earthy aromas. The palate is well concentrated with chocolatey, earthy, dried herb flavours, high-pitched acidity and dry sinewy tannins. The fruit is just holding up.

1996

★★★★★ / ★★★

A good soak in autumn and winter (the wettest July at Magill in 10 years) was followed by a generally dry and even growing season; a hot summer was followed by a damp March and a cool vintage; a great Grange year • Maturation of Grange and Bin wines discontinued at Magill • John Howard elected Prime Minister of Australia • The twin Petronas Towers in Kuala Lumpur become the tallest buildings in the world • The Port Arthur massacre in Tasmania.

Bin 95 Grange ▽

NOW ••• 2040

Deep crimson. Lovely classic Grange with intense dark fruit, dark chocolate, mocha aromas. Plush, generously flavoured palate with praline, blackberry, dark plum flavours, fine grainy/graphite tannins and underlying savoury, malty nuances. Finishes chocolatey firm, long and flavourful. A gorgeously seductive wine with lovely fruit power and richness. Still has decades to go. A great vintage. 94% shiraz, 6% cabernet sauvignon.

Kalimna Vineyard (Barossa Valley), Barossa Valley, McLaren Vale, Magill Estate (Adelaide).

'Fresh red fruits, sweet, dense, fleshy and lush. Possibly perfect.' (JR) 'Superb and complete, with powerful tannins, fresh acidity and deep earthy, fruity, spicy, wood notes. This wine has everything.' (MC) 'Tremendous blend of power, complexity and elegance.' (JC)

Bin 707 Cabernet Sauvignon ▽

NOW ••• 2030

Medium-deep crimson. Classic *crème de cassis*, praline, mocha aromas with some herb garden notes. Beautifully balanced wine with deep-set cassis, dark chocolate, herb garden flavours, fine plentiful silky tannins and new vanilla, malt oak. Some brambly, roasted earth notes. Finishes cedary firm and long. A richly concentrated wine with superb definition, vigour and complexity.

'Intense bundled-up cassis fruit on the nose and palate. Firm, fine structure. Freshness and harmony. So very young still. Even a cold November Beijing day cannot deny how truly fine a wine this is.' (CPT) 'Exotic nose with more spice and sweet oak. Intense, well-focused flavours with tight but silky tannins. A very clever style with freshness and vibrancy.' (LD) 'Pronounced smoky, coffee aromas. Masculine but restrained and still unapproachable. Full-bodied but explosive palate with lingering coffee flavours. Needs another ten years at least.' (CC)

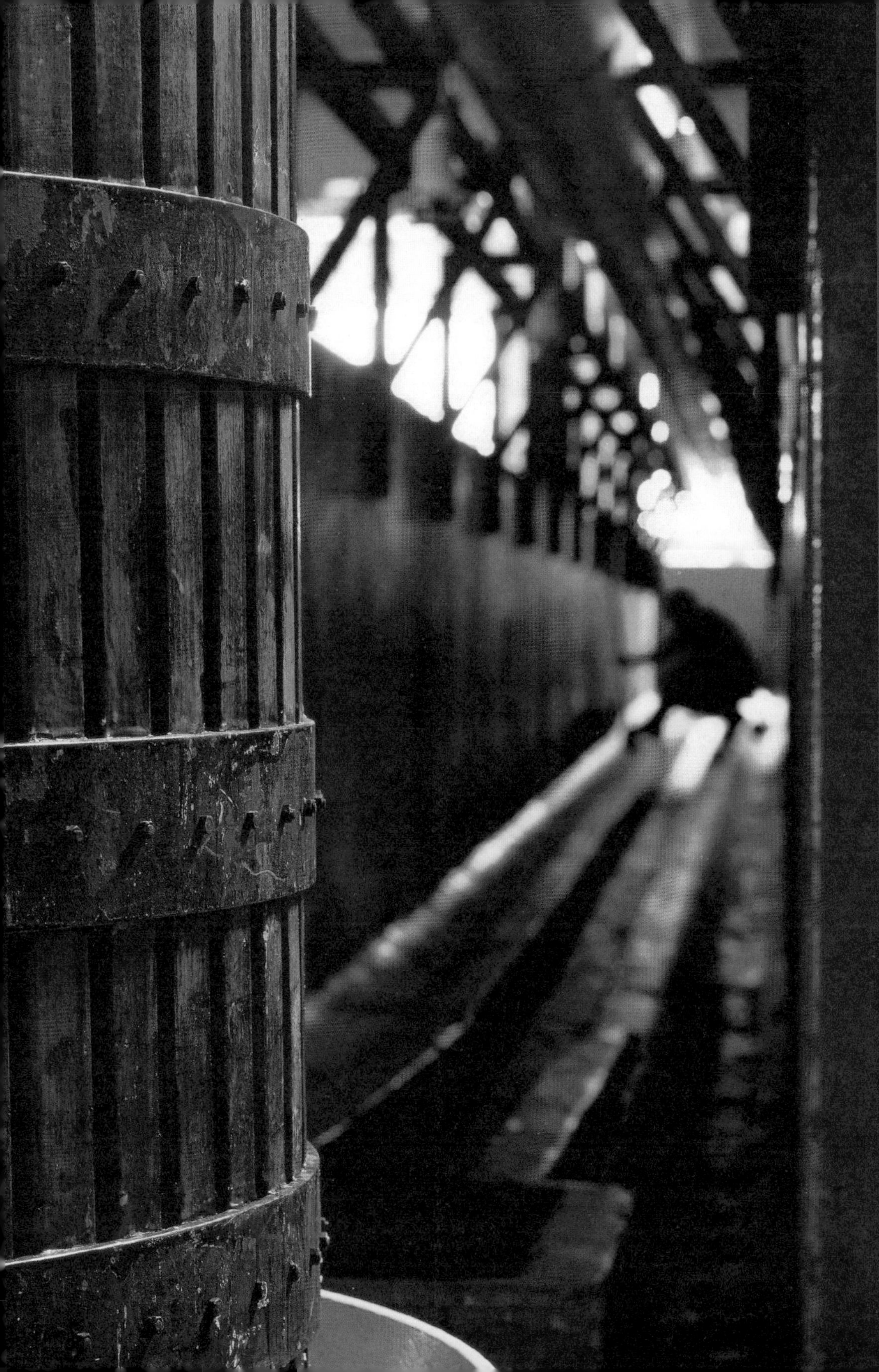

Magill Estate Shiraz ⬦

NOW ••• 2020

Medium-deep crimson. Very complex and aromatic with dark berry, herb garden, tobacco leaf, vanilla, malt aromas, generous sweet fruit flavours, plentiful lacy, chalky tannins and underlying savoury notes. Finishes leafy and long. A top vintage approaching full maturity. Should hold but probably not for more than five or six years.

'Lifted red and black berry fruits with savoury, tobacco nuances. A lovely round ripe luscious palate with balanced tannins and acidity.' (JF) 'A distinctive step up. Elegantly structured with a soft, gently sweet and juicy fruit profile.' (JH)

St Henri ⬦

NOW ••• 2025

Medium-deep crimson. Fragrant cassis, smoky, graphite, roasted chestnut aromas. Gracefully proportioned yet substantial with deep-set inky, cassis, smoky, liquorice, roasted chestnut flavours and slinky, long, graphite-textured tannins. Lovely sweet fruit drives over a chalky firm finish. Lovely power, density and length. A classic St Henri. Drinking well now, but will still improve. 90% shiraz, 10% cabernet sauvignon.

'A sensual bouquet of crushed violets infuses the vibrant red berry, confiture aromas. The palate is medium-bodied with supple tannins, a thread of fine acidity and satisfying harmony. There is an appealing openness and generosity to the finish reminiscent of a classic Rioja!' (NM) 'Youthful primary dark plum, cassis, cherry aromas with a dash of liquorice and spice. Dense, dark inky fruit with nicely firm tannic presence that adds freshness to the rather very good length. Balanced and attractive with a bright future.' (AL)

Kalimna Block 42 Cabernet Sauvignon ⬦

NOW ••• 2040

Medium-deep crimson. Intense blackcurrant, cedar, dark chocolate, mocha, liquorice aromas with apricot, gun smoke notes. Superb richness and generosity with blackberry, smoky, dark chocolate flavours, plenty of mocha oak and sweet ripe velvety tannins. A cedary, mocha finish with a long tail of fruit sweetness. Wonderful vinosity, freshness and balance. A great Penfolds wine. Deserves further cellaring. Best to keep.

'An impressive and distinguished wine with dusty/earth, charcuterie aromas and nuances of Moroccan spices. Powerful yet soft, smooth texture with liquorice, meat stock and dark plum flavours. Great wine, with thirty years ahead of it.' (HH) 'Incredibly pure blackcurrant essence fruit swallows up the spicy oak. It's plump and rich with chocolatey dark fruit notes and grippy tannins in perfect equilibrium. Another classic in the making.' (RKP)

Bin 389 Cabernet Shiraz ⬦

NOW ••• 2025

Medium-deep crimson. Intense praline, blackberry, panforte, malty aromas with lifted mint, aniseed notes. Plush, expansive and fresh with generous praline, chocolate/dark berry flavours, underlying vanilla, ginger, oak and fine plentiful chalky tannins. Finishes firm with plenty of flavour length. Rich and voluminous with superb tannin structure, power and fruit complexity. Drink now or keep.

'A well-defined core of secondary black fruits and dark chocolate oak encased in a bulletproof exoskeleton of beautifully poised tannins. A Bin 389 of consummate longevity and exacting definition.' (TS) 'Big-boned and shapely with a beautiful marriage of dark fruits, flavourful but powerful tannins and subtle oak notes.' (TL) 'A tough bastard but everything is in the right place with rich blackberry, slight eucalypt notes and dominating tannic structure.' (RKP)

Bin 407 Cabernet Sauvignon ⬦

NOW ••• 2016

Medium-deep crimson. Intense complex spearmint, herb garden, cassis, redcurrant, ginger, mocha aromas. Abundant ginger, spearmint, cassis, redcurrant flavours, underlying savoury notes and plentiful grainy, al dente tannins. Finishes leafy firm. An elegantly structured wine with lovely density and vinosity. A classic Bin 407.

'A wine of impressive stature and character. Layers of redcurrants, red pepper, tobacco and herb notes flow gracefully over finely textured tannins. Impeccably balanced; reflecting an outstanding vintage.' (TS) 'Vibrant cassis aromas complexed by balanced oak. Layered, hedonistic fruit-richness lazily flows across the palate. Oak and tannins are mere bystanders applauding the varietal fruit.' (JH)

Kalimna Bin 28 Shiraz ⬦

NOW ••• 2020

Medium-deep crimson. Intense complex cassis, black liquorice, mocha, earthy aromas with some flinty/smoky notes. Well concentrated and substantial, with classic dark plum, blackcurrant, black liquorice, dark chocolate flavours, lovely mid-palate richness and plentiful chalky tannins. Finishes sinewy firm and long. A beautifully balanced and fully mature Bin 28. It will further develop but it probably won't improve. Enjoy now or hold.

'A brooding toasty spicy liquorice nose. Concentrated liquorice, dark chocolate notes with almost tarry nuances, firm but ripe tannins and tight acidity.' (JB) 'Slight dustiness detracts, but otherwise a good core of black fruits, impressive persistence and nicely structured tannins.' (TS)

Bin 128 Shiraz ♦ NOW ••• 2018

Medium-deep crimson. Intense dark chocolate, black cherry aromas with a touch of mint and sage. Beautifully concentrated wine with dark cherry, praline, herb garden flavours and chalky, loose-knit tannins. Finishes leafy firm, buoyant and long.

'Deep, sound, ripe and rich on the bouquet. Walnut, savoury spices on the palate with moderate richness and fruit depth. A drying and quite firm finish.' (HH)

Old Vine Shiraz Grenache Mourvèdre NOW

Medium-deep crimson. Fragrant sandalwood, leather, polish, spicy aromas with a touch of mint. Quite tightly structured with fading earthy, sandalwood, leather notes, dusty tannins and marked acidity. Finishes savoury with some bitter-sweet notes. Drink up. 54% shiraz, 25% mataro, 21% grenache.

'Fresh lightly savoury, spicy nose. A juicy palate that straddles primary berry/cherry fruit and mature savoury spicy characters.' (JB) 'Animal/gamey notes with a core of earthy tertiary development and some persistence, though drying out and past its prime.' (TS)

Koonunga Hill Shiraz Cabernet ♦ NOW ••• 2018

Deep crimson. Developed blackberry, panforte aromas with herb garden aromas. Full-bodied wine with rich blackberry/liquorice, panforte, roasted chestnut flavours and plentiful chocolatey tannins.

'Glossy wine with sweet cherry, black fruits and still some tannins to shed.' (TG)

1997 ★★★/★★★★

Generally dry and cool conditions prevailed during October and November, then a hot burst of weather arrived during summer but cooler temperatures and a week of rain in February slowed down ripening; a warm dry period followed over vintage • RWT Barossa Valley Shiraz and Cellar Reserve Adelaide Hills Pinot Noir are first made • *Harry Potter and the Philosopher's Stone* by J.K. Rowling is published • Britain returns sovereignty of Hong Kong to the People's Republic of China • Diana, Princess of Wales, dies, aged thirty-six, in a car accident in Paris • Mother Teresa of Calcutta dies, at eighty-seven.

Bin 95 Grange NOW ••• 2025

Deep crimson. Lifted blackberry, panforte, leafy, smoky aromas. Well-concentrated blackberry, herb garden, sage, leafy flavours, fine-grained slinky tannins and plenty of smoky, mocha oak notes. Finishes chocolatey and long. An early-drinking Grange with lovely freshness, vinosity and drive. 96% shiraz, 4% cabernet sauvignon.

Kalimna Vineyard (Barossa Valley), Barossa Valley, McLaren Vale, Limestone Coast.

'Not as impactful as 1996 or 1998, but an excellent Grange with finesse and balance.' (MC) 'Roasted meat, sesame seed aromas with youthful brooding tannins.' (DM)

Bin 707 Cabernet Sauvignon NOW ••• 2018

Medium-deep crimson. Intense cassis, roasted walnut, cedar, herb aromas. Well-concentrated sweet fruit, redcurrant, walnut flavours, fine loose-knit sweet tannins and plenty of new malty oak notes. Finishes firm, chalky and long. Very expressive and giving. Probably best to drink soon.

'Attractive and open black cherry, black plum, cassis, red fruit aromas with cedar, tobacco notes. Rich and vibrant in fruit with strong tannin structure and very good length.' (ER) 'Polished wine with savoury sweet spicy notes and pronounced new oak.' (ST)

RWT Shiraz NOW

Deep crimson to brick. Intense dark cherry, mocha, panforte aromas with *sous bois*, mushroom, mint, sage notes. Well-concentrated, smooth wine with generous developed dark cherry, sage, mocha, earthy flavours, plenty of mid-palate richness and fine, silky tannins. Finishes long and sweet. 53% new French oak. Probably best to drink soon.

'A slightly diffuse bouquet with decayed scents of leather, scorched earth, mint and an almost feral, animally note. It expands nicely with aeration, with floral, violet notes emerging. The palate is soft on the entry with a touch of piquancy, good acidity and grainy texture. Noticeable volatility and fiery finish blows off and mellows in the glass.' (NM) 'Sweet fruit aromas, hints of liquorice and mature complex notes of incense, spice and leather (not too dissimilar from an old Hermitage). The palate is vigorous and ample with layers of dark fruit, soft tannins and a medium-to-long spicy finish.' (AL) 'Dried fruits, tobacco, a touch of sandalwood and liquorice. The palate is firm and taut with attractive ripe red fruits. Just entering its tertiary phase.' (MS)

Magill Estate Shiraz NOW

Medium-deep crimson. Fully mature chocolate, blackberry aromas with leafy, Vegemite, cedar notes. Well-concentrated but developed cassis, dark chocolate, sweet fruit flavours, complex tarry notes and sappy, leafy tannins. Finishes firm and lean. Beginning to fade.

'Dusty, earthy and callow with drying astringency and awkward imbalance.' (TS) 'The gentle juicy, sweet fruit characters roll on but without the conviction of the 1996.' (JH)

St Henri NOW ••• 2016

Medium-deep crimson. Developed redcurrant, prune, leather, leafy, liquorice aromas with a touch of grilled meats. Well-concentrated demi-glace, prune, dark chocolate, leather, herb garden flavours, plenty of mid-palate richness and pronounced chocolatey tannins. Finishes sappy dry, but long and sweet. Drink now or keep for a while. 92% shiraz, 8% cabernet sauvignon.

'A Bordeaux-like bouquet with touches of sea salt and hints of cooked meat/game, brettanomyces. The palate is medium-bodied with a lovely rich, creamy texture and a controlled, powerful long finish. A touch of alcohol blurs the delineation. Otherwise great potential.' (NM) 'Intense blackcurrant, leafy, slightly vegetal aromas. Gorgeous density of texture thanks to very fine-grained tannins. Round as a ball. A slight hot spot on the finish.' (NB)

Cellar Reserve Pinot Noir NOW

Medium-deep crimson. Leafy, forest floor, red cherry, roasted hazelnut aromas with silage, leather notes. Supple, red cherry, strawberry, silage, sage flavours and chalky dry tannins. Finishes chalky firm and savoury. Drink up.

'Old mellow leathery bouquet with slight raisin, stewy, beetroot notes. The palate is dry and lean with good depth, but the tannins dominate fading fruit.' (HH)

Bin 389 Cabernet Shiraz NOW ••• 2016

Medium-deep red. Fresh, redcurrant, leafy aromas with walnut, polished leather, mint and violet notes. Fleshy sweet redcurrant, plum flavours, firm gravelly tannins and underlying savoury nuances. Plenty of mid-palate richness but tannins muscle up towards the finish. Best to drink soon.

'Fumy dark chocolate, blackberry aromas with gentle spice notes. Smooth and lush with grippy tannins. Quite butch and firm, but lacks finesse.' (RKP) 'Fragrant cassis liqueur, exotic spice aromas with floral notes. Intense and concentrated with tight-knit fruit, firm velvety tannins and fresh acidity. A powerful, tightly coiled wine.' (JB)

Bin 407 Cabernet Sauvignon NOW ••• 2016

Medium-deep crimson. Fresh cranberry, blackcurrant aromas with some earthy/tobacco notes. The palate is

medium-bodied with developed red fruit, cassis, tobacco, sage flavours, underlying cedar notes and leafy dry but fine-grained tannins. Finishes grippy firm, long and savoury.

'A wine of modest dimensions with neatly judged oak and tannins balanced with fruit. Slightly dusty characters throughout.' (JH) 'A confectionary lift in aromatics. Primary redcurrant pastille flavours flow across the palate with underlying savoury tannins. Less intense and complex than some others in this line up.' (TL)

Kalimna Bin 28 Shiraz
NOW ••• 2015

Medium-deep crimson. Espresso/ground coffee, herb, redcurrant, dried fruit aromas with aniseed/liquorice, sweet fruit notes. Fresh, minerally and medium-bodied wine with espresso, redcurrant flavours, some leafy/silage notes and fine, loose-knit chalky, slightly grippy tannins. Finishes minerally and long. Best to drink soon.

'Lovely lifted smoky mocha aromas with blood orange notes. Medium-bodied, bold and lithe with ripe tannins, great texture and balanced acidity. A lively vibrant delicious wine with years ahead of it.' (JF) 'Exceeded expectations with its spicy red fruits and gentle tannins.' (JH)

Bin 128 Shiraz
NOW ••• 2015

Medium-deep crimson. Redcurrant, aniseed, slightly leafy, tobacco aromas with coffee and white pepper notes. Fresh white pepper, redcurrant flavours and fine slinky, dry tannins. Plenty of vinosity and freshness with a chalky firm finish.

'Sweeter, slightly raisin aromas with a lick of aniseed. Savoury and earthy with coarse rustic tannins and modest length.' (HH)

Old Vine Shiraz Grenache Mourvèdre
NOW

Medium-deep crimson. Fragrant dried roses, pot pourri, walnut, herb garden aromas with a touch of musky plum. Fully mature herb garden, walnut, leather flavours, sinewy/muscular, dry tannins and mineral acidity. The palate has dried out a touch. Best to drink up. 36% shiraz, 34% grenache, 30% mataro.

'Fragrant plum, cherry aromas with spicy, leather, menthol notes. The palate has plenty of fruit sweetness, grippy tannins and a long acid backbone.' (JF) 'Less of the ripe attributes of the 1996, with moderate concentration, loose-knit texture and slight iodine, pepper notes. Dried out a bit.' (RKP)

Koonunga Hill Shiraz Cabernet
NOW ••• 2020

Medium-deep crimson. Complex and developed espresso, roasted walnut, caramel, leafy aromas. A firm but substantial wine with roasted coffee, chinotto, leafy flavours, strong graphite tannins and noticeable alcohol. Still holding. Quite a surprise!

'Crisp acids and tannin provide definition and shape.' (JC) 'Warm fleshy wine with gorgeous meaty, spicy fruit and seamlessly integrated tannins. Finishes minty.' (DM) 'Attack is generous, four-square and dense. This has a long life ahead.' (TG)

1998
★★★★/★★★★★

A mild early growing season was followed by very hot, dry weather with virtually all damwater reserves exhausted; an exceptional vintage • Asian economic meltdown • Mark Taylor, Australian cricket captain, bats for two days against Pakistan and amasses 334 not out, equalling Sir Donald Bradman's Australian record set in 1930 • US President Bill Clinton admits in taped testimony that he had an 'improper physical relationship' with White House intern Monica Lewinsky • Disastrous Sydney to Hobart yacht race due to unpredicted tumultuous weather.

Bin 95 Grange ▽
NOW ••• 2045

Deep crimson. Very powerful, opulent and expressive Grange with intense blackberry, praline, liquorice aromas and hints of leather and spice. A sumptuous mouth-filling wine with plush blackberry, cassis, dark chocolate flavours, beautiful savoury, spicy new oak and plentiful velvety graphite tannins. Superb balance, length and richness of flavour. A classic year. 97% shiraz, 3% cabernet sauvignon.

Kalimna Vineyard (Barossa Valley), Barossa Valley, Magill Estate (Adelaide), Padthaway.

'Beautiful fruit and mouth-filling power.' (TG) 'Monumental with dark sweet fruit, chocolate, roasted wood notes and a mountain of fine tannins. One of the greatest Granges.' (MC) 'Tremendous opulence with juicy blackberry, boysenberry, superb depth and structure.' (RI)

Bin 707 Cabernet Sauvignon ◈   NOW ••• 2040

Medium-deep crimson. Ethereal and intense blackcurrant, mocha, vanilla, mint, sage aromas. Beautifully concentrated, rich and voluminous with smooth cassis, praline flavours, fine plentiful chocolatey tannins and underlying malty, roasted chestnut notes. Finishes velvety smooth. Superb fruit sweetness, vinosity and freshness. A powerful Bin 707 with the balance, richness and stuffing to age into its forties!

'Perfectly balanced and stunningly tasty with deep black fruits, mocha, tar, Chinese ink notes, big tannins and a very long finish.' (ST) 'A more dominant style with intense dark cherry, meaty, spicy aromas, rich powerful concentration and precision.' (LD)

RWT Shiraz ◈   NOW ••• 2020

Medium-deep crimson. Intense ripe blackberry, cherry stone, mocha, roasted chestnut aromas with balsamic notes. Really well concentrated and classic with rich blackberry, liquorice flavours, fine chocolatey tannins and underlying vanilla, malt oak. Finishes velvety and long with a sweet fruit flourish and some panforte notes. 100% new French oak. Progressing well. Drink now or keep.

'Fresh, elemental dark berry fruit with new oak and dark spices. More juicy, succulent and fleshy than 1997.' (MS) 'Exotic ripe fruits with complex earthy, peaty undertones. Gentle entry with compact sweet fruit, fine-grained tannins and a clear fresh, flourishing well-balanced finish.' (NB)

Magill Estate Shiraz ◈   NOW ••• 2020

Medium-deep crimson. Intense liquorice, dark fruits, black olive aromas with mint/spearmint and cedar notes. Concentrated dark fruits/mulberry/blackberry, black olive flavours, complex meaty nuances and sinewy, graphite textures. Well balanced with attractive fruit complexity and richness. Elegantly proportioned rather than powerful now. Approaching optimum drinking. Achieved great critical reviews at release.

'Aromatic wine with dark fruits and plenty of cedar, spice complexity. Medium-bodied with lovely depth of fruit and supple ripe tannins. A terrific wine.' (JF) 'Altogether more substance, yet still medium-bodied and soft with red and black fruits.' (JH)

St Henri ◈   NOW ••• 2028

Medium-deep crimson. Intense cassis, raspberry, redcurrant aromas with smoky, praline, herb garden, liquorice notes. Beautifully concentrated palate with dense juicy cassis, redcurrant, raspberry, praline flavours, fresh herb garden notes, seductive fruit sweetness and classic fine-grained tannins. Finishes velvety with superb length of flavour. Lovely buoyancy, volume and richness but will continue to develop. Classic St Henri vintage. Drink now or keep. 92% shiraz, 8% cabernet sauvignon.

'Fresh blackberry, spicy, smoky notes, young excellent tannins, plenty of sweet fruit and very good length. Finishes quite minerally. Great potential.' (PK) 'Intensely perfumed and alluring with violet, cassis, plum, cherry and mild spice aromas. The palate is immensely concentrated with masses of fruit and a very polished texture. Monolithic, persistent and resonant!' (AL)

Cellar Reserve Pinot Noir   NOW

Medium-deep crimson to brick red. Complex, earthy, roasted nuts, redcurrant, leafy aromas with some herb garden, barnyardy notes. Well-concentrated earthy, roasted nuts, redcurrant flavours with some spicy notes and plentiful but hard tannins. Finishes leafy dry. Drink now.

'A lean wine with animal notes, an overly firm structure and a slight hollowness.' (HH)

Cellar Reserve Sangiovese NOW ••• PAST

Medium-deep brick red. Developed coffee, toffee, roasted walnut, graphite aromas with sweet fruit notes. Developed roasted coffee, cherry stone, tobacco flavours with fine grainy tannins. But it finishes firm and sweet and sour. Old leather, slight maderised notes suggest this wine is towards the end of its days.

'Developed leather, lead pencil, cedar spice aromas with liquorice notes. It still has great presence and structure.' (TL) 'A pleasant light- to medium-bodied palate with tannins largely out of the picture. But spiced red fruits are still present. Drink – don't cellar.' (JH)

Bin 389 Cabernet Shiraz ▽ NOW ••• 2030

Medium-deep crimson. Powerful dark chocolate, dark berry aromas with cedar, malt, panforte and sweet fruit notes. Rich, voluminous and buoyant with superb rich sweet fruit, dark berry, praline notes, abundant fine chalky firm tannins and underlying savoury spicy notes. Finishes grainy firm, long and fruit-sweet. Lovely vinosity and generosity of flavour. An outstanding vintage with plenty of cellaring potential. Drink now but recommended to keep.

'The generosity of the vintage sings true with secondary berry fruits suspended in a brittle web of firm tannins. The fruit is just holding out against its structure, suggesting it is nearing its peak, but the youthful tannins declare that it is barely halfway there.' (TS) 'A full-bodied wine with multiple layers of blackberry, plum and blackcurrant fruits followed by well-balanced ripe tannins. A long future ahead.' (JH)

Bin 407 Cabernet Sauvignon ▽ NOW ••• 2020

Medium-deep crimson. Complex plum, roasted coffee/espresso/mocha aromas with underlying cedar, mint notes. Sweet buoyant plum, dark chocolate, roasted coffee/espresso flavours with sweet fruit/cassis notes and fine-grained tannins. Finishes firm, but the flavours are persistent. A classic Bin 407.

'Incredibly vibrant and fresh with pure cassis, leafy, mint aromas and mocha, camphor notes. It's medium-bodied and stylish with a flourish of balanced tannins and wonderful persistency.' (JF) 'Vibrant stewed berries and plums with porty, cedar notes. Finely textured but jammy with plenty of ripe tannins and great mouth-feel.' (RKP)

Kalimna Bin 28 Shiraz ▽ NOW ••• 2020

Medium-deep crimson. Intense blackberry, praline, walnut aromas with liquorice, bitumen notes. Rich voluminous wine with generous, but developed, dark berry, dark chocolate, roasted walnut, tobacco flavours and plentiful chocolatey dry tannins. Finishes fruit-sweet and long. A classic Penfolds vintage. Complex, generous and balanced. Drinking well now yet should hold for a while.

'Secondary red berry fruits and soft tannins are impressively intermeshed. Primary characters are giving way to secondary leather/savoury complexity. A touch of dryness on the finish suggests its life is beginning to draw to a close.' (TS) 'Intense savoury, meaty nose, slight stewed characters, obvious depth and concentration.' (JB)

Bin 128 Shiraz ▽ NOW ••• 2018

Medium-deep red to purple. Fresh red and black cherries with liquorice, nougat, roasted nut notes. Rich, voluminous, black cherry, praline, espresso flavours and fine chocolatey tannins with lovely mid-palate buoyancy. Finishes firm, long and tangy.

'Mellow and appealing spicy red-berry aromas with savoury notes, too. Palate is medium-bodied, dry and savoury with a hint of leanness. There's good depth and textural density. It will live for many more years.' (HH)

Bin 138 Shiraz Grenache Mourvèdre ▽ NOW ••• 2015

Medium-deep crimson. Fresh musky plum, fruit-sweet, dark chocolate aromas. Richly concentrated and buoyant with plenty of fruit sweetness, persistent dark chocolate, plum, prune flavours and loose-knit velvety tannins. Generous and voluminous. At the peak of development. Drink soon. 55% shiraz, 25% grenache, 20% mataro.

'Youthful raspberry, liquorice nose. Attractively concentrated, smooth and juicy with dark berry, raspberry, liquorice notes and plenty of sweet fruit characters. A pretty wine that embodies the wonderful freshness of a good Grenache.' (JB) 'Appealing raspberry/strawberry fruit with earthy, leathery notes. Seamlessly poised with velvety long tannins, impressive line and persistency.' (TS)

Koonunga Hill Shiraz Cabernet ⬥
NOW •• 2018

Medium-deep crimson. Intense plum, dark cherry, cedar aromas and flavours. Richly concentrated and buoyant palate with plenty of sweet plum, dark cherry, panforte flavours and chocolatey tannins. Finishes firm, long and sweet.

'Ripe round supple, excellent depth of fruit, fine cabernet tannins on the finish.' (JC) 'Smooth polished tannins and dense blackberry fruit.' (LM) 'Intense bitter cherry, dark berry fruit and plentiful gentle tannins.' (JR)

1999 ★★★★ / ★★★

Dry winter conditions were followed by intermittent rainfall during November and December, just enough to maintain healthy vines • The Barossa and McLaren Vale experienced heavy rains in March and ripening slowed, but despite this, vineyards with good drainage produced fruit of exceptional quality • Padthaway escaped the burden of March rain and experienced a great vintage • The euro currency is introduced • Australian swimmer Ian Thorpe breaks the first of thirteen individual long-course world records • Portugal returns Macau to China.

Bin 95 Grange ⬥
2016 •• 2045

Deep purple to crimson. Fragrant cedar, sandalwood, blackberry, espresso, malty aromas. Superbly concentrated wine with blackcurrant pastille, mulberry, tarry, wood spice, liquorice flavours, fine plentiful savoury tannins and mocha oak notes. Finishes chalky firm. Lovely fruit richness and volume. 100% shiraz.

Kalimna Vineyard (Barossa Valley), Barossa Valley, Magill Estate (Adelaide), McLaren Vale, Padthaway. Another 'sleeper Grange vintage'.

'Serious power with smoky, cassis, liquorice aromas. Vibrant, fleshy and spicy with rich red berry, mocha flavours and chewy tannins.' (JR) 'Brandied berry and chocolate aromas. Full-bodied, long and supple.' (JC)

Bin 707 Cabernet Sauvignon
NOW •• 2020

Medium-deep crimson. Violet, red cherry, blackcurrant, dark chocolate, malty, roasted chestnut aromas with some herb notes. Classically proportioned claret style with beautiful *crème de cassis*, praline flavours, underlying savoury, roasted walnut complexity, fine lacy, al dente tannins and mocha oak notes. Finishes firm and sweet. Delicious!

'Herbaceous nose with less obvious fruit. Lean and tight with grippy tannins, less richness of fruit and strong oaky finish.' (FW) 'Crème de cassis, *fine-herb notes with good balance and structure.' (LD)*

RWT Shiraz ⬥
NOW ••• 2020

Medium-deep crimson. Fragrant blackberry, iodine, cedar aromas with some liquorice, leather, mint notes. More developed than the 1998, with generous blackberry, cedar, leather, sandalwood, earthy flavours, round buoyant mid-palate, velvety tannins and underlying savoury oak nuances. Finishes chalky but long with an alcoholic kick and tannin plume. Complex with some rustic notes. 75% new French oak. Drink now or keep.

'The 1999 screams "Barossa" on the nose with delightful pure eucalyptus scents, succulent black fruits and tinges of brettanomyces. *The oak is well integrated and allows the fruit to express itself. The palate is full-bodied, opulent and viscous. A hedonistic yet well-balanced finish that glides to a long cassis, blueberry-tinged aftertaste. Smooth and sensual.' (NM) 'Meaty, minty aromas with some rustic notes. Juicy and richly flavoured with strong grippy tannins and very firm finish. It's full of character and elegance but not classic in any way.' (FK)*

Magill Estate Shiraz
NOW ••• 2018

Medium-deep crimson. Fresh, dried roses, pot pourri, aniseed/liquorice aromas with dark chocolate notes. Expressive cassis, espresso, herb garden flavours, underlying savoury, vanilla nuances and chalky, plentiful tannins. Fruit sweetness and mineral notes pervade across the palate and well past the finish.

'Black fruits, plums and blueberries slowly and assuredly become savoury and earthy. Lovely lingering blueberry, plum, sweet fruits are balanced neatly by fine, silky tannins. An excellent vintage.' (TS) 'Subdued fruit, firm bracing tannins and cleansing acidity. Good line and length.' (JF)

St Henri ▽ NOW ••• 2025

Deep crimson. Fresh, intense liquorice, blackberry, graphite aromas with roasted walnut, panforte notes. Deeply concentrated, fleshy and smooth palate with blackberry, dark chocolate, liquorice flavours and lovely supple, sweet polished tannins. Builds up firm and chocolatey at the finish with a long tannin plume. Very stylish. Drink now or keep. 89% shiraz, 11% cabernet sauvignon.

'A straight-laced bouquet of dark plum, damson fruit, orange blossom and mint notes. The palate is full-bodied, perhaps one of the richest St Henris thus far, although it does not possess the complexity and tension of a great Shiraz.' (NM) 'Intense blackberry, spicy, mineral notes. Elegantly structured with concentrated fruit and young ripe tannins.' (PK)

Cellar Reserve Pinot Noir NOW

Medium-deep crimson to brick red. Redcurrant, leafy, silage, vanilla aromas and flavours, chalky/fine-grained tannins and underlying savoury oak. Finishes firm with fruit-sweet, sappy, gamey notes. Drying out now. Drink soon.

'Mellow and aged with funky, animal aromas with vegetal/stalky notes. Palate is lean, narrow and a trifle tough. Not much fruit and charm.' (HH)

Cellar Reserve Sangiovese NOW

Medium-deep crimson to brick red. Mature leather, mushroom, red berry, spice aromas with some slightly oily/bitumen notes. Aged leather, black cherry, mushroom flavours with plentiful chalky tannins and underlying savoury nuances. Finishes chalky firm with a tannin plume. The tannins are taking over and the fruit is drying out. Drink soon.

'A very earthy nose with a touch of mustiness. Dried-out and scrawny on the palate without richness, fruit and depth. Fades away fast at the finish.' (HH) 'Smells of cherries, herbs and leather armchairs. Good mouth-feel, but it's drying out now with savoury flavours giving way to old-age characters.' (RKP)

Reserve Bin Eden Valley Riesling NOW

Medium-deep amber, brassy colour. Fresh lemon, toast, marmalade, oilskin aromas. Well-concentrated wine with rich lemon curd, tobacco, toasty flavours, plenty of mid-palate richness and crisp long acidity. Finishes long and minerally. At cusp of age now, but still delicious to drink.

'Very mature and developed with paraffin wax and elements of pineapple, poached peach and mango. Fairly savoury, dry and clean finishing. Starting to teeter on the brink, perhaps.' (HH)

Bin 389 Cabernet Shiraz NOW ••• 2020

Medium-deep crimson. Complex roasted chestnut, blackberry, graphite aromas with apricot, herb garden, grilled almond nuances. Richly concentrated with developed blackberry, graphite, dark chocolate, roasted chestnut, cola flavours, plentiful muscular tannins and underlying acidity. Salted liquorice, praline, bitter-sweet notes at finish. More advanced than 1998, but showing lovely fruit complexity. Drink now or keep.

'Plenty to like here with fresh cassis fruit, tight firm tannins and balanced acidity.' (JF) 'Deep blackberry, blackcurrant aromas with a savoury spicy undercurrent. A massively concentrated wine with good fruit weight, firm tannins and balanced acidity.' (JB)

Bin 407 Cabernet Sauvignon NOW ••• 2025

Medium-deep red. Fragrant cassis, dark chocolate aromas with black olive notes. Lovely *crème de cassis*, brambly flavours, a core of sweet fruit and some bitumen, roasted walnut notes. The tannins are sinewy firm and gradually take over the palate. It finishes bitter-sweet with some herb garden notes. A substantial wine, but will the tannins ever soften out?

'A riper expression of cabernet with plenty of dark fruits, and rich ripe flavours. Not in the least bit overblown.' (TL) 'Crunchy red fruits, blackcurrant, capsicum, leafy aromas. The tannins are still bracingly youthful and promise to sustain it for decades. Impressive persistence and style.' (TS)

Kalimna Bin 28 Shiraz
NOW ••• 2018

Medium-deep crimson. Dark chocolate, blackberry, herb aromas with dried leaf notes. Richly concentrated blackberry, dark chocolate, espresso flavours and fine loose-knit muscular tannins. Finishes cedary firm. More tannic than 1998, but still has generosity of fruit. Should further develop and improve. Drink now or keep.

'Svelte and elegantly structured but relatively subdued with an attractive spicy nose, dark chocolate, sour plum flavours, taut acidity and a spicy, mineral undercurrent.' (JB) 'Stacked to the rafters with ripe black fruits and tannins. Arguably needs a decade to decide which side of the Moon will win!' (JH)

Bin 128 Shiraz
NOW ••• 2018

Medium-deep red. Intense, liquorice all-sorts, sage, dark berry, malty aromas with a hint of rum 'n' raisin. Substantial wine with deep-set, dense liquorice, dark berry flavours, spicy notes and plentiful chocolatey tannins. Finishes firm, minerally and quite muscular.

'Big and full-bodied with savoury flavours, walnutty notes, and masses of firm drying tannins. A very solid wine with years left in it.' (HH)

Bin 138 Shiraz Grenache Mourvèdre
NOW ••• 2015

Medium-deep crimson. Fresh herb garden, liquorice, dark berry aromas with red cherry, chinotto/cola notes. Well-concentrated wine with herb garden, dark berry, raspberry, mulberry flavours, some brambly notes and fine loose-knit chalky tannins. Finishing leafy firm and long. Still quite juicy and complex but drink soon. 36% shiraz, 34% mataro, 30% grenache.

'Lifted wood varnish, cumin, turmeric spice aromas. The palate is fresh with plenty of fruit sweetness but the tannins and acidity dominate.' (JF) 'Full-bodied with earthy, gravelly notes, great palate tension and integration of flavours.' (TL)

Koonunga Hill Shiraz Cabernet ♦
NOW ••• 2020

Medium-deep crimson. Intense cherry stone, tobacco leaf aromas. Well-concentrated cherry, tobacco, espresso flavours with fine loose-knit grainy tannins. Finishes long and sweet. Still showing freshness and complexity. Drink soon.

'Perfumed, fresh and fruity with young tannins. Very good.' (MC) 'Simply delicious. Opulent now with perhaps ten years of maturation left.' (LM)

2000
★★ / ★★★

Dry, cool growing conditions with a hot spell in late January, although rain in late February and early March slowed ripening and vintage • The earliest vintage on record in Coonawarra • John Duval is the International Wine Challenge's 'Red Winemaker of the Year' in London • Y2K, the predicted Millennium Bug global technological crash, fails to materialise • A draft genetic blueprint of the human species is completed • Sydney hosts the Millennium Summer Olympics. At the closing ceremony, Juan Antonio Samaranch, chairman of the IOC, declares Sydney 'the best Olympic Games ever'.

Bin 95 Grange
NOW ••• 2025

Medium-deep crimson. Intense aniseed, redcurrant, rhubarb, blackberry, dark chocolate aromas with mocha, ginger oak. Sweet buoyant palate with redcurrant, mocha flavours and velvety, touch leafy tannins. Finishes bitter chocolatey firm but long and sweet-fruited. An atypical Grange but still expressive and delicious to drink. 100% Barossa Valley shiraz.

Kalimna Vineyard (Barossa Valley), Barossa Valley.

'An amazing blend of balance and power with luscious fruit and sappy textures.' (JR) 'Light, easy-drinking Grange with sweet warm blackberry, liquorice fruit.' (TG) 'Very pretty and pure with blackberry liqueur notes and plush, dense peppery tannins.' (RI) 'After a dozen years or so it seems to be losing the winter-green notes.' (PG)

Bin 144 Yattarna Chardonnay   NOW

Medium-deep straw gold. Intense lemon curd, verbena, nougat aromas with some amontillado notes. Much fresher than 2000 Bin 00A. Lemon curd, marzipan flavours with some creamy, nougat, caramel complexity, chalky, loose-knit textures and mineral acidity. Advanced in age but still drinking well for now. Drink very soon. 68.5% Tumbarumba, 31.5% Adelaide Hills. 13% new French oak.

'Rich butterscotch and roasted hazelnut aromas with chicken stock notes. Fully mature wine with voluminous flavour and absolutely perfect chardonnay character. Magnificent; a complete wine at its peak.' (HH) 'Still sound with mature chardonnay fruit, hints of vanilla bean and some maderised notes. It's beginning to lose complexity and real interest.' (RKP)

RWT Shiraz   NOW

Medium-deep crimson. Fresh redcurrant, praline, sweet-fruit aromas and flavours with apricot, roasted chestnut, black truffle, tomato leaf notes. The palate is medium-bodied but expressive with redcurrant, praline, spicy flavours, ripe loose-knit chocolatey tannins and plenty of cedar complexity. Finishes firm with redcurrant, tobacco notes. Generous and balanced but not particularly complex. 60% new oak. Drink now.

'Red fruit aromas with tomato notes. The palate is well saturated with sustained pure fruit flavours and refinement of tannins. Long and well balanced.' (NB) 'Blackberry confiture, smoke, mild spices and truffles. Lovely aromatic complexity. Northern Rhônish in style with plum, blackberry, pepper notes, plenty of freshness, backbone and serious length. Very nice drinkability and harmony but Syrah rather than Shiraz in style.' (AL)

Magill Estate Shiraz   NOW

Medium-deep crimson. Quite advanced confectionary, dried roses, apricot aromas with some hazelnut notes. Contrived red cherry, redcurrant, raspberry fruit, evolved apricot notes, vanilla oak flavours and chalky loose-knit tannins. Finishes a touch lacy but with plenty of fruit sweetness. Simple, lighter style but easy to drink now.

'Exuberant wild red fruits and smoky, woody fragrances. The palate is quite slender and compact; the fruit just holding up to the structure.' (TL) 'Medium-bodied wine with stewed fruit characters, mocha, vanillin oak notes and peppery notes. A touch hard.' (JF)

Reserve Bin 00A Chardonnay   NOW ••• PAST

Deep amber gold. Developed apricot, banana aromas with a touch of tobacco and lemon curd. Well concentrated but advanced with apricot, berry, tobacco flavours, fine cutting acidity with some creamy, butterscotch notes. Finishes chalky. Just about alive, but now reaching its end.

'Deep gold–amber. Significantly advanced with some marmalade oxidative, maderised characters creeping in. Lean and dry with mellow caramel, vanilla, butterscotch flavours. Still pleasant but drink up soon.' (HH) 'Overdeveloped for sure, but the structure is still minerally with lovely creamy mouth-feel and a chalky dry texture.' (TL)

St Henri   NOW

Medium-deep crimson. Fragrant blueberry, beetroot, tobacco aromas with black truffle notes. Medium-concentrated supple wine with fresh blueberry, red cherry, tobacco notes, fine lacy tannins and minerally elegant finish. Ready to drink. 100% shiraz.

'Dense and deep with truffles, black olive, gravelly notes, great fruit complexity and refreshing unobtrusive tannins.' (FK) 'Intense and exquisitely perfumed with crushed berry, violet aromas and earthy, black truffle notes. The palate is medium-bodied with elegant fruit, plenty of tannin backbone, freshness and finesse.' (AL)

Cellar Reserve Pinot Noir   NOW

Medium-deep crimson. Fresh lifted aniseed, tarry, leafy, dark chocolate aromas with polished leather notes. Sweet-fruited, slightly soupy wine with deep-set praline, red fruits, herbal/sappy flavours and pronounced chocolatey tannins. Finishes leafy and long. Drink up.

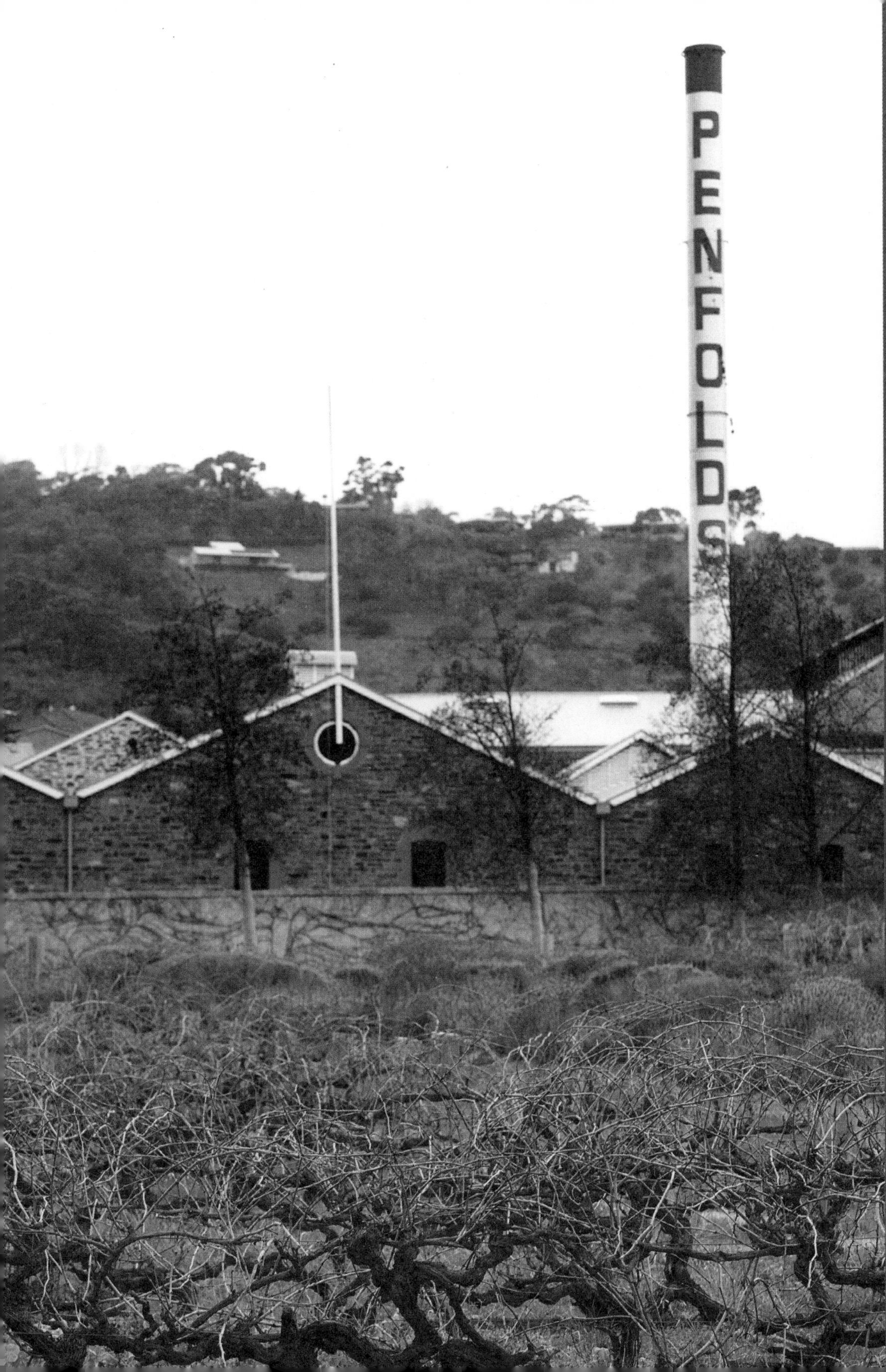
PENFOLDS

'Developed peppery, stemmy, slightly leathery aromas. Full-bodied with real density, but very dusty tannins. The palate has intensity, drive, line and length but tannins are winning the race.' (HH)

Cellar Reserve Sangiovese
NOW

Medium-deep crimson to brick red. Contrived cherry, raspberry, violet aromas. Sweet and flavourful, but simple with red cherry, ripe red raspberry flavours, plenty of sweet fruit notes and loose-knit, chalky tannins. Some herbal/touch sappy notes at the finish. Not hugely complex but fresh and buoyant. Not a wine to keep.
'The fruit has soured and the tannins remain astringent. An awkward wine that characterises a difficult season.' (TS) 'Already very developed in colour. Spice, leathery, earth notes are followed by an unexpected burst of sweet fruit on the finish.' (JH)

Reserve Bin Eden Valley Riesling
NOW ••• PAST

Medium amber, brassy colour. Intense baked lemon, nutty aromas with a touch of marzipan/butterscotch. Sweet brassy, but generous baked lemon, marzipan flavours and fine mineral cutting acidity. Already showing overdeveloped notes.
'Over-mature smoky burnt-toast aromas. Rich and smooth with additional pineapple notes and a very soft texture. Still drinking, but certainly has seen better days.' (HH)

Bin 389 Cabernet Shiraz
NOW ••• 2016

Medium-deep crimson. Fragrant redcurrant, cranberry aromas with geranium, herb garden notes. The palate is sweet and fleshy with dried fruits, plum, dark chocolate flavours, underlying savoury notes and slinky dry, loose-knit tannins. Finishes long and tangy. Pleasant to drink but it doesn't have the density or energy for long-term ageing. Best to drink soon.
'A stewed fruit, berry compote wine of fruit-sweetness and simplicity, overlaid on a firm web of tannins.' (TS) 'Fresh and youthful but the fruit struggles to provide the necessary mid-palate richness and finish.' (JH)

Bin 407 Cabernet Sauvignon
NOW ••• 2014

Medium-deep crimson. Fragrant redcurrant, chocolate, sage, dried herb aromas. Well-concentrated redcurrant, herb, praline flavours and slinky loose-knit dry tannins. A touch underpowered and constrained but still minerally and fresh. Ready to drink.
'Some mulchy threads. A touch one-dimensional in flavour and texture.' (RKP) 'Beautiful blackcurrant, violet aromas with a porty note. Pronounced tannins on the palate are unlikely to soften as the fruit further ages.' (TL)

Kalimna Bin 28 Shiraz
NOW

Medium crimson. Intense red cherry, blueberry, redcurrant, herb, pot pourri aromas. A juicy, easy-drinking wine with smooth blueberry, red cherry, redcurrant, red liquorice, plum flavours, and loose-knit, lacy tannins. Finishes silky with roasted walnut, mint notes. Not a wine to keep.
'Very savoury, dried-herb aromas; typical of this oddball year. Rich, fleshy, full and generous with plenty of sweet-fruit, touches of liquorice and firm tannins. A fair spike of acid lifts and carries the palate along.' (HH) 'Resin and chewing gum aromas. Awkward green tannins. Some mint too!' (JH)

Bin 128 Shiraz
NOW

Medium-deep red. Fresh redcurrant, dark berry, spearmint, leafy aromas. Lightweight wine with confectionary, redcurrant, red cherry, minty flavours, and chalky, leafy tannins. Finishes bitter-sweet and peppery. Drink soon.
'Dry, dusty, earthy aromas with touches of savoury, roasted meats. Not unattractive but lacks generosity and depth of fruit. Nonetheless, good for this vintage.' (HH)

Koonunga Hill Shiraz Cabernet
NOW

Medium-deep crimson. Fragrant cassis, herb garden aromas with some cedar savoury notes. Lightly concentrated wine with cassis, boysenberry, herb flavours, fine slinky loose-knit tannins and mineral notes. Surprisingly fresh and buoyant. Best to drink now.

'Lively and elegant with fresh fruit and well-delineated fine tannins.' (MC) 'Well done but not remarkable.'
(TG) 'Zinfandel-like nose. Jazzy and vibrant.' (LM)

| 2001   ★★★★ / ★★

Winter rains replenished soil moistures, which sustained vineyards during a very hot, dry summer; by late February to March cool, dry conditions prevailed, producing a very good red vintage • *The Sydney Morning Herald* named Max Schubert in its top 100 most influential Australians of the twentieth century • National Trust of South Australia lists Penfolds Grange as a Heritage icon • A bottle of 1951 Grange fetches AU$52 211 at auction • The London Stock Exchange becomes a public company • Saudi Osama bin Laden orchestrates al-Qaeda terrorist attacks on the US, bringing down New York's World Trade Center, war in Afghanistan begins • Ansett Australia, Australia's first commercial interstate airline, collapses leaving 10 000 people unemployed • Sir Donald George Bradman, 'The Don', Australia's greatest cricketer with a test-batting average of 99.94, dies aged ninety-two.

Bin 95 Grange   2016 ••• 2035

Deep crimson. Fresh praline, plum, blackcurrant aromas with cedar, herb garden notes. Very tightly structured wine with red plum, blackcurrant, roasted coffee, liquorice flavours, fine dense muscular tannins and plenty of mocha oak. Finishes chalky firm but fruit persists. Generously proportioned with a solid structure. Still needs more time. One of the hottest growing seasons since the 1920s. McLaren Vale had its highest mean January temperature since 1915. 99% shiraz, 1% cabernet sauvignon.

    50% Kalimna Vineyard (Barossa Valley), 50% Barossa Valley.

    'Deep black fruit, espresso, toast aromas with big tannins.' (LM) 'Terrific complexity of dark fruit and spice. Tannins are slightly coarse.' (JC)

Bin 144 Yattarna Chardonnay   NOW

Pale-medium straw gold. Developed, brassy, lemon peel, stewed pear aromas with silage and malt nuances. Tight lemon curd, vanilla flavours, concentrated acidity and chalky, al dente/hard textures. Finishes chalky and minerally with persistent fruit complexity. Tobacco notes suggestive of old age. Drink now. 13.5% alcohol. 83% Tumbarumba, 17% Piccadilly (Adelaide Hills). 60% new French oak.

    'Behind the detracting notes of cork wood is an impressive flow of succulent primary white peach and lemon zest fruit. The palate is long and honed with a perfect balance between youthful fruit profile and the roasted nut/toasty complexity of development.' (TS) 'Highly floral with slight cork wood notes. The palate is linear with strong acidity but the flavours are disjointed.' (TL)

Bin 707 Cabernet Sauvignon   NOW ••• 2020

Medium-deep crimson. Fragrant dark berry, praline, mocha aromas with a touch of mint, star anise. Rich and solid with upfront blackberry, dark cherry, redcurrant flavours, strong muscular firm tannins and spicy vanilla oak nuances. Finishes bitter-sweet, chocolatey and minerally. A hugely concentrated wine with substantial fruit and tannin power; 'a dark horse'.

    'Perfumed black cherry, cassis, dark plum, dark chocolate aromas with lovely cinnamon, vanilla, clove, liquorice notes. A robust and voluptuous palate with abundant sweet fruit flavours, chewy ripe tannins and lovely refreshing acidity. A superb vintage.' (ER) 'Evolved sweet earthy, herbal plum aromas, resolved tasty tannins, new oak notes and warm alcohol.' (ST) 'Blueberry, red cherry aromas with some meaty, mint notes. Huge concentration with muscular tannins.' (LD)

RWT Shiraz   NOW ••• 2020

Medium-deep crimson. Fresh brambly, musky, plum, mocha aromas with crushed ants, aniseed, vanilla, polished leather notes. Richly concentrated with developed plum, red berry fruit, underlying vanilla, spicy nuances and plentiful fine muscular tannins. Finishes al dente firm with plenty of flavour length. Already showing some advanced maturity. Will continue to develop but tannins' structure now prominent. 65% new French oak. Drink now or keep for a while.

'*Showing a surprising amount of maturity on the nose with little exuberance and presence; light, slightly rustic red berry fruit, undergrowth, singed leather and a touch of sandalwood. The palate is nicely balanced and clean, although comparatively simple. A conservative, rather four-square RWT with a tertiary finish. Pleasurable all the same but I cannot envisage this improving.*' (NM) '*Harmonious, rich and ripe with refreshing tangy orange zest notes. Round and soft on entry with cashmere-like tannins, impressive concentration and a flourishing finish.*' (NB)

Magill Estate Shiraz ♢ NOW ••• 2020

Medium-deep red to crimson. Intense black fruit, sandalwood aromas with vanilla, cedar notes. Tightly structured wine with powerful black fruits, roasted walnut, liquorice flavours and pronounced grainy, chalky tannins. Finishes muscular firm with lingering tangy, sweet fruit notes.

'*Generous sweet fruit with wood varnish, camphor notes and savoury, spicy complexity.*' (JF) '*Layers of blackberries, blueberries and succulent plums are laced with pepper, savoury spices of all kinds and leather notes. This is a full-bodied robust style with powerful fruit, fine firm tannins and outstanding line and length.*' (TS)

St Henri NOW ••• 2030

Medium-deep crimson to purple. Blackberry, dark cherry, dark chocolate aromas with some praline, salted liquorice, herb garden notes. Sturdy yet generous with plush sweet dark chocolate, dark cherry, plum flavours, spicy nuances and plentiful muscular dry tannins. Finishes grippy firm, yet the fruit follows through. A powerfully structured wine. 100% shiraz. Drink now or keep.

'*Provençal herbs, liquorice and rosemary. A medium-bodied palate with succulent tannins on the entry. The acidity is taut and deftly interwoven through the dense mass of black, liquorice-tinged fruit. A saline touch on the finish. It has great potential, but don't let the aromatics fool you into thinking that it is near ready.*' (NM) '*Intense blackberry, chocolate, tobacco, spicy flavours. Very young with opulent sweet fruit and powerful structure.*' (PK)

Cellar Reserve Pinot Noir ♢ NOW

Mature deep crimson. Very fragrant, complex aromas of strawberry, earth, apricot and herb garden with touches of dried roses/violet notes. Strawberry, cedar, spicy flavours and strong dry graphite tannins. Finishes savoury, sweet and long with some sappy notes. Very enjoyable to drink now.

'*Complex roasted meat aromas with stemmy, peppery, vegetal notes. Quite intense, deep and fairly tannic, with typical Penfolds structure. Has length and zip, aided by some slightly edgy acidity. Good characterful wine.*' (HH)

Cellar Reserve Sangiovese NOW

Medium crimson to brick red. Intense leather, amontillado, cranberry aromas with some roasted coffee, barnyard notes. Well-concentrated cranberry, leather, herb flavours and sinewy, dry tannins. It's drying out now but it still has a few years left.

'*Spicy earthy meaty nose with primary dark berry fruit still evident. The palate is developed with attractive spice, tea, dried herb, savoury flavours and fresh acidity. It's still drinking now but it's beginning to tip over.*' (JB) '*An earthy and gamey style with some barnyard notes and an astringency to its acidity.*' (TS)

Reserve Bin Eden Valley Riesling NOW

Pale medium, touch brassy colour. Grapefruit, lemon, oily, leafy aromas. Quite developed grapefruit, lemon curd, toasty, tobacco flavours, chalky loose-knit textures and crisp, almost sour acidity. A mouth-watering finish. It's beginning to show advanced development.

'*Toasty but lacks the detail and complexity of the best wines. It's lean, stern, taut, firm and very dry. Still quite sound but starting to maderise.*' (HH)

Bin 389 Cabernet Shiraz

NOW ••• 2020

Medium-deep crimson. Fresh cassis, cherry, redcurrant aromas with leafy, sage notes. Richly concentrated, substantial and buoyant with abundant sweet blackcurrant, dark chocolate flavours, fine slinky/velvety tannins and underlying malty, vanilla notes. Finishes chewy firm and fruit sweet. A generous wine but with a rigid structure. Drink now or keep.

'A rich warm velvety palate with an appealing ripe cherry, dark chocolate component, the tannins appropriately soft and pliant. The well of fruit will support it for at least a decade but it's an enjoyable drink now.' (JH) 'A forward style with berry, spice aromas and some mint, herb notes. The palate is tightly structured.' (RKP)

Bin 407 Cabernet Sauvignon

NOW ••• 2016

Medium-deep crimson to purple. Walnut, chestnut, black olive, meaty aromas with a touch of violet. Generously concentrated and voluminous with plenty of sweet fruit richness, blackcurrant, walnut, espresso flavours and granular/chunky dry tannins. It finishes muscular firm, long and sweet. The fruit is still ascendant, but drink soon.

'Red jelly, cassis aromas with loads of spicy, cedar notes. The palate is medium-bodied and persistent but the tannins are cumbersome.' (JF) 'Cherry, marzipan nose. More directly fruity with smooth red cherry, slight vegetal notes and juicy acidity.' (JB)

Kalimna Bin 28 Shiraz

NOW ••• 2018

Medium-deep crimson. Scented blackberry, plum, praline aromas with biscuity, savoury notes. Rich and buoyant with deep-set blackberry pastille, plummy, dark chocolate flavours and fine plentiful, muscular tannins. A long tannic, slightly leafy tail. Powerful and minerally but unlikely to soften out. Should hold for a while, but best to drink soon.

'A pretty fragrant wine with sweet plummy aromas and lovely density and fruit complexity on the palate. It's not overly dramatic but it has compelling drinkability.' (TL) 'Fragrant, ripe and black-fruited, with firm, slightly clunky tannins and a somewhat bulking demeanour.' (TS)

Bin 128 Shiraz

NOW ••• 2018

Medium-deep red to purple. Exotic, musky, violet, plum, rhubarb aromas with crushed ant nuances. Well-concentrated plum, blackcurrant, liquorice flavours, dense velvety tannins and underlying savoury oak notes. Finishes leafy firm and minerally. Best to drink soon although will hold.

'Spiced plum aromas with aniseed notes. Very ripe, slightly jammy, big, gutsy and tannic. Lacks the usual Coonawarra elegance.' (HH)

Bin 138 Shiraz Grenache Mourvèdre

NOW

Medium-deep crimson. Intense musky, roasted coffee/espresso aromas with walnut, dark plum notes. Fresh and juicy with deep-set roasted coffee, prune, dark chocolate flavours and lacy dry, touch al dente, tannins. Finishes leafy firm. Best to drink soon. 44% shiraz, 35% grenache, 21% mataro.

'Aromatic plum nose with incense, floral, gamey notes. A juicy palate with a lovely silky texture and acid/tannin balance. Finishes long and fragrant.' (JB) 'Another decade; another profile! Here the tannins are still doing their work by providing structure for the black and red berry fruits.' (JH)

Koonunga Hill Shiraz Cabernet

NOW ••• 2015

Medium-deep crimson. Fresh violet, musky, blueberry, cassis aromas with smoky nuances. The palate is supple and smooth with juicy cassis fruit, fine plentiful chalky tannins and underlying savoury nuances. Delicious wine.

'Confected berry fruit and dusty tannins on the finish.' (JC) 'Lovely ripe, penetrating dark fruits and refined well-managed tannins.' (TG) 'Pungent and intense with lovely florality and spiciness.' (JR)

2002

A long, cool summer with intermittent rains in the Barossa and Coonawarra was followed by a warm, dry autumn; a lovely vintage for both reds and whites • Components of Grange are once again vinified at Magill Estate for the first time since 1973 • A portion of the final Grange and Bin 389 blends are again matured at Magill, the first time since 1996 • John Duval steps down as Chief Winemaker • Peter Gago, Penfolds oenologist, is appointed Chief Winemaker • Michael Jackson receives the Artist of the Century award at the American Music Awards • Steve Fossett becomes the first person to fly solo around the world non-stop in a hot-air balloon, completing the journey in fifteen days • Terrorist atrocity in Bali by fundamentalist group Jemaah Islamiyah kills 202 and injures 209 people, including many Australians.

Bin 95 Grange ▽
2018 ••• 2045

Deep crimson to purple. Intense elemental inky, elderberry, blackberry aromas with meaty, vanilla, liquorice nuances. Beautifully balanced and powerful with elemental dark berry, elderberry, liquorice, aniseed flavours, fine plentiful grainy, al dente tannins and well-balanced malt, vanilla oak. A great Grange with huge substance and energy. Continue to hold! 98.5% shiraz, 1.5% cabernet sauvignon.

50% Kalimna Vineyard (Barossa Valley), 50% Barossa Valley.

'A Grange to be reckoned with! Intense black fruits, dark chocolate aromas and dense fine-grained tannins. Beautiful poise and length.' (TG) 'Fantastic structure and focus with blackberry, blueberry aromas and sage/ eucalyptus notes.' (DM)

Bin 144 Yattarna Chardonnay ▽
NOW

Medium straw-gold. Intense lime, grapefruit, herb garden, camomile, pineapple, tropical fruit aromas. Well-concentrated pure pineapple, grapefruit flavours with some savoury, nougat, grilled nut, vanilla oak nuances. Chalky textures with plenty of mineral acidity. It still has vinosity, freshness and drive but drink soon. 13.5% alcohol. 76% Adelaide Hills, 24% Tumbarumba. 55% new French oak.

'Complex citrus, mineral, savoury aromas with incense-like spice. Wonderfully balanced and integrated wine with elegant layers of mineral, white fruits, citrus, subtle oak underpinned by fine silky acidity. A great vintage.' (JB) 'Complex liqueur peach, ginger, nutty aromas with vanilla bean notes. Silky, fine-textured and long-flavoured with a slightly hard edge that detracts from the finish.' (RKP)

Bin 707 Cabernet Sauvignon ▽
NOW ••• 2040

Medium-deep crimson to purple. Ethereal and complex blackberry, roasted chestnut, praline, spice aromas with herb, mint notes. Richly concentrated, classically proportioned, smooth and generous with red cherry, redcurrant, dark chocolate flavours, all integrated perfectly with fine-grained polished tannins and toasty, roasted chestnut oak. Finishes firm with sweet buoyant fruit and a tail of savoury tannins. A sensational vintage.

'Quite restrained with crunchy black fruit, fresh sweet spices and overlying toasty notes. A solid core of sweet fruit but this one will need bags of time to come around.' (FW) 'Tighten your seatbelt for waves of rich dark cherry, ripe raspberry fruit with equal measures of firm yet very classy tannins. Layered, nuanced and persistent. Full-bodied Aston Martin finish!' (CPT)

RWT Shiraz ▽
NOW ••• 2028

Medium-deep crimson. Beautifully defined and exuberant dark cherry, dried fruits, praline, vanilla aromas with hints of liquorice, spice and sage. Generously proportioned, rich and voluminous palate with dense dark cherry, dark berry, praline, espresso flavours, chocolatey tannins and underlying cedar notes. Finishes chalky firm with superb fruit persistency and buoyancy. Seductive and perfectly balanced. A great RWT vintage.

'Fresh cherry, spicy, mint aromas and flavours with perfectly structured tannins. All the elements of fruit, tannins, alcohol and acidity are in harmony. This is a lovely RWT with superb cellaring potential.' (FK) 'Intense and perfumed with vanilla, smoke, dried figs, dates and prune notes. A seriously concentrated and vigorous palate with broad mouth-feel, polished texture, a plethora of dark fruits and a very long finish. Still fairly young and tight but deserves some more bottle age.' (AL)

Magill Estate Shiraz ◇ NOW ••• 2025

Medium-deep crimson. Scented violet, red cherry, earthy aromas with dried roses, garrigue notes. Well-concentrated, tight chalky wine with dried herb, cassis, leafy, spicy flavours, cedar oak complexity and al dente, peppery tannins. Finishes firm and long. A cool, expressive wine.

'Greater precision, focus and definition than previous vintages. Here the sweet fruit is balanced by dark berry notes and hints of chocolate. Good line and length.' (JH) 'Beautiful power and persistency. Has everything going for it.' (TL)

St Henri ◇ NOW ••• 2030

Medium-deep purple to crimson. Lovely fresh cassis, dark plum, redcurrant, praline aromas with floral, violet notes. Beautifully balanced with fresh juicy cassis, plum, redcurrant flavours and plentiful cedary tannins. Finishes chalky firm yet fruit sweetness pervades. A delicious expansive wine with superb vinosity, richness and persistence. Drink now, but best to keep. A classic St Henri. 90% shiraz, 10% cabernet sauvignon.

'Wonderful definition and femininity with raspberry, rose petal notes and an appealing Rioja-like richness and texture. It's very well balanced (but more backward than the nose suggests) with very good backbone and poise, a hint of white pepper and allspice towards the controlled, focused finish. One of the finest releases in recent years. Perhaps that cabernet is missed after all?' (NM) 'Soft subtle blueberry fruit with gentle earthy eucalypt notes and persistent tannins. Not the greatest length but very well balanced.' (NB)

Cellar Reserve Grenache NOW

Medium-deep crimson. Fresh, violet, musky, redcurrant, plum, sage aromas with some panforte notes. Sweet, fleshy wine with redcurrant, red cherry jam, musky, praline flavours and loose-knit, supple tannins. Generous and mouth-filling with a kick of peppery alcohol at the finish. Ready to drink, but should hold for a few years.

'In great condition for its ten years. Sweet vanilla, jammy, chocolate aromas. Soft, fleshy and round with an unexpected tannin firmness. An atypical grenache wine that should continue to age very well.' (HH) 'Lovely mature grenache nose with exotic spice, musky florals and rich plum/prune, raspberry, eau-de-vie notes. Velvety, mouth-coating and lush with ripe balanced tannins.' (RKP)

Cellar Reserve Pinot Noir NOW

Medium-deep crimson. Fragrant herb garden, leafy, aniseed, red fruit aromas with cut roses, espresso notes. Quite stalky flavoured, but it has its charm with dried roses, herb notes, plenty of mid-palate richness and fine plentiful, chalky tannins. Builds up leafy, dry and minerally at the finish. Interesting rather than beautiful. Drink soon.

'Sappy, slightly stewed, possibly a little raisined with some green-pea vegetal notes. Grippy structured with slight dead-fruit suggestions.' (HH)

Cellar Reserve Sangiovese ◇ NOW ••• 2016

Medium-deep crimson to brick red. Fragrant cherry stone, herb aromas with a touch of leather and praline. A lovely wine with sour cherry, sweet juicy fruit flavours, dark chocolate nuances, loose-knit velvety tannins and plenty of savoury undertones. The line is now reaching its true varietal expression. Drink soon.

'There's a primary poise to this wine, with its confident strawberry and plum fruit and classic Penfolds backdrop of firmly structured tannins. Oak is well handled and the vibrant acidity of the season upholds a fresh, lively and persistent finish. Tannins need some years yet to relinquish their grip.' (TS) 'It has plenty of sweet fruit but the palate is dominated by sour acidity, a hard tannin structure and finish.' (JF)

Reserve Bin Eden Valley Riesling ◇ NOW

Pale, medium straw. Fresh lemon curd, lanolin, dried leafy aromas with bottle-aged toasty notes. Well-concentrated lemon curd, toasty, lime flavours and crisp quartz-like acidity. Finishes dry yet long and minerally. Lovely complexity and flavour development. Should continue to hold for a while. Screwcap closures introduced with this vintage.

'Complex, bright, fresh and deliciously inviting with lovely buttered-toast aromas. Tight, fine, nervy/ energetic on the palate with excellent zestiness, flavour and depth. A benchmark style with youthful intensity and excellent aged characters.' (HH)

Bin 389 Cabernet Shiraz ▽   NOW ••• 2030

Medium-deep crimson. Intense dark chocolate, blackcurrant, mulberry, vanilla aromas with herb garden, mint notes. The palate is well concentrated and balanced, with deep-set cassis, mulberry, plum, praline flavours, an underlay of mocha, savoury notes and fine-grained tannins. It finishes classically firm with lingering savoury, sweet fruit nuances. A beautifully structured wine. Drink now or keep.

'*Super-concentrated and dense with deep dark fruits and solid structure without heftiness or excessive extraction. Seems like a directional shift in style.*' (TL) '*Has the mark of a cool vintage with primary black fruits and savoury, earthy, minty undertones. The tannins are surprisingly ripe.*' (JH)

Bin 407 Cabernet Sauvignon ▽   NOW ••• 2025

Medium-deep crimson. Intense blackcurrant, praline, herb garden, leafy aromas with cedar, spice, violet and Turkish Delight notes. Classically proportioned and superbly balanced with cassis, garrigue, dried herb flavours, some espresso notes and fine-grained tannins. Finishes firm with a lovely bitter-sweet aftertaste. Gorgeous vinosity and length. A brilliant vintage with lovely fruit complexity, power and balance. Drinking perfectly now, but should hold for a long time.

'*This has immense presence. It's dark, dense and weighty with big chunky/chewy tannins. They are in play now, but will they dominate in the future?*' (TL) '*A fragrant aromatic bouquet with red berry, blackcurrant fruit and a touch of forest floor. A finely etched palate is balanced with very good line and length.*' (JH)

Kalimna Bin 28 Shiraz ▽   NOW ••• 2025

Medium-deep crimson. Lovely intense liquorice, dark berry, dark chocolate aromas. Generously proportioned but elegantly styled with dark berry, liquorice, chocolatey flavours, superb mid-palate density and fine ripe velvety tannins. A delicious sweet fruit finish with minerally, salted liquorice, aniseed notes. This should further develop. Drink now or keep.

'*Attractive developed notes of spice, meaty, leather and dark fruits with a slightly peppery edge. The palate is quite restrained and tight with meaty, leather, chocolate flavours, moderate tannin presence and drying finish.*' (JB) '*Medium-weighted, ripe and generous with lush fruit and subtle spice/savoury complexity.*' (RKP)

Bin 128 Shiraz ▽   NOW ••• 2020

Medium-deep red to purple. Intense and expressive blueberry, sage, leafy aromas with a touch of spearmint. Buoyant but sinewy wine with blueberry, blackberry fruit, white pepper notes and touches of sage. Fine, graphite tannins build up sinewy firm at the finish, but the flavours are persistent. Drink now or keep for a while.

'*Earthy and subdued with herbal, pepper notes. Taut, lean, fine and grippy firm in structure. A serious long-term wine just starting to hit its straps now.*' (HH)

Bin 138 Shiraz Grenache Mourvèdre   NOW ••• 2016

Medium-deep crimson. Fresh but evolved herb, violet, praline, musky plum aromas with a touch of white pepper. Elegantly proportioned wine with praline, plum flavours, plentiful chalky/chocolatey tannins and juicy/tangy notes. Still has vinosity and generosity. Best to drink soon. 44% grenache, 30% shiraz, 26% mataro.

'*Very fresh typical sweet Barossa fruit notes. Rich and bold with plenty of ripe tannins and pronounced acidity. Still some grip at the finish.*' (JF) '*Raspberry, wild chocolate bouquet and palate with dried herb notes. Firm and robust with the most prominent tannins thus far.*' (TS)

Koonunga Hill Shiraz Cabernet ▽   NOW ••• 2018

Deep crimson. Intense blackberry, damson, herb garden aromas and flavours with some star anise notes. The palate is well concentrated with plenty of fruit sweetness, a touch of tobacco and ripe chocolatey tannins.

'*Well proportioned and balanced with the structure for a long life.*' (MC) '*Four-square, rich and meaty with glossy tannins.*' (TG)

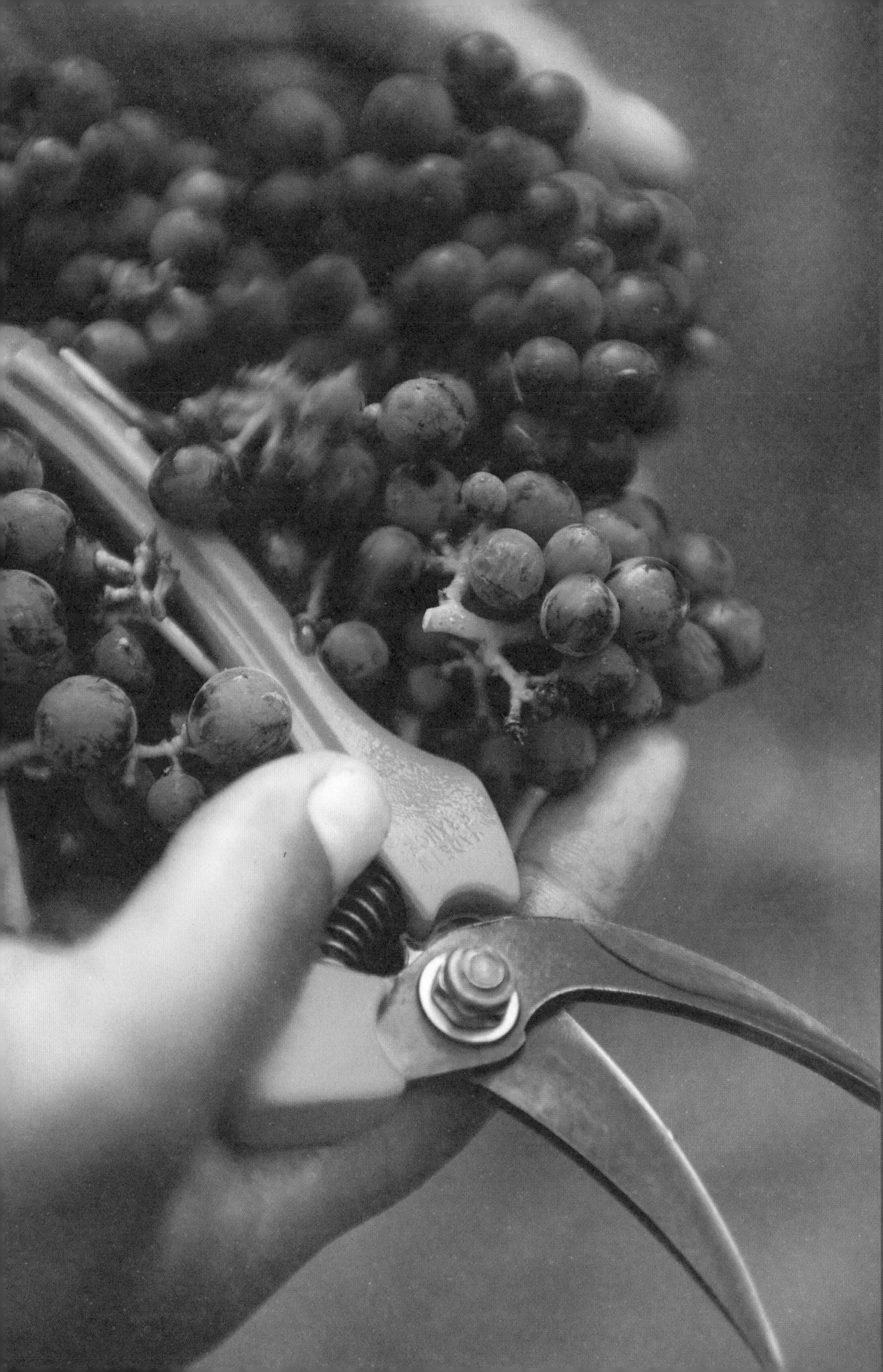

2003

A season of extremes: drought conditions prevailed over spring and summer but the hot dry weather was interrupted by heavy rainfall in February; vines struggled to reach phenolic ripeness • Coonawarra enjoyed better conditions with cool summer breezes moderating the growing season • Space shuttle *Columbia* disintegrates upon re-entry into the Earth's atmosphere, killing all seven astronauts on board • The SARS virus becomes pandemic, affecting people in more than twenty-five countries and eventually killing 774 people worldwide • The US Human Genome Project is completed, mapping the genetic blueprint of the human species • Saddam Hussein found and captured in a tiny cellar in a farmhouse in Iraq.

Bin 95 Grange
2016 •• 2030

Deep purple to crimson. Intense blackberry, praline, herb garden, malty aromas with graphite, leafy notes. Ample blackberry, brambly flavours with fine grippy tannins, plenty of mid-palate richness and mocha, malty oak. Finishes muscular, grippy. A robust yet balanced wine with plenty of vigour and freshness. 97% shiraz, 3% cabernet sauvignon.

Barossa Valley, McLaren Vale, Langhorne Creek, Coonawarra, Magill Estate (Adelaide).

'Pungent cherry, anise, herb aromas with spicy, black cardamom notes and ripe smooth tannins. Energy and focus.' (JR) 'Attractive violet, black fruit aromas. Powerfully tannic with roasted coffee and ripe black plum flavours.' (LM) 'Very fresh with balsamic notes, flowers, new oak, cedar and sweet spice. Very young and nervous with massive tannins.' (MC)

Bin 144 Yattarna Chardonnay
NOW

Pale straw-gold. Lemon, roasted hazelnut, tonic water aromas. A soft creamy wine, quite supple and smooth with lemon curd, hazelnut flavours and fine mineral acidity. Finishes minerally with plenty of flavour length. Best to drink soon. 13.5% alcohol. 100% Adelaide Hills. 69% new French oak.

'Slightly funky, wet-dog matchstick, marzipan notes. The palate lacks the purity of line of most Yattarnas.' (HH) 'A surprise after the 2002. Delectable fruit drives a supple smooth and very pure wine. A long lingering finish and aftertaste. Almost freakish!' (JH)

RWT Shiraz
NOW ••• 2016

Medium-deep crimson. Fragrant blackberry, mocha, marzipan, roasted chestnut aromas with some violet, wildflower notes. Medium-bodied and quite forward with mouth-filling blackberry, red cherry fruit flavours and touches of tobacco, roasted chestnut. Fine muscular, dry but plentiful tannins. Finishes firm, long and savoury. Complex and sinewy but probably won't improve greatly. 70% new French oak. Drink now or soon.

'Complex cherry, plum, pepper aromas with ripe tannins and medium acidity.' (PK) 'Balanced and compact with granular texture but lacks real fruit power. Finishes slightly "hot".' (NB) 'Familiar dark fruits with coffee/ spice notes. Opulent and juicy with mouth-filling dark luscious fruit and powerful tannins.' (MS)

Magill Estate Shiraz
NOW ••• 2015

Deep crimson. Intense blackberry, cedar aromas with some prune notes. Rich soupy wine with blackberry, prune, dark chocolate flavours and gritty, leafy tannins. Builds up sinewy and dry at the finish. Quite rustic but still has its appeal. Best to drink soon.

'Animal, barnyard notes gallop through from start to finish.' (TS) 'While the fruit is very ripe and complex, there are no porty, jammy notes; just a diminished structure.' (JH)

Reserve Bin 03A Chardonnay
NOW

Pale–medium yellow. Intense lemon curd, verbena, pear-drop aromas with a touch of matchstick. Creamy, concentrated and supple wine with plenty of lemon curd, juicy pear, apricot flavours, underlying vanilla and mineral acidity. Past the mid-point of its drinking plateau, but should last for a while. 88% new oak.

'Fragrant grapefruit and spice nose. Medium-weighted, but broad and developed, with grapefruit, spice, flinty aromas and flavours, some funky notes, an undercurrent of subtle toasty richness and a savoury finish.' (JB)

'An utterly delicious Chardonnay with rich ripe figs, stone fruit flavours, savoury nuances and a tight mineral acid line. It's generous, luscious and flavoursome yet fine and reined in.' (JF)

St Henri

NOW ••• 2025

Medium-deep crimson. Intense blackberry, praline, spicy aromas with violets, graphite, herb garden notes. Exuberant dark chocolate, dark berry, spicy flavours, abundant mid-palate fruit sweetness and rich chalky graphite tannins. Finishes grippy dry and long. A substantial wine that still needs time to unravel. Drink now but best to keep. 100% shiraz.

'Ripe stewed dark fruit, liquorice aromas. A rich and muscular palate with strong tannins and subdued fruit.' (FK) 'Intense blackberry, spice, minty notes with juicy flavours, massive tannins and fresh long finish.' (PK) 'Quite sweet, ripe and floral with cassis, plum, liquorice and spice. The palate is more restrained with strong tannin presence, a touch of dryness and nice Northern Rhônish black pepper/clove spiciness.' (AL)

Cellar Reserve Pinot Noir

NOW

Medium crimson to brick red. Developed earthy, pond water, prune, red cherry aromas. Well-concentrated, slightly soupy wine with earthy, prune flavours, some leafy notes and chocolatey, touch grippy tannins. Acidity pokes out at the finish but long and flavourful.

'Developed earthy, savoury bouquet with some leather, meat stock, Vegemite, charcuterie, raisin characters. Fairly dry, savoury flavours, with soft powdery tannins and a note of Vegemite that runs throughout.' (HH)

Cellar Reserve Sangiovese

NOW ••• 2018

Medium-deep crimson to brick red. Fresh cherry, plum, mocha aromas with a touch of sage. Ripe, fleshy wine with sweet cherry, plummy flavours, hints of herb and loose-knit, fine chalky, touch al dente tannins. Finishes chalky firm, long and fresh. A surprisingly expressive wine.

'This wine has all the elements in magical balance. The shape of the tannins, the dusty mineral notes, earthiness and acidity all meld together with captivating poise.' (TL) 'Excessive dry tannins and woody/briary notes. It won't soften before the fruit dies.' (JH)

Reserve Bin Eden Valley Riesling

NOW

Medium straw-gold. Fresh pear, oilskin, flinty aromas with herb garden notes. Richly concentrated lemon curd, pear-drop flavours with mild chalky textures and strong indelible acidity. Finishes tangy and long. A more forward and broader style. Ready to drink.

'A lovely wine with beeswax, light paraffin wax aromas. A fresh, bright and refined palate with appropriate bottle-age complexity. Taut, tangy, youthful and energetic. A poor vintage for reds but a great year for Eden Valley Riesling.' (HH)

Bin 389 Cabernet Shiraz

NOW ••• 2016

Medium-deep red to purple. Intense dark berry, cassis, dark chocolate aromas with sage, spicy notes. Fleshy and concentrated with juicy blackberry, cassis, raspberry, dark chocolate flavours, spicy nuances and chalky firm tannins. Finishes grippy with a peppery, alcoholic kick. A solid wine with excellent fruit generosity and drive, but best to drink soon.

'A powerhouse of ripe fruit and firmly scaffolded tannins, retaining impressive, succulent fruit persistence, but will it live out those tannins?' (TS) 'Relatively subdued wine with blackcurrant liqueur, woody spice notes. A slightly bitter, tight and darkly fruited palate with herbal notes at the finish.' (JB)

Bin 407 Cabernet Sauvignon

NOW ••• 2016

Deep red crimson purple. Cassis, praline, sage aromas with sweet fruit, violet notes. Well concentrated but a touch soupy, with rich cassis, plum, sage flavours, cedar/chewy tannins and underlying savoury oak. Finishes sappy firm, long and sweet-fruited. Robust and powerful but unlikely to improve. It should hold, but best to drink now.

'Ripe, macerated berry fruits and chocolatey firm tannins are jostling for position. Give it plenty of time and hope for the best.' (TS) 'Hard tannins prevail without compensating for fruit weight.' (RKP)

Kalimna Bin 28 Shiraz

NOW

Medium-deep crimson. Complex yeast extract, meaty, black fruit, raisin, liquorice aromas. Soupy dense and muscular with concentrated espresso, dried fruit, silage flavours and grippy/leafy tannins. Finishes chocolatey firm. Awkward and difficult. Unlikely to improve. Drink up.

'Better than expected. Slick tannin management in the winery has been well rewarded. But don't delay too long.' (JH) 'A stewy, ripe style with tawny port, coffee, dark chocolate flavours counteracted by firm, clunky tannins. Much better than expected.' (TS)

Bin 128 Shiraz

NOW ••• 2020

Medium-deep crimson to purple. Intense blackberry, sage aromas with slight musky, leafy notes. Supple, sweet-fruited palate with black cherry, liquorice, sage flavours and some white pepper, leafy, roasted nut notes. Fine-grained, chalky tannins build up firm and leafy at the finish. Still holding up very well.

'Developed earthy, dusty, savoury bouquet with roasted chestnut notes and grippy tannins. A big and butch wine with many years left in it. Surprising for such a difficult year.' (HH)

Bin 138 Shiraz Grenache Mourvèdre

NOW

Medium-deep crimson. Raspberry, mulberry, musky, violet aromas with porty notes. Generous, slightly confected, persistent sweet fruit, raspberry, mulberry flavours, underlying spicy notes and loose-knit chalky textures. Finishes minerally, firm and long. 69% grenache, 16% shiraz, 15% mataro.

'Classic sweet fruits, spice, dried fruit aromas. Plenty of energy, richness and texture with crunchy acidity at the finish.' (TL) 'Meaty, mixed berries and spice aromas, medium intensity and slightly metallic in texture. Finishes dry and savoury.' (RKP)

Koonunga Hill Shiraz Cabernet

NOW

Deep crimson. Intense redcurrant, musky aromas with some mocha, vanilla notes. The palate is medium concentrated with juicy redcurrant, plum, brambly, sage flavours and fine slinky/al dente tannins. Finishes firm but long and sweet. Drink now. Screwcap.

'Rich, sweet and suave with dark chocolate flavours and great persistency. This overdelivers.' (TG) 'Plush textures. Easy to drink.' (LM) 'Bright spicy, floral aromas, juicy/bitter cherry flavours. Lovely clarity and vivacity.' (JR)

2004

★★★★★ / ★★★★

Beneficial winter and spring rains were followed by cool to mild conditions over summer; ripening accelerated through a warm Indian summer resulting in near-perfect fruit • Green harvesting kept yields down in Coonawarra • An exceptional Penfolds vintage • 2004 Block 42 Cabernet Sauvignon is made in recognition of a perfect growing season • It also signified the reintroduction of Bin 60A • Bin 144 Yattarna is bottled commercially under screwcap for the first time • Google introduces its Gmail product to the public • *Homo floresiensis*, 'hobbit', skeletal remains discovered in a cave on the Indonesian island of Flores • A 9.3 magnitude undersea earthquake off the coast of Indonesia triggers a tsunami, causing a widespread catastrophe around the rim of the Indian Ocean.

Bin 95 Grange

2016 ••• 2050

Deep purple to crimson. Beguiling dark berry, liquorice, nutty, camphor aromas with some savoury notes. The palate is smooth and plush with fleshy blackberry, chocolate, espresso flavours, plentiful chalky tannins and savoury, vanilla oak. Finishes firm and flavourful with a long tannin plume. Lovely density and richness. A great Grange vintage. 96% shiraz, 4% cabernet sauvignon.

Barossa Valley (including substantial proportion of Kalimna shiraz), McLaren Vale, Magill Estate (Adelaide).

'Enchanting vintage with great concentration of ripe fruit, solid structure and fine tannins. Very impressive, complex and deep.' (MC) 'Blackberry, boysenberry, sweet vanilla aromas with some floral notes. A powerful wine packed with tannins. A peppery, chalky close.' (RI)

Bin 144 Yattarna Chardonnay ◈

NOW ••• 2015

Pale medium straw. Intense fragrant, herb garden, lemon, flinty, fino aromas. Fleshy, lemon, tropical fruit flavours with a creamy texture and underlying savoury, grilled nut notes, chalky/minerally textures and lovely prolonged linear acidity. A very seductive wine at the zenith of its life. Drink soon. 13.5% alcohol. 92% Henty (Victoria), 8% Piccadilly (Adelaide Hills). 100% French oak barriques but none of it new.

'There's a ripe peach, apricot and bubble gum note to the bouquet. The palate presents the same core of ripeness, giving a stronger impression of sweetness than any other wine thus far. A wine of considerable power and persistence.' (TS) 'Fresh and zesty with lovely fruit definition, complexity, lift and power. Mature tobacco, roasted nut characters are nuanced through the wine.' (TL)

Bin 707 Cabernet Sauvignon ◈

2016 ••• 2040

Deep crimson purple. Intensely fresh blackberry, dark chocolate, liquorice aromas with mocha, vanilla notes. Beautifully balanced, generous and luxuriant with sustained blackcurrant pastille, liquorice, toasted almond, panforte flavours, dense chocolatey sweet tannins and mocha/malty new oak. Finishes long and sweet with a slick of savoury tannins. An incredibly opulent wine with wonderful structure and harmony. A great Penfolds year! Best to keep for a while.

'A profound wine with splendid depth of fruit. Needs time to harmonise and express, even though the finish is quite seductive right now.' (FW) 'Elegantly structured wine with blackcurrant aromas and tight bitter tannins. Very polished.' (LD)

RWT Shiraz ◈

NOW ••• 2025

Deep crimson. Intense, fresh and fragrant with blackcurrant, bay leaf, panforte aromas and underlying roasted chestnut notes. Deliciously rich and complex with black fruit, bay leaf, cedar, panforte flavours, fine grainy tannins and vanilla, spicy oak notes. Finishes firm, persistent and sweet with aniseed notes. A beautifully balanced wine with lovely density and fruit complexity. A classic RWT with great potential. 69% new French oak. Drink now or keep.

'A little muted at first with broody dark berry fruit, a touch of black truffle, sous bois and dried rose petals. The palate displays very attractive poise and fruit intensity, superb acidity and wonderful tension and vivacity towards the feminine finish. This is understated but offers great potential.' (NM) 'Fresh stewed black fruit aromas with black olive, beetroot, mint, herb notes, delicious juicy fruit and chewy tannin backbone. The finish is not fully evolved, but it's dense, complex and nicely balanced.' (FK)

Magill Estate Shiraz ◈

NOW ••• 2025

Medium crimson to purple. Fragrant redcurrant, plum, herb, malty chocolate aromas. Classic, beautifully proportioned wine with redcurrant, plum, cassis flavours, underlying ginger, savoury, biscuity oak and chalky loose-knit tannins. Finishes firm, long and savoury. Lovely vinosity and freshness. Magill Estate at its best. Drink now or keep.

'More structure and better oak selection with plenty of black fruits, plum, dark chocolate expression; very good vineyard site definition too.' (JH) 'In a lovely moment with super concentrated blackberry fruit, rich dense textures and cedar new oak notes. Still on the ascendancy.' (TL)

Reserve Bin 04A Chardonnay ◈

NOW ••• 2016

Pale-medium yellow. Lanolin, flinty, grapefruit, lemon curd aromas with a touch of herbs and hazelnut. Supple and creamy with fresh hazelnut, peach, apricot flavours, underlying grilled nut complexity and flowing mineral acid presence. Finishes crisp and long. A multi-layered wine with beautiful balance and presence. At the vertex of its development. 85% new French oak. Screwcap.

'A considerable presence of white peach, lemon fruit is layered with taut notes of lime, roast nuts and toast. The palate is consummately poised with balanced acidity and lingering grapefruit notes. A powerful wine of outstanding freshness and complexity.' (TS) 'Barrel ferment smokiness tends to overwhelm the lightly honeyed

citrus fruit. Long and seamless with smoky, grilled nut, barrel ferment flavours, underlying dry savoury fruit and soft phenolics.' (RKP)

St Henri ⬦

2015 ••• 2035

Deep crimson. A remarkably expressive wine with inky deep elderberry, blackcurrant, blackberry aromas and star anise, mocha notes. Velvety and voluminous with plush blackcurrant pastille, dark cherry, praline, herb flavours and dense chocolatey tannins. Finishes firm, long and sweet. Classic Penfolds style with superb richness, density and balance. Delicious to drink now but worth waiting longer. 96% shiraz, 4% cabernet sauvignon.

'A broody, almost saturnine wine with hints of dark chocolate, liquorice and crushed violet. Masculine in structure with a vaulted backbone that keeps the fruit linear and strict. It unreservedly veers to the more Bordeaux-like St Henri, but it is rather impenetrable now and needs at least a decade.' (NM) 'Complex blackcurrant, spicy, mint aromas with certain mineral notes. Powerful, young and fresh with sweet fruit, ripe tannins, medium acidity and plenty of length.' (PK)

Bin 60A Coonawarra Cabernet Kalimna Shiraz ⬦

2016 ••• 2050

Deep crimson. Intense blackberry, chinotto, herb, elderberry aromas with ginger oak notes. Elemental, massively concentrated and unbelievably youthful. Buoyant, sweet-fruited wine with deep-set juicy blackberry, espresso, chinotto flavours and underlying ginger, mocha nuances. Tannins are still elemental, chalky and strong. Finishes chocolatey firm with plenty of drive and fruit sweetness. An exceptional vintage with superb cellaring potential. Keep.

'Amazingly youthful considering its immense core of primary blackcurrant and blackberry fruit, dark chocolate, coffee notes, perfectly matched oak and waves of firmly structured, mouth-coating tannins. Its fruit presence and sheer persistence are relentless. It will take a lifetime to fully unravel.' (TS) 'Nearly a decade old but still unevolved. Clean and pure with exotic herb notes. Ripe, juicy, concentrated and lush with fine dry tannins. Yet to show its best.' (RKP)

Kalimna Block 42 Cabernet Sauvignon ⬦

2018 ••• 2060

Fragrant, ripe pure blackcurrant, sage, violet aromas with malt, vanilla oak and liquorice. Lovely richness and concentration on the palate with dense cassis, liquorice, sweet fruit, praline flavours, underlying new oak and classic fine-grained tannins. Finishes chalky firm, with a plume of rich cassis fruit. Plush and beautifully balanced with understated power. A modern classic. Keep.

'Dior black, rich and luscious, silky and powerful with a heady glug of cabernet cassis, dried herbs, peppery spice, subtle new oak and lacy fine tannins. Fragrant as a spice market. Beautiful wine. A real yin and yang sort of wine.' (JF) 'The most opulent and beautiful delight with extraordinary cassis, violet aromatics, superb fruit density, richness, fleshiness and tannin bite. All the elements are interwoven immaculately. One of the greatest Penfolds wines of all time.' (TL)

Cellar Reserve Pinot Noir

NOW

Medium crimson. Complex strawberry, coffee, tobacco aromas with a hint of apricot and chinotto. Developed strawberry, cola flavours with spicy, vanilla notes and loose-knit but chalky dry tannins. Finishes firm/al dente but with good flavour length. Very complex and interesting wine.

'Quite a youthful wine in all respects with appealing brightness and freshness. It's still clean and alive with primary cherry-like flavours and sappy notes. Still in good condition.' (HH)

Cellar Reserve Sangiovese ⬦

NOW ••• 2018

Medium-deep crimson. Intense red cherry, herb, mocha aromas with a touch of strawberry. Well-concentrated red cherry, plum, strawberry flavours with fine chalky/touch brambly tannins. Acidity pokes out at the finish but still quite juicy and long.

'A deeply coloured, fruit-driven wine with attractive ripe plum aromas and gentle spice/floral notes. A juicy richness and satin-like texture on the palate with noticeable fresh acidity. Nice depth and dimension.' (JB) 'Primary red berry fruits are just holding out against a firm, slightly astringent undercurrent of tannins.' (TS)

Reserve Bin Eden Valley Riesling ⬦

NOW ••• 2016

Pale-medium straw. Fragrant lime, herb garden, sweet fruit aromas with developed oilskin notes. Slightly developed toasty, lime juice, lemon curd, oilskin flavours, lacy chalky textured tannins and tart indelible acidity. Finishes long and minerally. Showing plenty of complexity, vinosity and freshness. Drink now or keep for a while.

'Pristine lime-juice aromas. Fresh, crisp, lean and taut, with lemons and limes galore. It even has some sneaky richness on the palate, in spite of its delicacy and refinement. Sensational stuff.' (HH)

Bin 389 Cabernet Shiraz ⬦

NOW ••• 2035

Medium-deep purple red. Fresh juicy cassis, blackberry, brambly aromas with some herb/leafy notes. Tightly structured, well concentrated and expressive with blackberry, graphite, brambly flavours and classic fine-grained tannins. Finishes firm with a long flavourful tail. Wonderful vinosity and energy. A top Penfolds vintage. Should continue to evolve for decades. Drink now but best to keep.

'Fragrant earthy, cassis aromas and complex floral notes. Sweet-fruited and grippy with spicy notes, reined-in richness and cleansing acidity.' (JF) 'Consistent flowing style with intense berry fruit, movement and drive.' (TL)

Bin 407 Cabernet Sauvignon ⬦

NOW ••• 2025

Deep crimson purple. Intense cassis, sage, herb garden, garrigue aromas with cedar, spice notes. Classic blackcurrant, black cherry, cedar, black olive flavours and fine-grained, al dente tannins. Finishes firm with lingering vanilla oak notes. Generously proportioned wine with the concentration, power and balance for further ageing.

'Earthy, new leather, mocha, dark chocolate aromas with lifted floral notes. Full-bodied, super-concentrated and with grainy ripe tannins. A big wine.' (JF) 'Pure, elegant and intense with dark fruits, spice, herb and mint. Ripe and sweet with juicy fruit flavours, silky textures and herb, mocha notes.' (JB)

Kalimna Bin 28 Shiraz

NOW ••• 2018

Medium-deep crimson. Fresh blackberry, sage, praline aromas with a hint of herb. Lovely concentrated wine with superb energy and richness. Abundant blackberry, sage, praline flavours, underlying spicy notes and chocolatey dry tannins. Finishes firm, long, minerally and fruit sweet. Lovely fruit complexity, substance and harmony.

'The fruit is brighter than ever before with a distinct tension between tannin and acidity. It's a solid, vivacious wine that needs more time to unfold.' (TL) 'Toasty, roasted nuts, dark fruit aromas with liquorice and spice. An elegantly structured yet balanced palate with savoury notes, firm tannins, good length and drive.' (HH)

Bin 128 Shiraz ⬦

NOW ••• 2020

Medium-deep crimson. Intense, beautifully defined dark cherry, praline, panforte aromas. Buoyant rich and voluminous with dark cherry, dark chocolate flavours, plentiful chewy/velvety tannins and underlying vanilla savoury, spicy notes. A firm solid finish with superb persistency of fruit.

'Clean bright walnut, blood-plum aromas. A big, bold wine with amplitude, dimension and lots of tannin too. A seriously concentrated and structured wine for the long haul. Clean and bright, not that complex nor showing significant development.' (HH)

Bin 138 Shiraz Grenache Mourvèdre ⬦

NOW ••• 2018

Medium-deep crimson. Dark chocolate, plum, praline aromas. Very buoyant and fresh with lovely deep-set musky, plum, praline flavours. Quite fleshy and fruit-sweet on the mid-palate with loose-knit chocolatey tannins. Finishes minerally and long. At the peak of development with some tertiary elements just beginning to appear. Optimum drinking now. 49% grenache, 40% shiraz, 11% mataro.

'Briary, savoury, spice nose with lifted raspberry notes. Dusty spice, berry fruits, chocolate nuances and juicy acidity.' (JB) 'Intense berry fruit/chocolate aromas and emerging leather nuances. Supple and approachable with persistent fruit, well-balanced tannins and a slick chocolatey finish.' (TS)

Koonunga Hill Shiraz Cabernet ▽

NOW ••• 2018

Deep crimson. Fresh liquorice, aniseed, red cherry, redcurrant aromas with a touch of herb. Very buoyant and expressive wine with delicious juicy red berry flavours and dense chalky tannins. Finishes al dente firm and fruit-sweet.

'Mixed berry fruit, supple tannins and crisp acid on the finish. Nicely balanced.' (JC) 'Smoke, flint, dark fruits with plush tannins.' (DM) 'Rich and stylish wine with plenty of black fruits and penetrating flavours.' (TG) 'Aromatic and concentrated with the textbook Penfolds fresh close.' (LM)

2005

★★★★/★★★

Regular rainfall through winter into early spring, establishing good soil moistures and dam levels, then mild conditions, followed by a dry, late summer and autumn, led to optimum fruit ripeness; a good Penfolds growing season • Ray Beckwith wins the prestigious Maurice O'Shea Award for his contribution to the Australian wine industry • YouTube, the popular internet video site, is launched • The superjumbo jet aircraft Airbus A380 makes its first flight • Hurricane Katrina, one of the worst storms in US history, hits New Orleans.

Bin 95 Grange

2018 ••• 2045

Deep crimson. Intensely fresh violet, blackberry, elderberry, liquorice, cedar, mocha aromas with herb garden notes. Plush, richly concentrated wine with blackberry, elderberry, herb flavours, firm chalky/velvety tannins and spicy ginger, vanilla oak. Generously proportioned, powerful and expressive. Still needs time to unfold. 96% shiraz, 4% cabernet sauvignon.

Barossa Valley (including substantial proportion of Kalimna shiraz), McLaren Vale, Coonawarra.

'A penetrating wine with juicy dark berry, cherry stone fruit and lingering spiciness.' (JR) 'Potent and concentrated with beautiful violet, blueberry fruit and structured tannins.' (LM) 'Open, flamboyant and effusive. Richly textured, deep and layered.' (JC)

Bin 144 Yattarna Chardonnay

NOW ••• 2016

Pale straw, yellow. Intense pear skin, white peach, flinty aromas. Precise, very buoyant and generous with concentrated pear skin, white peach, apricot flavours, plenty of flinty, amontillado notes, underlying savoury oak nuances and fine mineral/spring-water–like acidity. Finishes long and tangy with a slight al dente grip. A really delicious wine with a youthful tightness of structure. Has completely outperformed expectations. Another four or five years at least. Outstanding vintage. 13.5% alcohol. 100% Adelaide Hills. 55% new French oak.

'Reserved mineral quartz aromas. Taut and reticent, perhaps without the dimension of some vintages, but it's still ascending and will build more character and richness. A firm, tight finish.' (HH) 'Matchstick character adds dimension to white stone fruit aromas. Like the 2004, it has superb seamless integration with vanilla bean, grilled nut, buttery complexity. Mellow, smooth and long with a very slightly astringent edge.' (RKP)

Bin 707 Cabernet Sauvignon ▽

2018 ••• 2035

Deep crimson purple. Intense *crème de cassis*, blueberry, vanilla, malt aromas. Plush and richly concentrated with deep-set cassis, blueberry flavours, gravelly tannins and plenty of mocha, malty new oak. Finishes leafy/grippy firm. Still elemental with luscious young cabernet fruit, superb density and power.

'Confronted with unmitigated greatness, we are sometimes reduced to just a few words. So, here goes. Scented oak, sweet cedary, blueberry fruit. Fine tannins. Rich and full-bodied. Aristocratic. I hope six stars are adequate.' (CPT) 'A very fleshy wine with blueberry, spicy, oaky aromas, refined ripe tannins. Lovely balance, elegance and texture.' (LD) 'Lush, generous and rich with delicious spiciness and well-polished tannins.' (YSL)

RWT Shiraz

NOW ••• 2025

Medium-deep crimson. Scented redcurrant, red cherry, musky, rhubarb aromas with cedar, tobacco, mint notes. A loose-knit, forward style with redcurrant, mocha flavours, hints of tobacco and toffee, fine plentiful chocolatey – slightly grippy – tannins and underlying vanilla notes. Finishes velvety and long with some mineral, spicy notes. A lovely wine with elements starting to harmonise and evolve. 70% new French oak. Drink now or keep.

'Bright red cherries, kirsch, vanilla and a touch of blueberry jam with underlying wild mint notes that should become more prominent with age. The palate is well balanced with a crisp, taut, slightly spicy, savoury opening. It builds nicely in the mouth; the acidity ever-present and lending a taut, citric thread towards the lively, slightly peppery finish. The first of this series that really impressed me.' (NM) 'Discreet and shy with elegant, slightly "cooler" red fruit characteristics, floral notes and fine spices. Young and unresolved at first, it then expands, and gives a lot more on the finish. There's freshness, subtle oak and youthful tannins.' (AL)

Magill Estate Shiraz   NOW ••• 2025

Medium-deep crimson. Ginger, plum, cedar aromas with savoury/spice notes. Dense, generous plum, cedar flavours, fine-grained, cedary textures and ginger/malt oak. Finishes firm, grippy tight, and a touch sappy. Drinking well now, but should hold for a decade at least.

'Vibrant red berry, vanilla aromas with plenty of spicy notes. The palate is dense and generous with rich sweet fruit complexity, loads of spice and structured, slightly rasping tannins. Delicious.' (JF) 'Macerated red berries, a slightly sludgy palate, supple fruit persistence and fine, tense, grippy tannins.' (TS)

Reserve Bin 05A Chardonnay   NOW

Pale yellow. Intense pear skin, white peach aromas with flinty notes and hints of grilled nuts/vanilla. Lemon curd, white peach, apricot flavours with underlying vanilla, spicy oak, chalky textures and fine indelible, mouth-watering acidity. Finishes long and tangy. Still youthful and bright with just a touch of maturity. 85% new French oak.

'Bright light–medium yellow-green. Restrained, waxy aromas, attractive toasted-nut and faintly smoky nuances. Refined and restrained all round; delicate and light on its feet. Some alcohol warmth? Has spring-watery lightness and delicacy. Lovely wine.' (HH) 'Fractionally more developed colour. Very tight and distinctly complex with some edgy notes. The fruit richness finally appears with a bit of coaxing. Very good but not great.' (JH)

St Henri   2015 ••• 2030

Medium-deep crimson. Fresh herb garden, mulberry, redcurrant, brambly aromas with dried fruit, liquorice notes. Dense and weighty palate with redcurrant, mulberry, brambly, roasted walnut flavours, plenty of mid-palate richness and fine lacy dry, slightly leafy tannins. Finishes minerally and long. A 'dark horse' vintage. Developing really well, but will it become a classic? Worth keeping to find out! 89% shiraz, 11% cabernet sauvignon.

'Ebullient notes of mulberry, raspberry, wild hedgerow, black truffle and tobacco with scents of mandarin and dried mango. The palate opens with that mulberry fruit profile, underpinned by crisp taut acidity. There is great structure and authority in this 2005, with impressive weight and persistency. This is a worthy follow-up to the great 2004.' (NM) 'Composed and vivid with highly polished tannins. Massive but so harmonious and seamless. It will be very exciting to follow this.' (NB)

Cellar Reserve Pinot Noir   NOW

Medium-deep crimson. Fragrant red cherry, chocolate aromas with espresso and toasted chestnut notes. Red cherry, roasted chestnut, leafy flavours, fine, touch al dente, chalky tannins and marked acidity. Finishes minerally and long. Quite chunky and robust in structure. Drink now.

'Dusty, vanilla aromas with some stem influence. Palate is dry and savoury with a certain hollowness. Not much pinot charm, but plenty of tannin grunt. Strange vermouth-like spices at the finish.' (HH)

Cellar Reserve Sangiovese   NOW

Medium-deep crimson. Fresh developed walnut, redcurrant aromas with a touch of graphite. Quite lean and tart with redcurrant, walnut flavours, grainy tannins and marked acidity. Finishes firm and tight. A savoury wine with a tight structure and only modest fruit intensity. It will keep, but it won't improve.

'A savoury, deep, mouth-filling wine with walnutty notes, some richness and flesh, but quite pervasive

chalky tannins.' (HH) 'The wine freshens with air revealing blackberry, raspberry fruit, gentle spice and dry grainy, but tight structure. A savoury finish.' (RKP)

Cellar Reserve Barossa Cabernet Sauvignon

NOW ••• 2018

Medium-deep colour. Ripe blackcurrant, dark plum, herb garden aromas with panforte, malty oak notes. Sweet, buoyant and fruity with blackcurrant, chocolatey flavours, balanced cedar/vanilla oak notes, a sappy freshness and attractive mid-palate richness. Tannins are plentiful and velvety. Finishes long and sweet.

'Touches of green-pea and bean in the aromas. Savoury flavours and drying tannins which tend to dominate the palate.' (HH) 'Strong cassis/berry fruit with balanced oak. Solid tannins carry the wine forward.' (JH)

Cellar Reserve Coonawarra Kalimna Shiraz

NOW ••• 2025

Medium-deep crimson. Fresh, blackcurrant, toasty, panforte, dark chocolate aromas with a touch of aniseed. Well concentrated and generous with blackcurrant, cassis, panforte flavours, balanced toasty, malty oak and complex walnut/hazelnut notes. The palate is velvety textured with plentiful fine tannins and prolonged sweet fruit notes.

'Mocha oak slightly dominates well-poised red berry/black berry fruits. The cabernet leads the structure and the shiraz fills out the mid-palate. Excellent, fine, taut tannins promise a long future.' (TS) 'Surprisingly, though pleasantly developed. A very elegant wine with good fruit and oak balance.' (JH)

Reserve Eden Valley Riesling ♦

NOW ••• 2016

Pale medium straw. Aromatic pear skin, mandarin aromas with flinty, oilskin nuances. Tangy, slightly lean with pear skin, apple, flinty flavours, fresh oilskin/developed notes, balanced, lacy tannins and mineral acidity. Finishes long and tangy. An elegant style with plenty of freshness, fruit complexity and harmony. Drink now or soon.

'A lighter-weighted wine with mineral aromas and toast/straw notes. Dry, lean and savoury with a tautness and line that is very good. Has a future ahead. Spring-water lightness and freshness.' (HH)

Bin 311 Chardonnay

NOW

Pale–medium yellow. Fresh, pear skin, white peach, apricot aromas with a touch of lanolin. Well-concentrated, creamy textured wine with white peach, apricot flavours, mid-palate buoyancy and crunchy mineral acidity. Finishes long and crisp. Still fresh and delicious with lovely fruit complexity and vitality. Best to drink soon. 100% Tumbarumba (NSW).

'Medium intense grapefruit, grilled nut aromas. Fine, fresh and highly toned with a nice line of mineral acidity driving through the wine.' (JB) 'Still youthful and inviting with lemony, creamy, toasted nut aromas. Fresh crisp, taut, refined, subtle and smooth. Lovely fruit-driven wine with many years left in it.' (HH)

Bin 389 Cabernet Shiraz

NOW ••• 2025

Medium-deep purple red. *Crème de cassis*, mulberry, praline aromas with a hint of herb, cedar/spice. Fresh seductive mulberry, blackcurrant, praline, herb flavours, plentiful, chocolatey tannins and underlying malty nuances. Finishes firm and cedary with superb fruit sweetness. Lovely density and richness of flavour with a sinuous structure. Compelling potential. Drink now, but best to keep.

'A delicious wine with jammy raspberries, blackberries and gentle spice. A plummy richness with terrific fruit intensity, velvety tannins and subtle oak. Ripe and fluent.' (RKP) 'Interesting just how much this wine is left in the shade of both the 2004 and 2006. The minty characters are accentuated, as is the relative lack of structure. A good Bin 389 but not a great one.' (JH)

Bin 407 Cabernet Sauvignon

NOW ••• 2020

Deep crimson purple. Intense blackcurrant, praline aromas with dried herb, chinotto/cola notes. Plush and concentrated with abundant cassis, chinotto, dried fruit flavours, plenty of mid-palate richness, underlying vanilla notes and dense chalky/velvety tannins. Finishes chocolatey firm. Lovely volume, balance and persistency. Still relatively youthful with many years ahead of it.

'A big-boned, big-bodied, black-fruited wine with solid form and structure. Yet it has finesse and elegance in its line and gait.' (TL) 'Still very primary with layers of persistent dark berry fruit and dark chocolate oak; finessed by well-poised tannins. Bin 407 is on the rise, and it's a captivating spectacle.' (TS)

Kalimna Bin 28 Shiraz
NOW ••• 2020

Medium-deep crimson. Intense dried fruits, blackberry, sage aromas with leafy *sous bois* notes. Luscious, almost elemental, redcurrant, dried fruit, blackberry, spicy flavours and fine chalky dry tannins. A rich chocolatey but muscular firm finish. Drink now or hold. Will last for the decade, at least.

'A rich, ripe, full-bodied wine with dense fruit, intense concentration, chalky ripe tannins and cleansing acidity. Finishes long and a touch hot. The Bin 28 style seems to possess more richness, weight and alcohol heat than in previous decades.' (JF) 'Sumptuous, fresh and alive with luscious fruit and velvety long grippy tannins. More vibrant than previous 2000s.' (RKP)

Bin 128 Shiraz
NOW ••• 2020

Medium-deep purple to crimson. Fragrant mint, camomile, aniseed, white pepper, red berry aromas. Medium-concentrated redcurrant, white pepper flavours and chalky, slightly sappy, tannins. Quite minerally and tannic at the finish. Drinking well now but cellaring outlook is probably medium term.

'A bit shy but clean and correct spiced plum aromas. Firm, tight and lean with latent fruit and softness of ripe tannins.' (HH)

Bin 138 Shiraz Grenache Mourvèdre
NOW ••• 2015

Medium-deep crimson. Fresh dark plum, chocolate, vanilla, raspberry, mulberry aromas. Raspberry, mulberry flavours, underlying sandalwood notes, bitter, chalky graphite textures and marked acidity. The fruit is still generous and fleshy, but the tannins build up quite muscularly firm at the finish. 71% grenache, 15% shiraz, 14% mataro.

'A luscious, full-bodied wine with real presence and definition. The palate is rich and ripe with grippy tannins and acid backbone in balance.' (JF) 'Builds on the impression of the 2004 with its freshness, red berry fruits and spicy backdrop.' (JH)

Koonunga Hill Shiraz Cabernet
NOW ••• 2016

Deep crimson to purple. Fragrant black cherry, plummy, chinotto aromas with some savoury, nutty, spice notes. Generously proportioned wine with black cherry, plummy flavours and fine velvety tannins. Finishes gritty firm and long. Not as substantial as top vintages, but holding up well. Drink or keep.

'Full-bodied and robust.' (JC) 'Lovely dry, elegant, medium-bodied wine with young tannins.' (MC) 'Cellaring is highly advised!' (LM)

2006
★★★★★/★★★★

Good winter and spring rainfalls were followed by mild to warm conditions over summer; 50 mm of rain at the end of February and a warm burst of weather accelerated ripening • While it rained intermittently during vintage, the overall quality of the fruit was excellent • A near-perfect vintage • Twitter is launched • Scotland Yard exposes major terrorist plot to destroy aircraft travelling from the UK to the US; all toiletries are banned from commercial aeroplanes • *No. 5, 1948*, a painting by abstract impressionist artist Jackson Pollock, is sold privately for a record price of US$140 million.

Bin 95 Grange ♢
2020 ••• 2050

Deep crimson. Intensely fragrant dark berry, praline, liquorice aromas with roasted chestnut notes. Very powerful and concentrated palate with blackberry, dark chocolate flavours, pronounced fine tannins and savoury, ginger oak. Finishes oaky firm but long and sweet. Lovely vinosity, energy and freshness. A great Grange. 98% shiraz, 2% cabernet sauvignon.

98% Barossa Valley (including significant contributions from Kalimna and Koonunga Hill vineyards), 2% Magill Estate (Adelaide).

'Rich, round, sweet, voluptuous wine with chalky tannins and lovely oak integration.' (TG) 'Hugely concentrated with powerful tannins. This is a monster! Needs bottle time.' (MC)

Bin 144 Yattarna Chardonnay ⬙

NOW ••• 2018

Pale yellow. Fresh grapefruit, yoghurt, cashew nut aromas with some flinty, chalky notes. A leaner style with grapefruit, lemon curd, cashew nut flavours, underlying vanilla, toasty notes, al dente texture and fine persistent, but marked acidity. Tonic water, bitter lemon notes at the finish. A mouth-watering wine with excellent fruit complexity, freshness and balance. Drink soon. 37% Derwent (Tasmania), 36% Adelaide Hills, 27% Henty (Victoria). 45% new French oak.

'A super-tight, stunning wine with everything beautifully balanced. It's on the lighter side of medium-bodied yet it's complex and focused with intense pure lemon, grapefruit characters, yeasty notes, toasty nuances and mineral drive.' (JF) 'Right in the main frame of the Yattarna style. The initial impact is supple and smooth with stone fruit characters. Then citrus/mineral notes tighten and lengthen the finish.' (JH)

Bin 707 Cabernet Sauvignon ⬙

2018 ••• 2040

Deep crimson purple. Fragrant, intense plum, liquorice, dark chocolate, cedar aromas with touch of leaf/herb garden. Gorgeously seductive wine with plum, dark chocolate, violet, graphite flavours. Slightly muscular, fine-grain tannins. Finishes chalky firm yet long and sweet. Delicious!

Barossa Valley, Coonawarra, Adelaide Hills.

'Intense, profoundly oaky wine with black cherry, cassis, herbaceous aromas and vanilla, cinnamon, clove notes. Some smoky, liquorice nuances. Abundant fruit and massive super-ripe tannins. A great 707 vintage in the making.' (ER) 'A ripe cabernet with big oak and mocha, tar notes. Fresh and lively with sweet tannins.' (ST) 'A solid wine with rich cherry fruits, fleshy texture and firm tannins. Good balance.' (LD)

RWT Shiraz ⬙

NOW ••• 2035

Medium-deep crimson. Very precise and undeveloped but beautifully proportioned and intensely flavoured. Classic dark cherry, blackberry, praline aromas with sage, vanilla, liquorice notes. The palate is lavishly concentrated with a palimpsest of rich ripe fruits, velvety smooth, almost satin-like tannins, mocha, vanilla oak and roasted chestnut, cedar complexity. It finishes grainy firm, with superb flavour length. A wine of great dimension, youthful substance and balance. Will need several years to reach its vertex. A great Penfolds vintage. Drink now but best to keep. 70% new French oak.

'Young, fresh, intense and complex with black cherry, spice, pepper aromas, ripe tannins and balanced acidity. Long finish.' (PK) 'We're back to textbook Barossa here, with sweet dark fruit, plum, cassis, liquorice and a whiff of vanilla. Great mouth-feel and presence, the wine displays a very nice concentration and firmness. Fresh and balanced without heaviness, excellent fruit and a very long finish. Will develop more complexity with bottle age.' (AL) 'Very fine wine with intense black fruits, hints of tobacco, sandalwood and spice. Plenty of minerality.' (MS)

Magill Estate Shiraz ⬙

2015 ••• 2026

Medium crimson. Fragrant, red cherry, plum, black fruit aromas with cedar/vanilla notes. The palate is still elemental with deep-set dark cherry, blackberry, cassis flavours, plenty of roasted chestnut/cedar notes, fine chalky fruit tannins and underlying savoury nuances. Finishes firm and long with a lovely roasted chestnut, graphite plume. Very seductive and classic in structure. The epitome of the Magill Estate style.

'Plenty of dark fruits, underlying new oak and lovely spicy complexity. Richly concentrated, dense and chalky textured.' (TL) 'Ripe fruit, sweet oak aromas with spice nuances. Dense, almost creamy textured wine with dark berry fruits, dried leaf notes and ripe tannins. It's still a touch raw, but the elements are coming together.' (JF)

Reserve Bin 06A Chardonnay

NOW ••• 2014

Pale yellow. Intense grapefruit, lime, flinty aromas with vanilla, roasted chestnut notes. Understated but superb grapefruit, lime flavours, nuances of savoury new oak, grilled nuts and plenty of flinty complexity.

Finishes chalky and minerally. A classic contemporary style. Surprisingly linear and tight but with wonderful vinosity. Should hold for at least a few more years but lovely to drink now.

'Very attractive white stone fruit, citrus, baked apple aromas with quiet yeasty complexity. It's tight, dry and long with fine grainy, savoury mouth-feel and a fine tapering finish.' (RKP) 'Complex and expressive wine with the palate fulfilling the promise of the aromas. A vibrant mix of white stone fruits and a garnish of citrus acidity. The new oak is not in the least bit assertive.' (JH)

St Henri ♢

2016 ••• 2035

Medium-deep colour. Beautiful blackberry, plum, liquorice aromas with praline, cedar, spice notes. Sweet, supple, voluminous wine with saturated inky blackberry, liquorice, praline flavours, some brambly/spicy tones and fine grainy/graphite tannins. Finishes chalky, al dente firm. Fruit sweetness and underlying mineral freshness lengthen the palate. Seductive and superbly balanced with stature, density and vinosity for the long term. 89% shiraz, 11% cabernet sauvignon. A classic St Henri vintage. Lovely to drink now but best to keep.

'Floral and feminine, with dried rose petals, vanilla, liquorice and wild strawberry notes. The palate is silky smooth and sensual with a cheeky tang of salted liquorice and grippy dry tannins dominating the finish. This has great potential. It completely and unreservedly adheres to the St Henri style.' (NM) 'Nice freshness, elegance and transparency. Already drinking so well.' (FK)

Cellar Reserve Pinot Noir

NOW ••• 2015

Medium-deep crimson. Red cherry, leafy aromas with a touch of spice, vanilla oak. Lovely fresh, generous and complex with red cherry, strawberry flavours and underlying spicy vanilla nuances. A rich and impactful mid-palate with plentiful slinky, touch gritty, tannins and a long sweet, slightly sappy finish. Drink now or soon.

'Sappy stemmy/whole-bunch aromas with slightly stewed notes. Quite youthful and inviting with more vibrant pinot fruit on the mid-palate than most of the mature wines. Still bears the structural hallmarks of Penfolds, with plenty of tannins.' (HH)

Cellar Reserve Sangiovese ♢

NOW ••• 2020

Medium-deep crimson. Roasted coffee, plum, cherry stone aromas with fresh walnut notes. Well-concentrated, plum, dark fruit flavours with pippy, cherry stone notes and fine pronounced chalky, dry tannins. Finishes brambly with plenty of savoury notes. A solid wine with plenty of fruit density and grip.

'Refined and harmonious, dense and concentrated with fresh cherry, tobacco, fruits, savoury oak complexity and a fine coat of tannins. Top notch.' (TL) 'Savoury, mellow and complex aromas; not fruit-forward at all. Deep, brooding, concentrated and solidly built with deep-seated black fruits, anise flavours, excellent grip and backbone, which extends the long palate. Very satisfying.' (HH)

Cellar Reserve Barossa Cabernet Sauvignon

NOW ••• 2025

Medium-deep colour. Intense cassis, praline, mocha aromas with some graphite notes. Substantial wine with concentrated cassis, dark chocolate flavours, rich velvety/chocolatey tannins and underlying malty oak. Finishes grippy and long. A classic Barossa Cabernet Sauvignon with strong varietal definition, generosity and robustness. Should develop really well. Drink now or keep for a while.

'Ripe cabernet aromas with beautiful structure, generosity of fruit and weight. The tannins give the wine a velvety texture and lasting quality on the palate.' (TL) 'Restrained dusty blackcurrant, lead pencil aromas. But jammy, lush, ripe and complete on the palate with a fine tannic grip.' (RKP)

Bin 51 Eden Valley Riesling

NOW

Pale–medium gold. Intense flinty, lemon curd aromas with a touch of grapefruit and oilskin. Well-concentrated and developed wine with deep-set lemon curd, grapefruit flavours and some toasty, nutty, flinty, whetstone notes. Al dente, chalky texture and pronounced mineral acidity. Finishes long and sweet with some oilskin/whetstone notes. A richer style with excellent fruit intensity, volume and balance. Ready.

'Quite complex with lemon zest, toasty aromas. Clean, dry, tangy and fairly light with a slight grip to the back palate and finish. An almost thick textural dryness that persists on the aftertaste. Nevertheless, a rich, flavoursome and harmonious wine.' (HH)

Bin 311 Chardonnay

Pale–medium yellow. Fragrant, tropical fruit, lemon curd aromas with passionfruit notes and grilled nut complexity. Tangy lemon curd, tropical fruit, grapefruit flavours, underlying fino notes, lacy/chalky textures and fine cutting acidity. Finishes long and sweet. Delicious to drink now.

100% Tumbarumba (NSW).

'Pure grapefruit, preserved lemon aromas immaculately knit with grilled nut complexity. Fruit cascades across a minerally, chalky and finely textured palate. This exacting, seamless wine is marvellously reflective of Tumbarumba's cool-climate alpine foothill position.' (TS) 'Fine and subtle with lemon peel, toasty nuances. It's still super-tight and not overworked in any way with delicious clear primary fruit, bright crisp acidity and a long flavourful finish.' (JF)

Bin 389 Cabernet Shiraz ▽

Medium-deep crimson. Intensely perfumed *crème de cassis*, dark cherry, vanilla, malt aromas with garrigue notes. Generous, expressive and superbly balanced with blackcurrant, dark cherry, praline, vanilla, malt flavours, lovely mid-palate juiciness and fine cedary tannins. Finishes firm with a lingering plume of fruit sweetness. Beautiful wine with great structural integrity and balance. 'This is one of the greatest Bin 389s of all.' You can drink this now but best to keep.

'A pristine and tightly wound Bin 389 with primary blackcurrant, mulberry, black plum and black cherry fruit tightly encased in a cage of beautifully fine tannins and classy dark chocolate oak.' (TS) 'A personal-best Bin 389 with multi-layered black fruits, built-in tannins and a stab of dark bitter chocolate. Seductive and imperious.' (JH)

Bin 407 Cabernet Sauvignon ▽

Medium-deep crimson. Blackberry, praline, herb, spice, cedar aromas with a hint of sage. Buoyant, concentrated wine with rich, voluminous blackberry, herb, brambly, dark chocolate flavours and supple, dry leafy tannins. Finishes gritty/sappy with savoury/malt nuances. Superbly balanced with a long future.

'Fragrant black fruits on nose with dark chocolate, violet notes. Powerful and plush with velvet textures supporting some really lovely concentrated cassis fruit.' (JB) 'Everything is here in abundance, but time is definitely needed to resolve and unify the components.' (JH)

Kalimna Bin 28 Shiraz ▽

Medium-deep crimson purple. Beautifully pitched Turkish Delight, dark berry, liquorice aromas with some roasted chestnut, cedar notes. Voluminous, ripe and smooth with musky, dark berry, liquorice, praline flavours and fine loose-knit chalky tannins. Mineral acidity emerges at the finish. Lovely richness and harmony.

'Fragrant nose combining dark fruit, liquorice, dusty spice notes. A juicy bright palate lifted by bright acidity with dark cherry, black berry fruit, dark chocolate toasty notes.' (JB) 'Dark fruits, Rhône spices and star anise notes. Dry, savoury and concentrated with densely packed supple tannins and a long finish.' (HH)

Bin 128 Shiraz ▽

Medium-deep crimson purple. Intense, almost elemental, blackberry, blueberry aromas with sage, leafy, musky notes. Concentrated, generous blueberry, blackberry pastille, liquorice/aniseed flavours with leafy notes. Tannins are quite muscular with a leafy, dry edge but the fruit is rich and buoyant. Finishes firm and a touch sappy but the volume, density and persistency of fruit give the wine superb power and balance. One of the greatest Bin 128s. Worth keeping for a while.

'Black plums, spices, pepper, dusty, savoury aromas. Powerfully structured with terrific depth of flavour and concentration. A long, long finish with satisfying weight and muscle. Impressive wine built for the long haul.' (HH)

Bin 138 Shiraz Grenache Mourvèdre

Medium-deep crimson. Fresh, intense, pure mulberry, raspberry, brambly aromas with strawberry, dark chocolate notes. Well-concentrated mulberry, plum, herb garden flavours with supple sweet fruit notes and chocolatey tannins. Finishes brambly and juicy. A classic year. 39% shiraz, 32% grenache, 29% mataro.

'Beautiful fresh red fruits and floral notes. Very juicy, expressive, generous and fresh.' (TL) 'Fresh red berry, plum fruits are perfectly framed by firm, finely textured tannins. An impressively crafted wine with beautiful youthful poise, outstanding persistence and drive.' (TS)

Thomas Hyland Chardonnay ◈ NOW

Medium yellow. Fresh lemon, grapefruit, flinty, oyster shell aromas. Lovely complex wine with intense lemon, grapefruit, tonic water flavours, hints of vanilla, malt, new oak, fine chalky textures and mineral acidity. This has developed more slowly than expected! It's still fresh, complex and balanced with plenty of liveliness. A remarkable wine. Drink now.

'Full yellow, bright and clear: excellent colour for a six-year-old Chardonnay. Beautifully mellow and mature bouquet of toasted hazelnuts, stone fruits and oatmeal with malty/vanillin elements. It's attractively light- to medium-bodied and superbly balanced without heaviness or oaky clumsiness, yet it has richness and fleshy density. Captivating!' (HH) 'This wine challenges the drink-now proposition.' (PG)

Koonunga Hill *Seventy-Six* Shiraz Cabernet ◈ NOW ••• 2020

Medium-deep crimson. Vibrant dark chocolate, dark cherry, ginger, aniseed aromas and flavours with fine grainy tannins and plenty of mid-palate richness. Finishes chalky firm, long and sweet.

'Violets and black fruits balance the heft and density of the wine.' (LM) 'Super-refined and powerful wine.' (TG) 'Rich and concentrated with sweet ripe fruit and young fine tannins.' (MC)

Koonunga Hill Shiraz Cabernet ◈ NOW ••• 2020

Medium-deep crimson. Exuberant dark cherry, juicy cassis aromas with hints of sage, mint and aniseed. The palate is beautifully concentrated with deep-set choco-berry fruit and plentiful velvety tannins. A wine with superb richness and flavour length. Delicious to drink now or keep for a while.

'My excitement is returning here; pretty, almost floral aromas, plush sweet fruit flavours, mint/eucalyptus notes and a clean refreshing finish.' (DM) 'Dense, rich, deep, more serious style with lovely weight and structure.' (TG)

2007 ★★★ / ★★★★

Drought conditions prevailed over winter and through the growing season; around a third of average annual rainfall, causing soil moistures to dry up and old vines and unirrigated vineyards struggled through some of the toughest conditions in living memory • Barons of the Barossa award Grange winemaking team (comprising Peter Gago, Steve Lienert, Andrew Baldwin and Paul Georgiadis) 'Winemaker of the Year' • Sydney hosts the first of a global series of Live Earth benefit concerts to raise awareness and combat global warming • Beginning of the Global Financial Crisis • UK and Australia ban smoking in all public places.

Bin 95 Grange 2018 ••• 2040

Deep crimson. Dark chocolate, elderberry, blackberry aromas with leafy, aniseed notes. Elemental and juicy with concentrated blackberry, elderberry fruit, muscular, al dente tannins and mocha, ginger oak. Finishes gravelly dry and firm. Very powerful wine. 98% shiraz, 2% cabernet sauvignon.

Barossa Valley, McLaren Vale, Magill Estate (Adelaide).

'Polished wine with beautiful lifted blackberry, sweet fruit aromas and a spicy, refreshing, supple palate.' (LM) 'Very perfumed and elegant, yet powerful and balanced.' (RI)

Bin 144 Yattarna Chardonnay ◈ NOW ••• 2015

Pale yellow. Fragrant camomile, honeysuckle, waxy, lemon aromas with biscuity nuances. Upfront, generous creamy lemon curd, pear, peach flavours, subtle herb garden notes, fine chalky, touch grippy textures and underlying biscuity oak characters. Finishes chalky firm with plenty of sweet fruit follow through.

A well-balanced wine with understated power and grace. Drink soon, though. 49% Derwent River (Tasmania), 37% Adelaide Hills, 14% Henty (Victoria). 35% new French oak.

'Lime-leaf aromas with honey beeswax. Generous, full and smooth with a rich fruit-sweet middle palate. Gorgeous.' (HH) 'A savoury nose of chalky minerals, white fruits, nuts and cream gives way to an almost lush palate of great length and savoury presence. It has a lovely creamy texture and finishes long and fine. It's broader and more obvious than 2006 and 2008 but it has similar purity and charm.' (RKP)

Bin 707 Cabernet Sauvignon
2016 ••• 2040

Deep inky purple. Intense blackberry, redcurrant, red cherry, malty aromas with herb, mint, aniseed notes. Concentrated and lush with deep-set cassis, redcurrant, herb flavours, plenty of sweet fruit notes, chocolatey, leafy tannins and luxuriant spicy/malty new American oak. Finishes chewy firm, long and flavourful. Substantial and muscular wine with superb fruit density and tannin richness. Padthaway, Barossa Valley, Coonawarra.

'Pungent blackberry, blueberry, black pepper, herbaceous aromas. Very rich and lush with plenty of new oak and fruit puissance.' (FW) 'Intense cassis, black plum, black cherry, eucalyptus, herbaceous aromas. Wonderfully ripe and chewy with a tremendous core of black fruits and well-integrated acidity. Very young and appealing.' (ER)

RWT Shiraz
2014 ••• 2022

Deep crimson. Fresh blackberry, dark cherry, bay leaf, cedar, mint aromas. Richly flavoured and full-bodied with juicy blackberry, dark chocolate, minty flavours, ample fine dry muscular, slightly leafy tannins and underlying savoury oak. Finishes firm and dry with some vanilla notes. Youthful and tight but should develop well over the medium term. 71% new French oak.

'Balanced and relatively restrained with balanced oak and the promise of perfume to come. Fleshy texture and black fruit flavours overpowered by assertive tannins. Should eventually come around.' (NB) 'Young blackberry, spice aromas with lovely concentration, plentiful tannins and mineral acidity.' (PK)

Magill Estate Shiraz
NOW ••• 2018

Medium-deep crimson. Fragrant ginger, black fruits, soft red berry, miso aromas. Densely concentrated with attractive sweet blackberry, mulberry fruit, brambly textures, pronounced cedar oak and leafy tannins. Finishes sappy, dry and long. Drinking well now. Unlikely to improve.

'Lifted pure dark berry fruit with violet nuances. Tannins are firmly composed, slightly drying and promise a long future.' (TS) 'The toughness of the vintage is less evident here. The wine is remarkably restrained, long and balanced.' (JH)

Reserve Bin 07A Chardonnay
NOW ••• 2016

Pale yellow. Fragrant, lemon curd aromas with herb, lime, verbena notes. Medium intense lime/lemon, tonic water flavours with bitter textures and marked acidity. It still has line and length but herb garden and hard edges prevail. A modest vintage with limited potential.

'Settled and balanced with flinty mineral notes, hints of oak and fresh acidity.' (TL) 'A seamless progression of fine white fruitiness and very light nuttiness.' (RKP)

St Henri
2016 ••• 2035

Medium-deep colour. Fragrant elderberry, cassis, liquorice aromas with fresh bay leaf, forest floor notes. Lovely juicy elderberry, cassis flavours with espresso/dark chocolate notes and vigorous, fine leafy tannins. Finishes chalky firm and juicy/brambly. Well concentrated and substantial with underlying muscular structure. Still needs time to meld. 100% shiraz.

'A Bordeaux-like persona with blackberry, cedar, a touch of truffle oil and sous bois, but without the intensity of recent vintages. The palate is medium-bodied with chewy, ripe tannins. Very primal and a little closed, but nicely balanced with a slight austerity towards the finish.' (NM) 'Open dark berry, crème de cassis, liquorice aromas with a whiff of truffle. The palate has very good substance with salty liquorice, crème de cassis notes, appetising freshness and good length. Restrained and elegant finish.' (AL)

Cellar Reserve Chardonnay

NOW ••• 2015

Pale–medium yellow. Fresh matchstick, lemon curd, flinty, verbena aromas. Smooth creamy lemon curd, flinty flavours with fine chalky textures and crisp, pronounced acidity. Finishes long and tangy. Drinking really well now, but should keep for a while.

'Quite youthful and tight for its age with smoky, toasty savoury notes. Palate offers much more in terms of character. It's soft, rich, silky textured and smooth, dry but not austere with discreet barrel-ferment nuances.' (HH) 'A New World style with plenty of barrel, lees and malolactic complexity. It's fuller and richer than the Bin 311 wines but not as fine as post-2008 vintages.' (RKP) 'A single vineyard wine.' (PG)

Cellar Reserve Pinot Noir

NOW ••• 2016

Medium crimson. Classic red cherry, violet, sappy, meaty aromas. Very complex and tightly structured with red cherry, woody flavours, plentiful but hard tannins and fresh but striking acidity. Finishes firm and tight with some strawberry, brambly notes.

'Developed dusty, earthy, stemmy aromas. Palate is lean and plain with low fruit intensity but lots of spicy, sappy, savoury flavours, deep and ripe powdery, but assertive, tannins.' (HH)

Cellar Reserve Sangiovese

NOW ••• 2016

Medium-deep crimson. Fresh blueberry, praline, sage aromas with plenty of sweet fruit. A fruit-dominant wine with blueberry, praline flavours and velvety/touch al dente tannins; the acidity carries the flavours across the palate. Finishes long and sweet, but quite grippy/leafy.

'Supple, ripe, dark berry fruits contrast with firm, dusty, drying tannins. It reflects the vine stress of this drought year, and it's not going to improve.' (TS) 'It's sound enough with liquorice, spice notes, fruit richness and dry grippy tannins. But it lacks vibrancy and balance.' (RKP)

Bin 51 Eden Valley Riesling

NOW ••• 2018

Pale–medium gold. Fresh toasty, lemon curd, flinty aromas with a touch of verbena. Lovely purity of fruit and complexity on the palate with developed lemon curd, toasty, flinty flavours and some leafy notes. Al dente texture and fine but pronounced acidity give balance and vinosity. Finishes long and fruit sweet. Showing classic Eden Valley Riesling expression. Ready to drink, but it should also keep for a while longer.

'A mineral, earthy/flinty style. Light, lean and without great line. A clean, dry finish. Doesn't possess the fruit depth or seamless quality of its brethren.' (HH)

Bin 311 Chardonnay

NOW

Pale yellow. Fresh oyster shell, lime, lemongrass aromas. A high-tensile palate with oyster shell, gun flint, bitter lemon flavours, al dente textures and fresh sharp indelible acidity. Very up-tight and mono-dimensional. Could further develop but best to drink up. 100% Orange (NSW).

'A completely different shape and personality with powerful white peach fruit and phenolic texture. It showcases the power of Orange, but lacks the refinement, minerality and structural energy of Tumbarumba fruit. A lesser wine in this flight.' (TS) 'A tightly structured wine with lemon mineral aromas, a touch oak and excellent chalky mouth-feel. The fruit is just starting to unfold at the finish.' (TL)

Bin 389 Cabernet Shiraz

NOW ••• 2020

Medium-deep crimson. Fragrant blackberry pastille, white pepper aromas with a touch of violet. Medium-bodied with gentle white pepper, blackberry, spice flavours and loose-knit chalky al dente tannins. It doesn't have the precision, weight or overall balance for longevity. Drink now or keep for a while.

'Some forward dark chocolate, mint notes and medium-bodied palate, yet there is structure to burn.' (JH) 'Lovely ripe fruits with lifted florals and spice. Concentrated and fresh with grippy tannins and fresh acidity.' (JF)

Bin 407 Cabernet Sauvignon

NOW ••• 2020

Medium-deep crimson. Blackcurrant, mocha, herb garden aromas with roses, ironstone notes. Classic wine with black olive, cassis/blackcurrant fruit flavours and fine, muscular, leafy tannins. Finishes chalky firm and tight. An early-drinking style but should hold for a while.

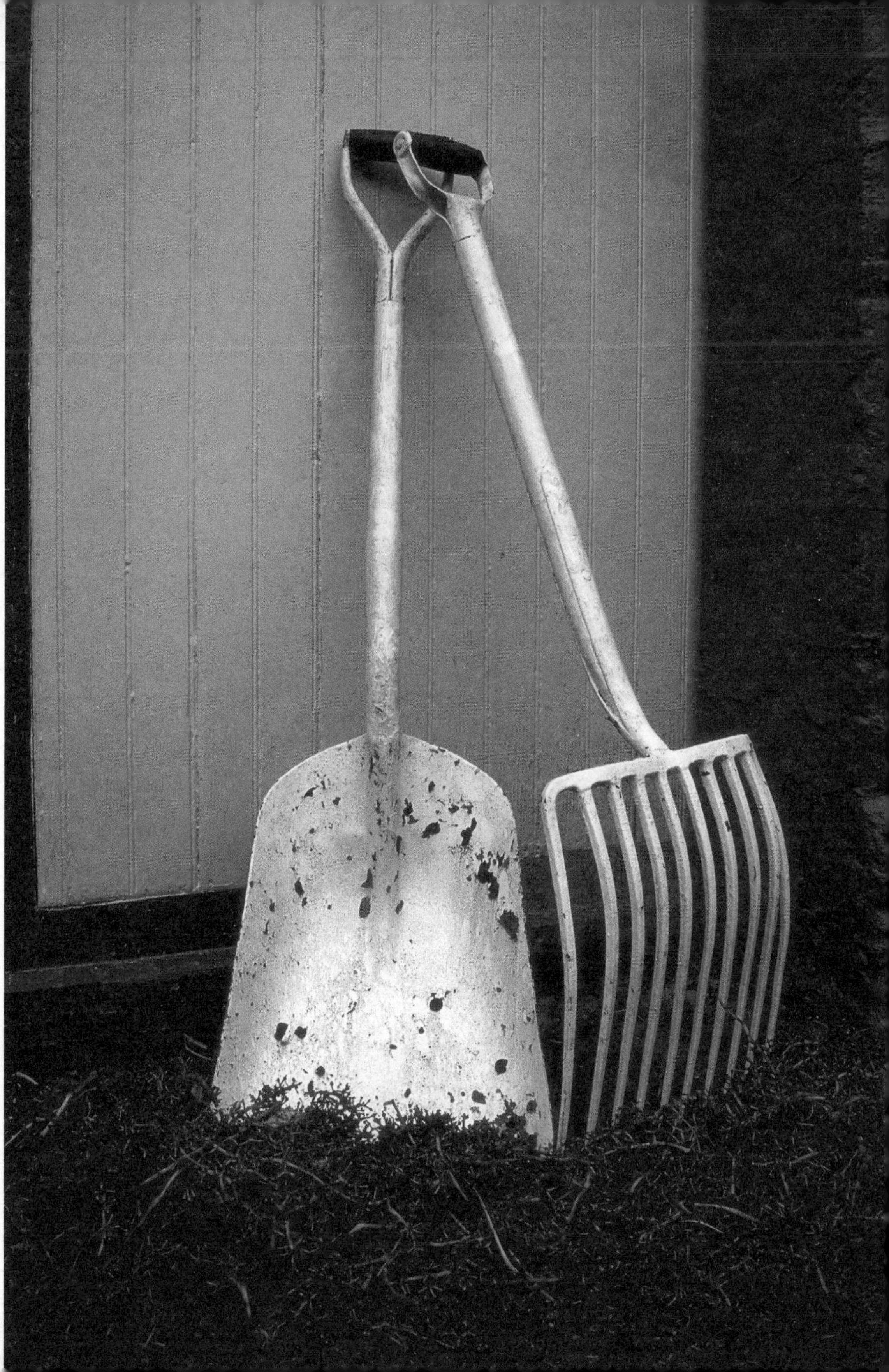

'Beautifully aromatic with lifted floral, cassis freshness, earthy mocha, savoury notes, plenty of sweet fruit, ripe tannins and a persistent long finish.' (JF) 'Juicy blackcurrant aromas with a nice spicy lick of oak. Smooth and forward in style with grippy tannins. Drink soon.' (RKP)

Kalimna Bin 28 Shiraz

NOW ••• 2015

Medium-deep colour. Intense redcurrant, sage, mint aromas with a hint of tobacco. Densely concentrated wine with sweet redcurrant, dark plum, tobacco flavours and plentiful chalky firm, almost solid, tannins. Finishes leafy with bitter-sweet notes. This will hold, but best to drink while the going is good.

'A big brute of a wine with deep rich colours, beautiful fragrance and dense rich powerful fruit. It's open and giving, but solid with plenty of tannin grip.' (TL) 'Impressively concentrated black fruits are foiled against a wall of oak tannins. Will the fruit ultimately hold out?' (TS)

Bin 128 Shiraz

NOW ••• 2020

Medium-deep crimson to purple. Intense red cherry, aniseed, chestnut aromas with some mint, sage notes. Medium-concentrated, juicy red cherry, aniseed, touch leafy flavours and chalky, dry tannins. Finishes muscular dry but still has sweetness of fruit.

'Rich, deep, fleshy and generously flavoured with firm authoritative tannins and fumy alcohol. Some meaty complexity and spicy, bay leaf notes. Built for the long haul.' (HH)

Bin 138 Shiraz Grenache Mourvèdre

NOW ••• 2016

Medium-deep crimson. Fresh raspberry, herb garden aromas with notes of dried roses. Medium-concentrated but ample with raspberry pastille, herb flavours, complex praline notes and cedary dry tannins. Finishes supple, sweet and long. Enjoy while the fruit is ascendant. Drink soon. 39% shiraz, 32% grenache, 29% mataro.

'Berry compote fruits with milk chocolate notes. Impressive length and drying, robust tannins.' (TS) 'Fresh ripe dark berry aromas. Fruit sweetness on the palate is offset by drying tannins and sour notes on the finish.' (JF)

Bin 2 Shiraz Mourvèdre

NOW ••• 2015

Medium-deep crimson. Fragrant redcurrant, blueberry, plum aromas with a touch of herb garden. Supple redcurrant, blackberry, dark plum, aniseed flavours and plentiful fine chalky tannins. Finishes firm and long with noticeable fresh acidity. Nice fruit complexity and balance.

'Fresh, clean and true with spicy star anise, blueberry and blackberry aromas and hints of black pepper. Elegant, medium-bodied and fruit-driven, with appealing softness of tannin and balance.' (HH)

Thomas Hyland Chardonnay

NOW

Medium yellow. Intense lemon curd, pear skin, white peach, light tobacco aromas. Well-concentrated lemon peel, grapefruit flavours with creamy sweet fruit, tobacco notes, chalky, al dente textures and minerally, long acidity. Still fresh but showing some bottle age development. Drink soon.

'Slight burnt toast notes with some poached peach, lemon juice characters. The palate is soft and round, nicely formed and balanced with attractive bottle age. Of modest complexity and character.' (HH)

Koonunga Hill Seventy-Six Shiraz Cabernet

NOW ••• 2018

Medium-deep crimson. Dark plum, dark cherry, praline aromas with some leafy notes. Palate is richly concentrated and powerful with sweet dark berry fruit and plush chalky dry tannins. Finishes firm and tight.

'Powerful dark berry, mocha, liquorice aromas and expansive sappiness.' (JR) 'Big, silky, supremely polished wine.' (LM) 'Lush with pronounced tannins.' (DM) 'Discordant with syrupy versus chalky hard tannin notes.' (JC)

Koonunga Hill Shiraz Cabernet

NOW ••• 2018

Medium-deep crimson. Intense blackberry, elderberry, liquorice aromas and flavours. A very opulent, solid, juicy and richly flavoured wine with generous fruit sweetness and pronounced pippy/leafy tannins. Acidity pokes out at the finish. A medium-distance runner. Drinks really well now but tannins will muscle in.

'An early maturing wine with generous fruit.' (MC) 'Huge wine yet keenly balanced.' (LM)

A difficult but remarkable year marked by the longest heatwave ever recorded in South Australia: a fifteen-day run of over 35°C between 3 and 17 March made it a vintage of two halves • In Coonawarra, cool Southern Ocean breezes moderated temperatures • A vineyard management year with outstanding results: a great Penfolds vintage • Reintroduction of Bin 169 Coonawarra Cabernet Sauvignon and Bin 620 Cabernet Sauvignon Shiraz • Ray Beckwith is awarded the Medal of the Order of Australia • Black Monday as world-wide stock markets report their biggest ever one-day drop • Australian Prime Minister Kevin Rudd formally apologises to the 'Stolen Generations' • Barack Obama becomes the first African American to be elected President of the United States.

Bin 95 Grange ◈

2020 ••• 2055

Deep crimson. Plush blackberry, dark chocolate, vanilla, mocha aromas. Rich and voluminous with fresh powerful blackberry, elderberry, liquorice fruit, plentiful dense chalky tannins and underlying new malty, mocha oak. A superbly concentrated wine with incredible fruit density, power and balance. A classic Penfolds year. Receives 100 points from Robert Parker's *Wine Advocate* (Lisa Perrotti-Brown, MW) and *Wine Spectator* 2013. 98% shiraz, 2% cabernet sauvignon.

Barossa Valley, Clare Valley, Magill Estate (Adelaide). Difficult growing conditions were overcome by meticulous vineyard management, fruit selection from physiologically earlier-ripening older vines.

'Very refined and elegant with a similar profile to 2010, with big tannins, purity and strength.' (MC) 'Massive with a brilliant core of fruit, palate-coating tannins and vibrant acidity. The inherent quality of this wine shines through.' (DM)

Bin 144 Yattarna Chardonnay ◈

NOW ••• 2020

Pale yellow gold. Wonderfully complex, fresh and fragrant, flinty, brine, grapefruit, lime, camomile aromas. Beautifully focused, tight and linear with plenty of energy and richness. Grapefruit, camomile, flinty, matchstick flavours and underlying savoury oak. Finishes chalky, dry and incredibly long. A sublime wine with superb vinosity, volume and minerally persistency. Drink now or keep for a while. 89% Derwent River (Tasmania), 11% Adelaide Hills. 49% new French oak.

'High-toned and flinty with aniseed/fennel, smoky notes. Complex and intense with lovely weight, richness and balance. A really good line of acidity supports and lengthens the palate. Excellent wine with plenty of potential.' (JB) 'Super-tight and mouth-watering with plenty of barrel ferment notes and yeasty complexity. It's enticing, wild and yet so fine with lovely palate weight, length and drive.' (JF)

Bin 707 Cabernet Sauvignon ◈

2020 ••• 2045

Deep crimson purple. Intense violet, inky, cassis, plum, espresso aromas with vanilla/mocha notes. Very approachable, plush and vivacious with dense, ripe, juicy blackcurrant, mulberry, aniseed flavours, ripe chocolatey tannins and beautifully balanced vanilla/mocha oak. Superb flavour length. Immensely concentrated, gorgeously seductive wine with remarkable balance, generosity and power. Will live on and develop for decades. A great Penfolds vintage! Coonawarra, Barossa Valley, Wrattonbully, Padthaway.

'Very young and tight with ripe black fruits, the sweet spice of vanilla and cloves and plenty of toasty oak. Plush and plump with grippy structure and lovely length. Needs time to mesh.' (FW) 'Smoky mineral sweet fruit notes. Concentrated and round with sweet tannins.' (LD) 'Generous sweet fruit characters with plenty of freshness, new oak and some spice notes. It's like an Old World meets New World style.' (CC)

RWT Shiraz ◈

2016 ••• 2035

Deep crimson. Intense blackberry, mulberry aromas with mocha, dark chocolate, spicy notes with a warm ethereal lift. Rich, voluminous and dense with intense blackberry, dark chocolate, spicy, cedar flavours, plush velvety tannins and underlying roasted chestnut, vanilla notes. Finishes chocolatey firm, slightly hot and very long. Still quite elemental but balanced and round. Will last decades. 83% new French oak. Best to keep.

'Primal and surly at first: tightly coiled macerated dark cherries, dark plum, blackcurrants and blueberry shrouded in a fug of alcohol. The palate is very smooth and voluptuous. It's packed full of sweet, opulent red fruit that needs several years to develop complexity. The constituent parts are all here for a great RWT.' (NM) 'Pure fruit, blackberry, crushed berries, violet aromas. Attractive well-rounded structure with vibrant dark fruit, fine integration of oak and an elegant, long aftertaste.' (AL)

Bin 169 Coonawarra Cabernet Sauvignon ▽ 2016 •••• 2040

Medium-deep crimson. Fragrant praline/mocha aromas with vanilla oak, roasted chestnut notes. Well-concentrated, powerful wine with praline, mocha, blackberry, sage flavours, fine-grained, chalky tannins and integrated ginger, oak nuances. Lovely volume and richness on the mid-palate. Finishes chalky, al dente firm, with superb fruit sweetness and length. Very classic in structure. Keep for a while.

'Starting to show the first signs of opening up with cedar notes on the bouquet and light and shade on the palate. Utterly beautiful texture, and varietal expression. Will always drink well.' (JH) 'Lifted floral, crushed blackcurrant aromas with briar notes and a whisper of tar. A completely unevolved palate with brawny structure, tangy flavours and oaky notes. Very Penfolds in stature.' (RKP)

Magill Estate Shiraz ▽ 2016 •••• 2030

Deep colour. Intense aniseed/liquorice, blackberry, blueberry aromas with some herb notes. Very lifted, concentrated, densely packed wine with blackberry, panforte, brambly flavours, beautiful *crème de cassis* mid-palate juiciness, plentiful chalky, ripe tannins and lovely savoury, roasted chestnut, oak notes. Finishes firm, fruit-sweet and long. Exuberant and elemental yet balanced and delicious. The best is yet to come. A great Magill Estate. The fruit was picked in early February, a good month before the heatwave.

'Rich, powerful and not at all overripe with superb line of flavour and fine tannins. Finishes bitter-sweet.' (TL) 'Defined and poised with intense black berry fruits and plush velvety textures. It finishes grippy firm but long. Balanced and delicious.' (JF)

Reserve Bin 08A Chardonnay 2014 •••• 2018

Pale yellow. Flinty, matchstick aromas with toasty, lemongrass notes. A well-concentrated, medium-weighted wine with lemon, flinty flavours, pronounced linear acidity and underlying savoury nuances. A crunchy dry finish with a bitter lemon/tonic water aftertaste. Should develop more richness and complexity with bottle age.

'Fresh attractive burnt matchstick nose with hints of toast. A fragrant flinty palate with richness balanced by a tight line of citrus and mineral notes. Creditable, different and interesting.' (JB) 'Charcuterie, bacon fat, figs and gun flint over a core of white peach and grapefruit pith. The wine is beginning to flesh out and harmonise. Great persistence, acid line and energy.' (TS)

St Henri ▽ 2018 •••• 2040

Deep purple to crimson. Fresh, deep blueberry, dark plum, blackberry aromas with dark chocolate, panforte, roasted nut notes. Rich, voluminous, layered and dense with gorgeous seductive blueberry, dark plum, blackberry, praline flavours and plentiful chocolatey fine tannins. Finishes chocolatey and sweet. Superb opulence, vigour and length. A great Penfolds vintage. 90% shiraz, 10% cabernet sauvignon.

'Gloriously opulent. Huge wine but satin smooth and harmonious with extraordinary length.' (NB) 'Intense blackberry, spicy notes with excellent concentration, fruit complexity and young present tannins. There is real freshness, power and length.' (PK)

Bin 620 Coonawarra Cabernet Shiraz 2020 •••• 2055

Deep crimson to purple. Intense inky, blackcurrant, herb garden aromas with notes of dried roses, pot pourri and savoury oak. An elemental yet infinitely complex wine with voluminous sweet fruit, blackcurrant, herb garden flavours, superb fleshiness, underlying ginger, malty oak and fine, sinewy but long tannins. Finishes firm, juicy and brambly with a touch of sage. Follows confidently in the footsteps of a classic. Keep.

'Fabulously concentrated blueberry and blackberry aromas, enormous fruit sweetness, decadent, lush fruit and texture. A totally seductive and harmonious wine of great power, monumental depth and persistence.

A very great wine in the making. Astonishing.' (HH) 'Fragrant ethereal blackcurrant, lavender, mint notes. Superb, concentrated and faultless with a lovely core of dark fruits and mocha. Velvety, seamless and long. More forward than the 2004 Block 42; a great wine nonetheless.' (RKP)

Cellar Reserve Pinot Noir
NOW ••• 2016

Medium-deep crimson. Fragrant pure strawberry aromas with sage, spice, praline notes. Well-concentrated strawberry flavours with varnish, leafy, spicy nuances, plenty of sweet fruit and plentiful lacy dry tannins. Finishes long, sweet and supple.

'Spicy, herbal, stalky vegetal notes tend to dominate the fruit characters. A twinge of acidity and ample tannins. Not a bad wine.' (HH)

Cellar Reserve Sangiovese ▽
NOW ••• 2020

Medium-deep colour. Intense dark cherry, malty, sweet fruit aromas with a touch of dark chocolate. Lovely buoyant, generous wine with dark cherry, malty, sweet fruit flavours and supple, savoury, satin tannins. Finishes chalky, long and sweet.

'Ripe dark fruits and liquorice on the nose. The palate is infused with dark cherry fruit, clove, liquorice notes and hints of dried fig/tobacco. A very youthful satin-textured wine. Some earth, spicy characters at the finish.' (JB) 'Attractive dark berry, almost plummy aromas. Unashamedly full-bodied but it does have balance in all departments.' (JH)

Cellar Reserve McLaren Vale Tempranillo
NOW ••• 2018

Medium crimson. Intense cherry stone, ginger, chinotto/cola aromas with mocha, graphite, earthy notes. Quite supple with sinuous cherry stone, oily, liquorice flavours and fine, bitter-sweet tannins. Builds up chalky and dry.

'Earthy, spicy, tobacco notes with a beautiful floral, violet lift. The palate is structured but fleshy and concentrated with fine mineral acidity that lengthens the flavours.' (TL) 'Rich spicy nose with plum/prune, anise aromas and dominating vanilla oak. Fleshy and gentle, medium-bodied and rounded with soft powdery tannins. Easy-going and balanced with good length.' (HH)

Cellar Reserve Barossa Cabernet Sauvignon ▽
NOW ••• 2023

Deep crimson purple. Very seductive, musky, dark berry, blackcurrant aromas with roasted chestnut, vanilla/malt notes. The palate is still elemental but beautifully concentrated with musky, dark berry, blackcurrant, graphite flavours, intense malty, ginger, oak and abundant sweet and savoury tannins. It finishes firm and chocolatey with plenty of fruit sweetness. A classic Penfolds vintage.

'A mouth-filling wine. Less varietal than the 2006, but it will develop over 10–15 years. It will always speak more of place than variety.' (JH) 'Squishy berry fruit, light florals and lead pencil notes are followed by a soft, long, juicy and perfectly balanced palate. It's refined and tight with great cellaring potential.' (RKP)

Bin 51 Eden Valley Riesling ▽
NOW ••• 2016

Pale–medium colour. Fragrant orange, mandarin, lemon aromas with a touch of petrol! Mandarin, oily flavours and a crisp, touch sour acidity. Quite lean on the mid-palate but builds energy at the finish.

'A very youthful-looking wine with mineral, dusty, earthy, flinty aromas and the beginnings of toast. Taut, reserved, nervy and lean, with a tight focus, great line and length. It's all ahead of this one.' (HH)

Bin 311 Chardonnay ▽
NOW ••• 2014

Pale yellow. Flinty, oyster shell, lemon curd aromas with a touch of pineapple. Well-concentrated, minerally wine with intense lemon curd, tonic water flavours, emerging honey/toasty complexity and crisp, linear acidity. A leaner style but shows plenty of vinosity and freshness. Drink now or keep for a few years. 100% Tumbarumba (NSW).

'Very attractive wine with intense oyster shell, grapefruit, green plum notes and a fusion of bright mineral acidity. Strong fresh flinty characters run beautifully through the wine.' (JB) 'Sherbetty lemon, lime aromas, flinty complexity and honeyed notes. Lovely flowing mouth-feel; complete, fine and silky with a dry austere end. A worked style that results in more richness and complexity.' (RKP)

Bin 389 Cabernet Shiraz 2016 ••• 2050

Medium-deep crimson. Fresh, powerful and elemental with intense inky, mulberry, blackberry, aniseed, ginger, malt aromas. The palate is immensely concentrated, fleshy and vibrant with plush blackberry, cassis, mulberry fruit, underlying ginger, malty, vanilla oak notes and plentiful chocolatey sweet tannins. It finishes velvety firm, sweet-fruited and long. Lovely density, richness and length. Easy to drink now, yet still best to keep.

'A powerful core of deep blackcurrant, cassis and mulberry fruit concentration ripples and tumbles from start to an impressively long inky finish.' (TS) 'Lifted ripe primary dark berry fruits. Exuberantly juicy with concentrated mouth-filling fruit infused with violet and exotic spice notes. The tannins are quite muscular but it has the fruit weight to support them.' (JB)

Bin 150 Marananga Shiraz ◈ NOW ••• 2030

Deep crimson. Intense, dark berry, dark chocolate aromas with roasted chestnut, vanilla notes. Well-concentrated dark berry, espresso, roasted chestnut flavours with chewy firm tannins and vanilla, malty oak complexity. Finishing firm, chalky and fruit-sweet. Plenty of volume and richness but still elemental. Drink now or keep for optimum enjoyment.

'Subtly fragrant nose of dark fruits and spice. Plush and soft with dark fruit, cedar, roasted spice, dark chocolate flavours and ripe velvety tannins. A hedonistic style.' (JB) 'A full-bodied yet relatively soft, pillowy wine with fairly high extract. I have always liked it within the context of the Penfolds red wine style.' (JH)

Bin 407 Cabernet Sauvignon ◈ 2016 ••• 2030

Deep crimson. Beautiful, musky, cassis aromas with a touch of sage, vanilla and roasted chestnut notes. Rich and buoyant, fruit-driven wine with seductive cassis, plum flavours, superb vanilla, malt, roasted chestnut notes and velvety graphite tannins. Finishes grainy firm with aniseed notes. A very generous style with great balance, volume and weight. Easy to drink now but the best is yet to come.

'A definite shift up in ripeness with powerful aromatics and fruit presence.' (TL) 'Abundant blackcurrant, mulberry, dark chocolate flavours and ripe tannins. An excellent outcome.' (JH)

Kalimna Bin 28 Shiraz NOW ••• 2025

Medium-deep crimson. Intense liquorice/aniseed, dark berry, praline aromas. Still elemental, but very seductive with incredibly concentrated, sweet buoyant dark berry, plum, liquorice flavours and brambly firm tannins. Finishes juicy and sweet. Balanced and delicious, but the best is surely to come?

'Ripe dark plum and Morello cherry nose and flavours. A voluptuous/velvety palate with plenty of fruit intensity and ripe powdery tannins. Very true to type.' (JB) 'Luscious juicy wine. Not in the least porty. Surely picked at the right time in various regions; balanced, supple and long.' (JH)

Bin 128 Shiraz ◈ 2015 ••• 2025

Deep crimson to purple. Intense, classic blueberry, blackberry pastille aromas with aniseed, dark chocolate, herb garden, pot pourri notes. Superbly buoyant wine with rich blueberry, blackberry flavours, some aniseed, white pepper notes, underlying savoury oak and loose-knit but firm, chalky tannins. Lovely persistency at the finish. Completely articulates the modern style. Delicious to drink now but will get better with more time.

'Very attractive dried spices, walnut, bay leaf, garrigue aromas with hints of star anise. Big, solid wine with a grippy tannin finish. Forceful, emphatic and long. Not a wine of particular finesse at the moment.' (HH)

Bin 138 Shiraz Grenache Mourvèdre NOW ••• 2020

Deep crimson. Intense, smooth, blackberry, musky aromas with praline notes. Rich voluminous, seductive palate with abundant sweet blackberry/cassis, praline flavours and plentiful chocolatey, velvety tannins. Finishes firm but sweet-fruited. Lovely balance and weight. Drink or keep for a while. 42% shiraz, 30% mataro, 28% grenache.

'Full-bodied, rich and ripe with plenty of savoury, spice notes and menthol nuances. The palate is deeply concentrated with ripe black fruit, spicy flavours, grippy tannins and balanced acidity. A touch hard on the finish.' (JF) 'Lovely floral, primary red berry fruit aromas and flavours. Soft and inviting; lovely.' (RKP)

Bin 2 Shiraz Mourvèdre ⬦

NOW ••• 2018

Deep purple to crimson. Fresh, dark berry/dark chocolate aromas with lifted panforte, roasted walnut notes. Buoyant and fresh with sweet prune, dark berry, praline flavours and plentiful, chalky tannins. Finishes al dente firm and juicy. A well-balanced wine with lovely fruit density and complexity. But drink soon.

'Lovely fresh black fruit, floral, spicy aromas. Steps up a gear with tremendously deep, concentrated dark fruit and spice flavours, a smooth tannin backbone and excellent length. Quite powerful.' (HH)

Thomas Hyland Chardonnay

NOW

Medium colour. Fresh grilled nuts, marzipan, white peach, pear skin aromas with creamy notes. Fresh, white peach, marzipan flavours with chalky, al dente textures, plenty of mid-palate richness and volume, underlying oak and marked indelible/cutting acidity. Lovely flavour length. Generous creamy style but ready to drink.

'Plenty of toasty smoky oak and low-level sulphides, which somewhat dominate the fruit. Rich, juicy palate with attractive flavour and softness; sulphides are present again but do not cloud the fruit. Very pleasant drinking right now.' (HH)

Koonunga Hill Autumn Riesling ⬦

NOW ••• 2016

Pale green-gold. Beginning to show some bottle age development with brown lime, lemon curd, burnt sugar, toasty aromas. Well concentrated and fleshy with plenty of lime, lemon curd, oilskin flavours and fresh steely acidity.

'Still has a young greenish hue. Lean, tight, lime, kerosene aromas with hint of honey and toast. Crisp acid on close.' (JC) 'Spicy, ginger, lime with mineral oil, petrol. Slight sweetness is balanced by the acidity.' (DM)

Koonunga Hill *Seventy-Six* Shiraz Cabernet ⬦

NOW ••• 2022

Deep crimson. Primary elderberry, dark plum, aniseed, liquorice aromas and flavours. The palate is supple and sweet with superb fruit density and ripe tannins. Finishes hot, firm and long.

'A hot and spicy wine accentuated by alcohol.' (TG) 'Intense bold berry fruit with classic dry tannins at the close.' (JC)

Koonunga Hill Shiraz Cabernet ⬦

NOW ••• 2025

Deep crimson to purple. Very bright dark cherry, boysenberry aromas with some savoury roasted chestnut notes. A gorgeously seductive wine with plenty of sweet fruit, dark cherry, *crème de cassis*, plum, boysenberry flavours, savoury complexity and dense velvety firm tannins. Finishes with a sinuous sweet-fruited flourish. Plenty of cellaring potential, but sensational to drink now.

'Full-bodied, intensely fruited wine with briary, blackberry spice fruit.' (JC) 'Rich, big and promising.' (TG) 'Intensely floral with primary dark fruits and long sappy textures.' (JR)

2009

★★★★★/★★★★★

The ten-day heatwave during the end of January and early February caused some concern, but overall mild conditions in the Barossa and Coonawarra leading up to vintage were near perfect • A relatively small crop but very high quality vintage • The earliest start of vintage on record at Magill Estate (3 February) • Horrific bushfires in Victoria leave 173 dead in the worst bushfire disaster in Australian history • 'Swine flu' pandemic declared by the World Health Organization • The film *Slumdog Millionaire* wins top prizes at the US Academy Awards.

Bin 95 Grange

2018 ••• 2048

Deep crimson to purple. Very fragrant elderberry, dark cherry, liquorice aromas with underlying ginger, mocha, malty nuances. Lovely concentrated palate with deep-set dark cherry, elderberry flavours, fine dense leafy tannins and ginger, malty oak. Finishes bitter-sweet with massive tannins and plume of extract. Impressive, solid wine with extraordinary density, robustness and lasting power. 98% shiraz, 2% cabernet sauvignon.

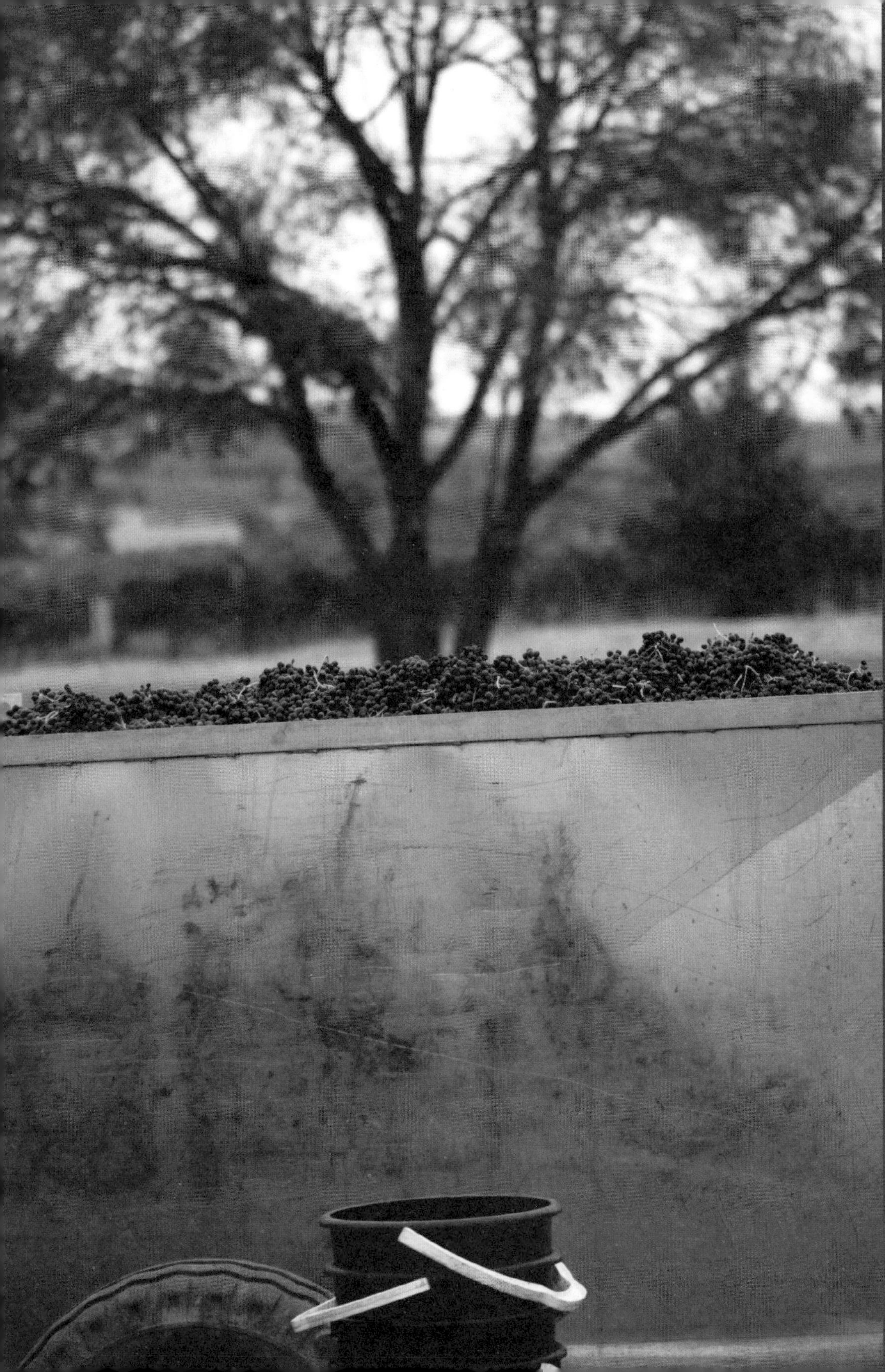

Barossa Valley, McLaren Vale, Clare Valley, Magill Estate (Adelaide).

'Inky purple. Deep powerful wine with blueberry, gingerbread, liquorice aromas and flavours. A rich expansive and powerful wine.' (JR) 'Surprisingly accessible with fragrant pomegranate, boysenberry, violet aromas and ripe velvety tannins.' (LM)

Bin 144 Yattarna Chardonnay ♦ NOW ••• 2018

Pale yellow-gold. Classic lemon, grapefruit, grilled nuts, yeasty aromas with underlying savoury oak. Well-concentrated grapefruit, grilled nut, yeasty flavours with fine cutting mineral acidity. Quite tightly structured but developing fruit complexity and creamy richness. Finishes chalky, minerally and long. Lovely to drink now but also has future potential. 51% Derwent River (Tasmania), 44% Henty (Victoria), 5% Adelaide Hills. 40% new French oak.

'More forward than 2008 and 2010 but lacks nothing in complexity and richness. A smooth creamy wine with stone fruit, pastry, grilled nut, subtle spice notes. Seamless and complete with a long dry finish.' (RKP) 'The Siamese twins of finesse and vinosity interplay with grapefruit juice, white flowers and white stone fruits. Like the 2008 and 2010, it is beautifully restrained and convincing with lovely mouth-feel and drive. There is no break in the line. Like a grand cru *Burgundy, it totally absorbs the oak and effortlessly carries the malolactic notes.' (JH)*

Bin 707 Cabernet Sauvignon 2020 ••• 2045

Deep inky purple. Elemental with musky, elderberry, black fruit, plum, aniseed aromas and vanilla, savoury oak notes. Richly concentrated, juicy and powerful with lashings of ripe blackcurrant, musky, dark cherry fruit, ample fine-grained 'caressing' tannins and fresh mocha, malty new oak. Finishes sappy firm, bitter-sweet and long. A massive wine which has yet to show its true colours. Barossa Valley, Padthaway, Coonawarra.

'Perfumed fresh cassis, plum, graphite, herb aromas with Chinese ink and medicine notes. Sweet, rich and heavy but completely seamless.' (ST) 'Loads and loads of rich ripe blackcurrant, black berry fruit, very pleasant minty touches and nicely judged toasty oak. Forceful and unbearably young with a ton of gorgeous juicy fruit.' (FW)

RWT Shiraz 2018 ••• 2038

Deep crimson. Elemental dark berry, elderberry, liquorice, aniseed aromas with vanilla, dark chocolate, cedar, spice notes. Lovely concentrated, generous and buoyant palate with juicy elderberry, dark berry, chinotto/cola notes, rich chocolatey tannins and integrated vanilla, ginger notes. It finishes chalky firm, with abundant fruit sweetness and flavour length. Ample and seductive with impressive vinosity, density and richness. A classic 'iron fist in velvet glove' style. Great potential. Only time will tell. 60% new French oak. Best to keep.

'Lovely floating blueberry, boysenberry, kirsch-like aromas. Beautifully controlled but no shortage of density of fruit intensity, exquisitely balanced tannins. Succulent and seductive and very natural.' (NB) 'Black, sweet fruit aromas with vanilla-oak notes. Very young and expressive with sweet tannins, medium body and balanced acidity.' (PK)

Bin 169 Coonawarra Cabernet Sauvignon 2016 ••• 2035

Deep crimson to purple. Intense blackcurrant, ginger, mint aromas with underlying savoury notes. Generous and substantial wine with cassis, herb garden flavours, slinky dry, loose-knit tannins and savoury, mocha oak nuances. Finishes chalky firm with a long tail of cassis, herb notes. Elemental, potent and true to its origins. A lovely wine with great potential. Keep for a while.

'A very smart wine with dark fruits, hints of violet/herbs and a subtle spicy savoury undertone. Fragrant and juicy with an angular tannin structure and marked acidity.' (JB) 'The bouquet has primary cabernet definition of tight precision with black- and redcurrants, capsicum, cedar and a lift of violets. The palate is tightly coiled with a wall of tannins. Monumentally structured with enduring potential.' (TS)

Magill Estate Shiraz ♦ 2016 ••• 2030

Medium crimson to purple. Intense redcurrant, plum aromas with camomile, ginger, peppermint, savoury notes. Fresh, seductive and fleshy with plum, redcurrant, aniseed flavours, underlying ginger, oak nuances and plentiful, chocolatey textures. Finishes firm and long. Plenty of fruit sweetness and vinosity. The second in a trio of great vintages.

'Beautifully pristine and focused black plum fruit and violets. A seductive core of juicy, pure sweet fruit and finely textured tannins. Luscious and simultaneously assertive. One of the greats.' (TS) 'Richly layered plum, black cherry, blackberry fruits. Supple soft with harmonious fruit/oak flavours and lingering tannins.' (JH)

Reserve Bin 09A Chardonnay ▽

Pale yellow. Very fresh, beautifully focused wine with flinty, matchstick, grapefruit aromas and underlying savoury notes. Attractive wine with grapefruit, lemon curd, flinty flavours and yeasty complexity, loose-knit, chalky textures and marked acidity. Finishes tangy and long. Very classic with elegance, vinosity and power. An impressive wine with an equally impressive Australian Wine Show record.

'Classical style with white stone fruits, citrus notes perfectly integrated with some funk, flinty nuances. Finely flowing, complex and long, tight and refined. Finishes dry and savoury.' (RKP) 'Super-elegant and refined. Lovely now or in a decade. Total command of the art and science of winemaking.' (JH)

St Henri

2016 ••• 2040

Deep inky purple. Perfumed cranberry, blackberry, liquorice aromas with winter green, savoury, violet notes. Well-concentrated and generously flavoured with juicy blackberry, cassis, raspberry, spicy notes and plentiful ripe granular tannins. Finishes chalky firm, long and sweet. Still elemental. 97% shiraz, 3% cabernet sauvignon.

'A ravishing, sensual, flamboyant bouquet of mulberry, raspberry, orange sorbet and strawberry jam. It is the kind of bouquet you could spread on toast! The palate is medium-bodied, well defined and shows great precision with succulent ripe tannins. It has a fleshy texture and great weight with a mineral-rich finish. This is another outstanding St Henri, one of the best!' (NM) 'Massive with purple cassis, blackberry aromas and aromatic herbs. Rich, full and concentrated with liquorice, sweet dark berry spicy notes and a long, seamless finish.' (AL)

Cellar Reserve Adelaide Hills Pinot Noir ▽

NOW ••• 2018

Medium-deep crimson purple. Fresh violet, dark cherry, praline, sage aromas with underlying vanilla. Smooth supple wine with dark cherry, praline, herb flavours and vanilla, spice nuances and fine-grained texture. Finishes chalky and long. Lovely weight and extract. Drink now or later.

'Shy, slightly closed wine with spicy clove, nutmeg notes and youthful firm tannin structure. A bold, tight, powerful pinot. Not a wine of charm, though.' (HH)

Cellar Reserve Barossa Valley Sangiovese

NOW ••• 2022

Medium-deep crimson. Dark cherry aromas with a hint of mint. Firmly structured palate with generous dark cherry, dark chocolate flavours and mid-palate richness but sinewy, muscular tannins. Flowing acidity at the finish. Will probably develop more complexity with age but tannins are unlikely to soften out.

'Lifted floral, red cherry, red liquorice aromas, slightly confectionary-like fruit on the palate with sour acidity and hard tannins. Finishes bitter-sweet. A medium-bodied wine with moderate depth and structure.' (JF) 'Straightforward cherry, almond kernel aromas. It's medium-bodied, but the high acidity knocks it off balance.' (RKP)

Cellar Reserve McLaren Vale Tempranillo

NOW ••• 2018

Medium crimson. Very fresh, juicy, plum, dark cherry aromas with some musky nuances. Sweet, juicy, pippy fruit flavours with dark cherry, raspberry notes, fine, chalky tannins and underlying brambly notes. Finishes chalky yet long and juicy. Nice but not refined. Best to drink soon.

'Earthy strawberry, red fruit nose with hints of leather. A fruit-driven wine with attractive juicy fruit and earthy undertones.' (JB) 'Luscious dark berry fruits meet an intensely spice-filled palate with nuances of Christmas pudding, stewed dried fruits. Firm, fine, taut tannins frame a long finish. A Penfolds red but not definitively Tempranillo.' (TS) 'Two years in oak. We won't do that again!' (PG)

Bin 51 Eden Valley Riesling

NOW ••• 2018

Pale colour. Intense flinty, oilskin, waxy, lime aromas. Well-concentrated wine with plenty of lime, sweet fruit flavours, some pithy grapefruit notes, mineral/salty complexity and long, fine indelible acidity. A flavourful finish with lovely fruit impact. A touch broader than 2010. Drink now or keep.

'*Remarkably youthful wine with taut, mineral, lemon pith aromas and a touch of waxiness. Dry and savoury throughout, with a tight, firm line and spring-water lightness. Very clean dry finish that rolls on and on.*' (HH)

Bin 311 Chardonnay ⊽

NOW ••• 2016

Pale yellow. Fragrant, white flower, camomile, pear skin, flinty/matchstick aromas. Buoyant, creamy and classical with camomile, pear, white peach, grilled nut flavours and fine fresh, searing acidity. Finishes minerally and long. 100% Tumbarumba (NSW).

'*Lean and flinty with citrus, toast, brioche notes and bracing acidity. A mostly savoury nuanced wine but plenty to like here.*' (JF) '*Tight, crisp, fresh energy and drive through the palate. Delicious pink grapefruit flavours with controlled oak.*' (JH)

Bin 389 Cabernet Shiraz

2018 ••• 2045

Medium-deep crimson to purple. Fresh, exuberant dark cherry, elderberry, star anise aromas accompanied by savoury, ginger oak. Richly concentrated, elemental and powerful wine packed with dark cherry, elderberry, liquorice, star anise flavours, assertive firm, almost brutish, tannins and plenty of new vanilla oak. Dense, almost impenetrable fruit is matched equally with a solid structure. Its full potential will take decades to realise. Best to keep.

'*Sensational crunchy blackcurrant, redcurrant, blueberry definition. Primary and concentrated with consummate power and strength. Has the balance and poise rarely seen at this young age. A compelling future ahead.*' (TS) '*A storm of wine with huge fruit and wonderful muscle power. It will take at least ten years to ease off and develop its truth.*' (TL)

Bin 23 Adelaide Hills Pinot Noir

NOW

Medium crimson to purple. Fresh violet, dark cherry, praline, sage aromas with underlying vanilla. Smooth, supple wine with dark cherry, praline, herb flavours, lovely extract, fine grainy texture and vanilla, spice nuances. Finishes chalky and long. Ready to drink but should keep for a while.

'*Sappy black cherry aromas with some bacon-fat notes. Soft, ample and mouth-filling, but leaner than the 2009 Cellar Reserve Pinot Noir.*' (HH)

Bin 150 Marananga Shiraz

NOW ••• 2025

Medium-deep crimson. Intense redcurrant, praline aromas with liquorice/aniseed notes. Plush, sweet fruit, redcurrant, praline flavours, plentiful chocolatey tannins and malty, ginger oak notes. Finishes firm and chalky with a peppery kick. The flavours flow across the palate with great persistency. Lovely density, richness and volume.

'*Concentrated black fruit aromas with slightly smoky/toasty oak. A tannic, firm, rather stern palate with an oaky hardness. A long, smooth finish.*' (HH) '*Wonderful primary black plum, black cherry aromas. The palate carries beautiful ripe fruit, without sacrificing anything in depth and concentration. Well-gauged dark chocolate oak and firm, fine tannins bring support and balance.*' (TS)

Bin 407 Cabernet Sauvignon ⊽

2018 ••• 2040

Medium-deep crimson to purple. Intense cassis, plum, floral, violet, vanilla aromas with aniseed/liquorice notes. Buoyant and sinuous with cassis, sweet fruit, prune, blackcurrant, aniseed, violet, cedar, plum flavours, underlying savoury oak and grainy, chewy tannins. Finishes chalky firm and inky. Has the substance and balance to last the distance.

'*A pristine expression of youthful Cabernet with intense blackcurrant, tobacco leaf, capsicum notes, waves of dark chocolate oak and mouth-coating tannins. Could this be one of the longest lived Bin 407s of all?*' (TS) '*Plummy, black fruits, cedar notes. Richly concentrated but tight with long grippy tannins.*' (RKP)

Kalimna Bin 28 Shiraz 2015 ••• 2030

Medium-deep colour. Fresh aromatic blueberry, blackberry, aniseed aromas with underlying ginger/ spice notes. Superbly concentrated, elemental and richly flavoured with voluminous blackberry, blueberry fruit, plenty of juicy/pippy notes and sinuous silky textures. Finishes velvety firm with remarkable length. Exuberant yet balanced and controlled.

'Intense lifted floral red liquorice notes with spicy, savoury nuances. Deliciously full-bodied with lovely concentration, ripe round tannins and balanced, refreshing acidity. A very sleek but powerful wine. Lots to like here.' (JF) 'Right on form with impressively poised black plum and blackberry fruit, finely structured tannins, beautiful persistence and balance. Give it plenty of time.' (TS)

Bin 128 Shiraz NOW ••• 2025

Medium-deep crimson to purple. Fresh gingerbread, red cherry, musky aromas. Well-concentrated red cherry, musky, mint flavours, underlying savoury, spicy notes and fine-grained, quite muscular tannins. Medium intensity at the finish but good length of flavour. Has good balance and fruit complexity. Drink now or keep.

'Intense star anise, gum leaf/mint high notes over meaty, clove-like aromas. Dense, concentrated and packed with flavour and guts. Impressive with long ageing potential.' (HH)

Bin 138 Shiraz Grenache Mourvèdre ▽ NOW ••• 2020

Medium-deep crimson. Fresh musky, blackberry, raspberry, plum aromas with brambly notes. Rich, ripe, concentrated palate with musky, blackberry, raspberry flavours, plenty of mid-palate richness and lovely chocolatey, graphite tannins. A brambly, juicy, minerally finish. Plush and exotic with lovely vinosity and freshness. Delicious drinking. 67% grenache, 21% shiraz, 12% mataro.

'Thrilling primary violet notes and youthful exuberance prevail over sizeable chunky tannins.' (TL) 'Firm, fresh and youthful flavours and texture. Lovely mix of juicy dark fruits and softer red berries.' (JH)

Bin 2 Shiraz Mourvèdre NOW ••• 2020

Deep purple to crimson. Intense blueberry, rhubarb, liquorice aromas with a touch of dark chocolate. Smooth glossy musky, blueberry, rhubarb, dark cherry, spicy flavours, chocolatey, slightly assertive tannins and underlying savoury nuances. Finishes chalky firm but long and sweet.

'Fresh, lifted floral, black pepper, black berry aromas and flavours. An authoritative spine of firm tannins, but only modest length compared to 2008 and 2010.' (HH)

Thomas Hyland Chardonnay NOW

Medium–pale yellow. Fragrant flinty, matchstick aromas with camomile, white peach, apricot notes. Beautiful wine with white peach and apricot flavours and some toasty/grilled nut notes and chalky/al dente textures. Finishes minerally. Ready to drink.

'Nutty, creamy wine with hints of toast and nougat. The palate is soft and round, smooth and easy-going but offers only modest depth of fruit and character.' (HH)

Koonunga Hill Autumn Riesling NOW ••• 2016

Pale green-gold. Fragrant verbena, pear skin, lemon curd, oilskin aromas. Very fresh and buoyant wine with lovely volume of fruit, al dente textures and strong indelible crisp acidity.

'Simple well-made riesling with white flowers, petrol characters, medium power and clean, fresh mouth-watering attack.' (TG) 'Sappy, pear skin, spicy aromas and flavours. Lovely clarity, precision and vivacity.' (JR)

Koonunga Hill *Seventy-Six* Shiraz Cabernet NOW ••• 2022

Deep crimson to purple. Fragrant red plum. Cherry aromas with some leafy musky notes. A juicy sweet, richly flavoured palate with plummy, red cherry, brambly notes and al dente tannins. Finishes sappy firm and long. Underlying muscular structure with plenty of baby fat.

'Smoky, cherry, cola aromas with a touch of liquorice. Good cut, focus and bite.' (JR) 'A voluptuous dark-fruited wine. An impressive winemaker-driven style.' (TG)

Koonunga Hill Shiraz Cabernet

NOW ••• 2020

Deep crimson to purple. Intense cassis, elderberry, liquorice aromas with a twist of mint. Palate is fresh, almost creamy with smooth, juicy blackberry, elderberry fruit and plentiful chalky tannins. Finishes brambly and long.

'Smooth and elegant with lush juicy fruit. Delish!' (DM) 'Rose petal "Guerlain" nose with youthful, bright, brooding palate. A serious wine.' (TG) 'Vivid black and blue fruits with flowery/lavender notes. Long and penetrating flavours, with a long peppery finish.' (JR)

2010

★★★★★/★★★★★

'The Millennium Vintage that arrived a decade late.' Peter Gago • After a wet and cold winter, generally mild conditions prevailed during the growing season with occasional heat spikes and cool periods; open-canopy management and a longer growing season allowed the fruit to ripen evenly with optimum flavour development, tannin ripeness and balanced acidities • A great Penfolds vintage • Reintroduction of Bin 170 Kalimna Shiraz • Australian Jessica Watson becomes the youngest person to sail around the world solo, unassisted and non-stop • Paul the Octopus becomes internationally famous for predicting the outcome of games in the FIFA World Cup • WikiLeaks releases a collection of over 250 000 American diplomatic cables, including 100 000 marked 'secret' or 'confidential'.

Bin 95 Grange ◈

2020 ••• 2060

Deep inky purple to crimson. A Grange of remarkable power and finesse. A classic 'iron fist in velvet glove' style, with seductive inky, elderberry, blackberry fruit, star anise, leafy nuances and mocha oak. An extravagant and expressive palate with saturated musky, inky, blackberry, elderberry fruit, plentiful satin-like tannins and underlying ginger, mocha oak. Superb fruit and tannin ripeness, wonderful concentration and balance. A classically structured and beautifully proportioned Grange with superb ageing potential. 96% shiraz, 4% cabernet sauvignon.

Barossa Valley, Clare Valley, Adelaide Hills (close to Magill), McLaren Vale, Magill Estate (Adelaide).

'A blockbuster wine; packed with fruit from front to back. Spectacular!' (TG) 'Ripe, full, plush with great length.' (JC) 'Earthy, mineral and dense as though drinking crushed stones saturated in blackcurrants.' (DM) 'Impressive purity and definition with the structure of tannins and acidity to last decades.' (MC)

Bin 144 Yattarna Chardonnay ◈

2015 ••• 2025

Pale yellow-gold. Fragrant camomile, quartz-like, lemon, herb aromas with matchstick, savoury notes. Quite restrained and lean, but vivacious and precise with delicate camomile, quartz, lemon flavours and striking, mouth-watering acidity. Finishes crisp and persistent. Superb underlying power and percussion. This really needs a few years to build composure, richness and volume. Undoubtedly a great vintage with superb potential. 96% Tasmania, 4% Adelaide Hills. 57% new French oak.

'Ultra-restrained, shy, reticent and elemental with lemon notes. The palate is tremendously tight, refined, intense and linear with great persistence and carry. Nothing sticks out; it's seamless. Hard to describe for that reason, but the theme, as with all Yattarnas, is citrusy fruit-driven characters, great purity, delicacy and line.' (HH) 'Gorgeously refined bouquet of elegant lemon blossom, crunchy pear, almond and vanilla. The palate is breathtakingly refined with honed mineral structure and high-tensile acidity that penetrates long into the finish. This is Yattarna at its zenith. A captivating wine with enduring cellaring potential.' (TS)

Bin 707 Cabernet Sauvignon ◈

2020 ••• 2050

Deep purple. Intense *crème de cassis*, blackberry, plum, malt, toasted oak aromas. Dense and elemental with exuberant dark cherry, bitter chocolate, elderberry, aniseed flavours, fine ripe chocolatey tannins and plenty of new vanilla, toasty oak. A ripe juicy finish with a plume of savoury tannins. Amazingly solid and concentrated yet luxuriant and balanced. A great vintage, possibly the greatest so far? Barossa Valley, Padthaway, Coonawarra, Wrattonbully, Adelaide Hills.

'Intense, gorgeous black cherry, blueberry, damson aromas with some floral, aniseed notes. Abundant, superbly made Cabernet with massive concentration and juicy ripe tannins. Great ageing potential.' (ER) 'Expressive and abundant violet, cassis, raspberry aromas with impressive concentration and wrap-around aristocratic tannins.' (CPT) 'Fresh and multi-layered with pronounced sweet fruit, great structure and tannin presence. Will evolve more complexity with age.' (CC)

RWT Shiraz ⬦

2020 ••• 2050

Deep crimson. Intense, elemental and luscious with expressive crushed blackberry, cranberry, elderberry aromas and underlying dark chocolate notes. Smooth buoyant, blackberry, cranberry, elderberry flavours with some mint, star anise notes, abundant chocolatey fine tannins and mocha, vanilla oak notes. Finishes long, sweet and velvety smooth. Extraordinary in its richness, power, weight and balance. Possesses all the hallmarks of a great Penfolds vintage. Great cellaring potential. 80% new French oak. Best to keep.

'Vivid, lush, plush and powerful.' (NB) 'Excellent purity, harmony and weight with good emphasis on fruit. Clean, vibrant and long but unresolved. Should become something very good in the future.' (AL) 'Potentially a great RWT with juicy blackberry, violet notes, fine tannins of perfect balance and finely knit acidity. Everything is there. Let's see how it goes.' (FK)

Bin 169 Coonawarra Cabernet Sauvignon ⬦

2018 ••• 2040

Deep crimson to purple. Fresh *crème de cassis*, praline aromas with a touch of sage, aniseed/star anise. Rich voluminous wine with deep-set cassis, dark chocolate flavours, fine-grained plentiful tannins and vanilla, spicy oak. Finishes firm and brambly with superb richness and vinosity. Superb regional and varietal expression. Keep it for a while.

'Incredibly vibrant, pure blackberry spice aromas, grainy firm tannins that will mellow out over time, lovely acidity and a super-long finish; as if there's no tomorrow. Put this away for several decades.' (JF) 'Very rich, toasted roasted aromas with intense cassis-like overtones. Rich and dense in the mouth with immense black fruit flavours, seamlessly interwoven oak and liquorice, black olive notes. A very powerful, complex and serious wine.' (HH)

Magill Estate Shiraz ⬦

2018 ••• 2035

Deep crimson to purple. Impressive pure blackberry pastille, musky plum, *crème de cassis* aromas with herb garden notes. Very juicy, elemental and sumptuous with beautiful, concentrated, blackberry pastille, plum flavours, lovely plentiful, graphite tannins and savoury vanilla oak. Finishes chalky firm and long with some brambly notes. Amazing density and energy for such a medium-weighted wine. A great Magill Estate Shiraz.

'Glorious lifted floral, ripe primary black berry aromas, lovely sweet fruit flavours and abundant ripe tannins. Fresh, youthful and exuberant but needs time to settle.' (JF) 'Everything is in perfect harmony with intense black fruits, spicy complexity and a velvety baseline of tannins.' (TL)

Reserve Bin 10A Chardonnay ⬦

NOW ••• 2020

Pale yellow. Beautifully intense, lemon rind, grapefruit, white peach aromas with mineral, flinty, chalky notes, grilled nuts and yeasty complexity. Well-concentrated white peach, camomile, flavours with superb yeasty, chalky, grapefruit textures, a touch of tonic water and vanilla oak notes. Finishes chalky firm. A delicious wine with potential now unfolding.

'Lovely funky nose with plentiful flinty struck-match characters. A rich and full palate with very generous flavours, smooth textures and tight acidity. An extended finish and aftertaste. A long-term future but surprisingly accessible at this age.' (HH) 'It has lovely exotic fragrance with spicy new oak notes, amazing fruit drive and length. Some shifting and balancing needs to happen before it settles into a whole entity.' (TL)

St Henri ⬦

2020 •••2050

Deep inky purple. Very expressive but elemental elderberry, liquorice, musky plum, praline, sweet fruit aromas. Voluminous, juicy and inviting with opulent elderberry, plum fruit flavours and dense sweet chocolatey tannins. Finishes velvety smooth with superb flavour length. All the elements are there in complete harmony, but several years are needed to reach optimum maturity and drinking window. A great vintage. A modern classic St Henri. 100% shiraz.

'Extremely promising with dark plum, beetroot, gamey, violet notes, juicy density and solid tannin backbone.' (FK) 'Incredibly youthful with masses of dark inky fruit. A very fine combination of concentration, dark fruits, freshness and fine tannins. Finishes long but still difficult to judge.' (AL)

Bin 170 Kalimna Shiraz
2020 ••• 2055

Deep crimson to purple. Fresh, elemental and powerful blueberry, blackberry aromas with star anise, herb notes. Massively concentrated with saturated blackberry, mulberry, blueberry fruit, lifted liquorice notes, underlying ginger spice, new oak and velvety, perfectly ripe tannins. An extraordinarily plush wine with remarkable percussion and energy. Finishes smooth and chocolatey. All the elements are in harmony, but this will take years to develop. A wonderful wine from a great Barossa Valley vintage. The fruit is sourced entirely from the Kalimna Vineyard's Block 3C.

'Exceptional richness and depth. The elegance is waiting to appear. This has all the magic of the remarkable 2010 vintage.' (JH) 'The bouquet is pristinely focused with lifted violets, pure blackberries and plums. The palate is a sheer epiphany of power, integrity and unremitting precision. Astonishingly generous with open fruit, beautifully fine pronounced tannins and integrated dark chocolate oak. Mesmerising for its approachability and potential longevity.' (TS)

Cellar Reserve Chardonnay ▽
NOW ••• 2018

Pale–medium yellow. Intense white peach, vanilla, grilled nut aromas with complex nutmeg, spice notes. Well-concentrated white peach, lemon curd, vanilla, grilled nut flavours, with malty, toffee, crème caramel notes, lovely mid-palate richness and yeasty complexity. Finishes crisp and long. Superb drinking now although, under screwcap, it should hold for a while.

'High-toned wine with youthful fruit intensity and plenty of toasty oak notes, creamy complexity, some smoky notes, elegant acidity and chalky textured minerality.' (JB) 'Fresh yet supple and mouth-filling with juicy overlay. Controlled winemaking is totally convincing.' (JH)

Cellar Reserve Adelaide Hills Pinot Noir ▽
NOW ••• 2022

Medium-deep purple to crimson. Fresh red cherry, praline, vanilla aromas with chinotto notes. Well-concentrated red cherry, praline flavours, vanilla, malty notes, plentiful chalky tannins and fresh mineral acidity. Finishes long and flavourful with some dried roses/herb nuances.

'Bright, spicy dark cherry, plummy aromas with dried herb overtones; not especially pinot-like but it's very good. Bright and fresh with excellent fruit depth and volume. A real step up for this wine.' (HH)

Cellar Reserve Barossa Valley Sangiovese ▽
2015 ••• 2025

Medium-deep crimson to purple. Lovely intense, musky, cherry aromas with a hint of vanilla and white pepper. Very attractive, juicy wine with red cherry, plum, elderberry flavours, ginger, aniseed nuances and fine-grained but firm tannins. Some brambly notes build up at the finish but richness and acidity persist. Still yet to harmonise. Should develop well. Best to keep for a while.

'A bouquet of mixed spice and dark berry fruit compote. There's considerable intensity of blackberry and mulberry fruits, with tart acidity and firm, finely poised tannins. Dark chocolate oak supports a long finish. There's promise here.' (TS) 'Dark berry, sour cherry in one department and tannins in another. Well matched but yet to come together. Good potential.' (JH)

Cellar Reserve McLaren Vale Tempranillo
NOW ••• 2020

Medium crimson. Fresh elderberry, sage, praline aromas with some herb garden, cola notes. Well concentrated with smooth elderberry, dark cherry, praline flavours and fine chocolatey, dry tannins. Finishes firm and a touch sappy.

'Lovely bright cherry fruits with chinotto/cola notes, savoury nuances, lively acidity and slightly drying tannins.' (JF) 'Appetising wine with stewed berries and savoury forest floor, earthy notes. A tight astringent palate.' (RKP)

Cellar Reserve Barossa Cabernet Sauvignon ▽
2015 ••• 2025

Deep crimson to purple. Intense liquorice, dark chocolate, elderberry, blackcurrant aromas with a touch of aniseed. Another substantially concentrated, voluminous wine with elderberry, blackcurrant flavours, ginger, spicy, vanilla/malt notes and plentiful chocolatey, slightly grippy tannins. Finishes firm and long.

'*Super-bright young Cabernet with fresh cassis aromas intermixed with cedar box and Moroccan spices. Ripe lively tannins and pronounced acidity on the palate. Will age for another 15 years plus.*' (JF) '*Rich fleshy and fruit-sweet, but no suggestion of over-ripeness. A delicious wine with a firm upright structure and spicy oak complexity.*' (HH)

Cellar Reserve Block 25 Kalimna Mataro ♥ 2014••• 2024

Medium-deep colour. Fragrant, aniseed, blackberry, cardamom, spicy aromas. Well balanced with dense luscious sweet blackberry, spicy, panforte flavours, plentiful firm grainy tannins with underlying new mocha, vanilla oak. Finishes firm and brambly with an alcohol kick. A rich, voluminous wine with superb fruit concentration and pronounced tannin frame. Its debut at the London Wine Trade Fair in 2012, where it won Best Red Wine of the Fair, created a mini sensation. Best to keep for a few years at least.

'*Fresh, fruit-driven plum, light herbal/spice aromas with a hint of pepper. Deep and plush with ample flavours, fruit-sweetness and grippy tannin follow-through.*' (HH) '*Classical lush blackberry jam, spice, minerally aromas and flavours. It's substantial and grainy with a particularly tannic structure. A long floral finish.*' (RKP)

Reserve Bin Adelaide Hills Merlot 2015 ••• 2025

Medium-deep colour. Very elemental with lifted spearmint, dried roses, plum, violet, black fruit aromas with a touch of graphite. Sinewy in texture, with juicy fruit, brambly, mint, herb flavours with robust tannins. Needs several years to unfold and soften.

'*Intense cassis aromas with savoury, herbal, medicinal notes. Restrained dark fruits, savoury herb, medicinal flavours, overt tannins and pronounced acidity. On a knife's edge. Needs a few more years to loosen up.*' (JB) '*Aromatic and varietal with plummy leafy notes. The palate is tight and slightly tannic but well balanced. Needs time.*' (RKP)

Bin 51 Eden Valley Riesling ♥ NOW ••• 2022

Pale colour. Classic lemon, lime, flinty, whetstone aromas. Fresh and delicious wine with intense lemon curd, lime flavours and a faint touch of toast. Lovely density and richness with a light chalky, grippy texture and superb mineral acidity. Finishes long and tangy. A really well-balanced wine.

'*Very fresh aromas of intense limes, lemons and grapefruit with an extra dimension of floral notes. Taut, fine and delicate with great line and persistence. Clean and bone-dry throughout. A pristine, tight, refined, seamless wine of lovely delicacy but also intensity and depth. Enormous potential here.*' (HH)

Bin 311 Chardonnay ♥ NOW ••• 2015

Pale yellow. Lemon curd, vanilla aromas with verbena/herb garden notes. Lemon curd, grapefruit, verbena flavours with some yeasty complexity, crisp mineral acidity and plenty of length. Some herb garden, al dente notes at the finish. A lovely wine with subtle fruit complexity, freshness and line. 100% Tumbarumba (NSW).

'*Pure white peach, fresh lemon fruit aromas with complex almond, grapefruit notes. A seamless textural presence with wonderful pure and persistent grapefruit characters and minerally poise.*' (TS) '*Tropical hints over grapefruit, mineral notes give a lush, seamless and subtly complex nose. It's a dry, austere type. It's not like Chablis per se but like an Aussie equivalent in weight, dynamics and style.*' (TL)

Bin 389 Cabernet Shiraz 2020 ••• 2050

Deep crimson purple. Impressively powerful with intense blueberry, blackberry, liquorice aromas and dark chocolate, ginger, marzipan, oak nuances. A substantial palate with remarkable richness and concentration. Saturated blackberry essence, juicy fruit, liquorice, aniseed flavours are balanced by abundant ripe generous tannins and plenty of vanilla, malt, ginger oak notes. It finishes chocolatey firm with tremendous drive and depth of flavour. All the elements are in harmony. It's delicious but ultimately this wine needs time to reveal its true potential. Keep.

'*A superb wine that encapsulates everything you could wish for. It's full-bodied with a cascade of flavours, yet it possesses a relaxed elegance and assuredness that every element is in perfect balance.*' (JH) '*Succulent fruit, plums, lemons, raspberries, earth and gentle spice. The palate is plush with a lovely seamless progression to a bite of ripe tannins.' (RKP)*

Bin 150 Marananga Shiraz ◇ 2016 ••• 2030

Deep crimson. Intense liquorice, elderberry, blackberry, meaty, ginger/spice aromas. Very fresh and balanced. Superbly concentrated wine with plenty of fleshy elderberry, blackberry fruit, chocolatey/spicy oak flavours, liquorice nuances, and dense satin-like tannins. Finishes firm and tight. A substantial wine with lovely density, fruit definition, power and complexity.

'A powerfully structured, impenetrable wine with full-bodied exuberant ripe fruit, spicy earthy notes and minerally mouth-feel. Impressive.' (JF) 'A magnificent wine with remarkable purity of fruit, meaty complexity and toasty/wooded notes. The palate is beautifully balanced with superb drive and intensity of flavour.' (TL)

Bin 407 Cabernet Sauvignon ◇ 2020 ••• 2040

Deep crimson. Intense dark berry, cassis, dark chocolate, violet, dried roses aromas. Beautifully concentrated, seductive, opulent wine with superb *crème de cassis*, plum, blueberry flavours, plenty of vanilla oak and chocolatey sweet tannins. Finishes leafy firm and persistent. Lovely fruit definition, vinosity and balance. Elemental yet showing brilliant potential.

'This is very drinkable now, but it still needs time to reach its full composure. The palate is supple, shapely and not at all heavy. Is this a style shift, I wonder?' (TL) 'Wonderfully perfumed wine with floral, inky, cassis aromas, perfectly ripe dark berry fruits, round ripe poised tannins and balanced acidity. Superb length and line. All the elements, including oak, are still integrating but nothing is poking out.' (JF)

Bin 23 Adelaide Hills Pinot Noir NOW

Medium-deep colour. Fresh red cherry, praline, vanilla aromas with a touch of chinotto/cola. Well-concentrated red cherry, praline, fruit flavours with underlying vanilla, malty notes, fresh grippy/chalky tannins and mineral acidity. Finishes long and flavourful with some dried roses, herb notes. Drink now or keep for a while.

'Earthy, forest floor, savoury aromas; it's not showing much primary fruit. Big, powerful and concentrated with grippy tannins. Impressive but not charming.' (HH)

Kalimna Bin 28 Shiraz ◇ 2015 ••• 2030

Medium-deep crimson to purple. Powerfully intense aniseed/liquorice, blueberry, mulberry, blackberry fruit aromas with ginger, mint notes. Vigorous, elemental and fleshy with generous blueberry, blackberry flavours, superb mid-palate richness and plentiful brambly, chocolatey tannins. All the elements are in perfect harmony. Impressive vinosity and finesse. A great Bin 28. Worth holding on to.

'Concentrated, floral and black-fruited. Tightly packed on the palate with ample supple tannins and a very long carry. An excellent wine with superb ageing potential.' (HH) 'Sumptuous wine with impeccable texture, balance and varietal expression. Impossible to ask for more at this price point.' (JH)

Bin 128 Shiraz ◇ 2015 ••• 2035

Deep colour. Intense aniseed, floral, camomile, black pepper, blackberry, elderberry aromas. Well-concentrated elemental wine with elderberry, black cherry flavour, plentiful chocolatey tannins and mocha, vanilla, roasted chestnut, oak complexity. Finishes chalky, firm and long. Elemental, powerful and beautifully balanced with superb cellaring potential.

'Bright and fresh with ripe, floral, cherry eau-de-vie *aromas and plenty of red and darker fruits. Fruit-driven with great intensity of flavour, marvellous fruit of pinpoint ripeness and perfect structure. A marvellous wine. This will last for 50 years, like the 1963.' (HH)*

Bin 138 Shiraz Grenache Mourvèdre ◇ NOW ••• 2025

Deep colour. Pristine blackberry, liquorice/aniseed aromas. Elemental, exuberant and seductive with beautiful sweet blackberry, liquorice/aniseed, herb flavours, pronounced chalky tannins and mineral acidity. All the elements are perfectly balanced. Superb density, richness and volume. Delicious drinking now, but will enjoy a second life! 50% grenache, 27% shiraz, 23% mataro.

'Voluptuously fruited and velvet-textured wine with notes of plums, cherries, marzipan, exotic spices and florals. A great wine; my favourite in this flight.' (JB) 'This wine is loaded with primary dark berries, mixed spice notes and pepper, anise, liquorice nuances. Considerable concentration and fruit persistence are supported by beautifully textured tannins and suede-like finesse. Impressive.' (TS)

Bin 2 Shiraz Mourvèdre ◈ NOW ••• 2025

Deep purple to crimson. Intense elemental blackberry, brambly, aniseed, herb garden aromas. Dense exuberant juicy blackberry, liquorice/brambly flavours with some panforte, dark chocolate notes and plentiful chalky dry tannins. Finishes al dente, long and sweet. Exuberant and expressive.

'Fruit-driven, almost raw young red-wine aromas with aniseed, pepper, spices and blue-fruit notes. A rich smooth and soft palate with ample but supple tannins, and a long farewell. Excellent drive, carry, length and line. The structure is impeccable.' (HH)

Thomas Hyland Chardonnay ◈ NOW ••• 2014

Pale yellow-gold. Intense pear skin, flinty aromas with a touch of white peach and lanolin. Superbly fresh and complex wine with flinty, grilled nuts, white peach flavours and fine indelible mouth-watering acidity. Finishes chalky and long.

'Youthful fresh nose with predominating cashew nut aromas. Clean and crisp with softness, balance and some savoury elements. A lighter-weighted, seamless and clean style. Great value really.' (HH)

Koonunga Hill Autumn Riesling NOW ••• 2018

Pale green-gold. Intense bright lemon curd, grapefruit, camomile, jasmine aromas. The palate is superbly balanced with sweet grapefruit, lemon curd flavours and fine minerally acidity. Finishes long and sweet.

'Elegant and ethereal with lime rind, citrus aromas, bright acidity and slightly bitter aftertaste.' (TG) 'More mid-palate weight than 2008 and 2009, perhaps slightly sweeter. Not as electric but more sumptuous.' (LM)

Koonunga Hill *Seventy-Six* Shiraz Cabernet ◈ NOW ••• 2025

Deep inky purple. Intense plush elderberry, musky, liquorice aromas. Very concentrated and powerful with deep-set juicy elderberry, dark plum, liquorice flavours, firm brambly tannins and underlying savoury complexity. Still very elemental but superbly balanced. A wolf in lamb's clothing.

'Very well proportioned. Will live for many years.' (MC) 'Plush and intense, a star in the making.' (DM)

Koonunga Hill Shiraz Cabernet ◈ NOW ••• 2025

Deep inky purple. Elemental and powerful with intense elderberry, cassis, praline, herb garden aromas and flavours with a touch of star anise. The palate is immensely concentrated, juicy and richly flavoured with dense chalky tannins and superb flavour length. A very substantial wine with great balance and vinosity.

'The missing ingredient is time!' (PG) 'Very good pure fruit and well-defined fine young tannins.' (MC) 'Super wine with primary/lifted fruit and soft, plush, delicious flavours.' (TG)

‖ 2011 ★★★ / ★★★

Cool wet conditions prevailed during the growing season, with significant disease pressure • This was a vineyard management year with open canopies and crop thinning to achieve optimum fruit ripeness; surprisingly good wines were made • Bin 707 and Bin 169 are not made • Launch of 2008 Penfolds Bin 620 Cabernet Sauvignon in Shanghai • A 9.3 magnitude earthquake off Tohoku, Japan, triggers powerful tsunami waves of up to 40.5 metres, causing widespread death and devastation • Osama bin Laden, founder of Al-Qaeda, is ambushed and killed in Pakistan by a CIA-led operation • The Royal Wedding of Prince William and Catherine Middleton at Westminster Abbey.

Bin 144 Yattarna Chardonnay

NOW ••• 2020

Pale colour. Lovely intense white peach, pear, grapefruit, camomile aromas with flinty, grilled almond notes. Crisp and vibrant on the palate with beautiful white peach, lemon curd, pear, flinty flavours, some yeasty complexity, superb mid-palate richness and crunchy, long, penetrating acidity. Wonderful line and length. An impressive result from a difficult year. Tasted in Sydney 2013. 95% Derwent Valley (Tasmania), 5% Adelaide Hills. 64% new French oak, 36% one-year-old barriques.

Cellar Reserve Adelaide Hills Pinot Noir

NOW ••• 2015

Medium-light purple to crimson. Fragrant, red cherry, leafy aromas with vanilla, ginger oak notes. Well-concentrated red cherry, vanilla, spicy flavours, fine, lacy, touch metallic tannins and tangy acidity. Finishes chalky with medium length and intensity. Drink now.

'Fresh, clean vibrant spicy aromas with definite stalky whole-bunch characters. Lively and lighter-bodied but good flavour and intensity. A surprisingly attractive wine for the vintage. An early-drinking style.' (HH)

Bin 23 Adelaide Hills Pinot Noir

NOW

Medium crimson. Fragrant, red cherry, leafy aromas with a touch of vanilla/ginger. Light- to medium-bodied with red cherry, vanilla flavours, fine lacy tannins and tangy acidity. Finishes chalky with medium length. A lighter style than previous vintages, reflecting a difficult growing season.

'Bright cherry, spice, pepper aromas with slightly green stalky notes. Palate is a touch lean, but attractive in a lighter, more fragrant style.' (HH)

Bin 51 Eden Valley Riesling

NOW ••• 2022

Pale colour. Lemon curd, camomile aromas with some flinty notes. Lean, lemon curd, flinty flavours and fine but strong acidity. Finishes quite sour. A very lean, austere Riesling but will develop more weight and richness with age.

'Lovely lifted floral aromas with jasmine, citrus blossom notes and subtle yeast/minerals background nuances. Tight, fine and minerally. Searing acidity contributes to great length and focus. Long-term potential.' (HH)

Bin 311 Chardonnay

2014 ••• 2017

Pale yellow. Fragrant flinty, oyster shell, pear skin aromas. Very lean, acidic wine with lemon, grapefruit, bitter tonic water flavours, tight linear structure and chalky, minerally finish. An austere style that should develop a touch more weight and richness with age. Keep for a year or so. 100% Henty (Victoria).

'Shy, reticent and under-developed with minerally, slaty aromas. It's taut, dry and lean with invisible oak. Better in a year or two?' (HH) *'Vibrantly fresh and tightly focused wine with crisp grapefruit, white peach notes and very good acidity. Will develop slowly but surely.' (JH)*

Bin 128 Shiraz

NOW ••• 2018

Medium crimson. Fresh red cherry, redcurrant, white pepper, roasted nut aromas with hints of vanilla. Fresh redcurrant, white pepper flavours, savoury, touch leafy tannins and underlying roasted almond/vanilla oak. Finishes sinewy firm with good flavour length. A relatively early-drinking style. Tasted at release in 2013.

Bin 138 Grenache Shiraz Mataro

NOW ••• 2018

Medium-deep colour. Fresh, vivid, floral, musky plum, cola, aniseed aromas. A well-concentrated palate with juicy plum, raspberry-jam flavours, fine bitter-sweet, slightly pippy tannins and fresh mineral acidity. Ready to drink, and delicious. Tasted at release in 2013. 65% shiraz, 20% grenache, 15% mataro.

Thomas Hyland Chardonnay

NOW ••• 2014

Pale yellow-gold. Fragrant white peach, apricot, muesli aromas with a touch of vanilla. Lean, tightly structured wine with white peach, pear skin flavours and marked acidity. Some chalky textures. Finishes quite lean and austere. More richness and volume will develop with bottle age.

'Fruit-driven with faint oatmeal notes, moderate intensity and a relatively short finish. A pared-back style reflecting the season.' (HH)

Koonunga Hill Autumn Riesling

NOW ••• 2018

Very pale green-gold. Very fresh lemon, tonic water, herb, violet aromas. A lean elemental wine with lemon curd, tonic water, grapefruit flavours and fine cutting acidity. Acidity flows across the palate like electricity!

'Really fresh, zippy, zest lime with great length and balance.' (JC) *'Lemon, lime focus with laser-like acidity. Hint of wet slate, lemon oil. Snappy finish.'* (LM)

Koonunga Hill Shiraz Cabernet

NOW ••• 2015

Medium crimson. Fresh blueberry, cranberry, raspberry aromas and flavours. Generous and supple on the palate with surprisingly seductive fruit and silky dry textures. The tannins build up firm and slightly sappy at the finish. An impressive result considering vintage conditions. Punches well above its body weight. An early-drinking style though. Tasted in Sydney 2013.

2012

★★★★ / ★★★★★

Ideal wet conditions over winter replenished soil moistures • Early budburst and flowering, followed by generally mild dry conditions, a burst of heat in January and an early cool vintage led to a low-yielding but high-quality crop: an excellent Penfolds vintage • Penfolds Nuriootpa Winery is 100 years old • Eleven ampoules of Penfolds 2004 Block 42 Cabernet Sauvignon are released and sold at $168 000 each • Peter Gago is the recipient of the Institute of Masters of Wine/*the drinks business* 'Winemakers' Winemaker Award' • Dr Ray Beckwith, OAM, dies aged 100 • John Bird completes his fifty-third vintage; Steve Lienert completes his thirty-fourth vintage • Diamond Jubilee of Queen Elizabeth II • The film *The Artist* wins five Academy Awards and becomes the first silent film to win since 1927 • Indian batsman Sachin Tendulkar becomes first cricketer to score 100 international centuries • Cyclist Lance Armstrong at the centre of a doping scandal is stripped of his seven Tour de France victories and banned from the sport for life.

Bin 144 Yattarna Chardonnay ◆

2015 ••• 2025

Pale colour. Fresh lime, apricot, verbena, vanilla aromas with flinty, matchstick notes. Concentrated lime, lemon curd, apricot, white peach fruit, underlying vanilla, grilled nut notes and plenty of creamy richness. The flavours are carried across the palate by long indelible acidity. A superbly balanced and expressive wine that will further develop in bottle. Tasted in Sydney 2013. 38% Derwent Valley (Tasmania), 37% Henty (Victoria), 25% Adelaide Hills. 45% new French oak, 55% one-year-old barriques.

Bin 51 Eden Valley Riesling ◆

NOW ••• 2025

Pale colour. Intense grapefruit, lemon curd aromas with a hint of verbena. Well-concentrated lemon curd, yeasty, grapefruit, ginger flavours, plenty of mid-palate richness and long, cutting acidity. Persistent and flavourful finish. Very classic with the vitality, line and flow to age really well.

'Fresh yeasty, light ester aromas and floral/blossom notes. Palate is taut, delicate, dry and firm. Nervy and latent, as expected in such a young wine. A very good Bin 51 still emerging from the chrysalis.' (HH)

Thomas Hyland Cool Climate Chardonnay

NOW ••• 2018

Pale colour. Fresh flinty lemon, herb, pink grapefruit, tonic water aromas. The palate is lean and minerally with fine clear acidity and chalky textures at the finish. Lovely energy and persistence of flavour. An elegantly structured wine that should develop more richness with age. Tasted in Sydney 2013.

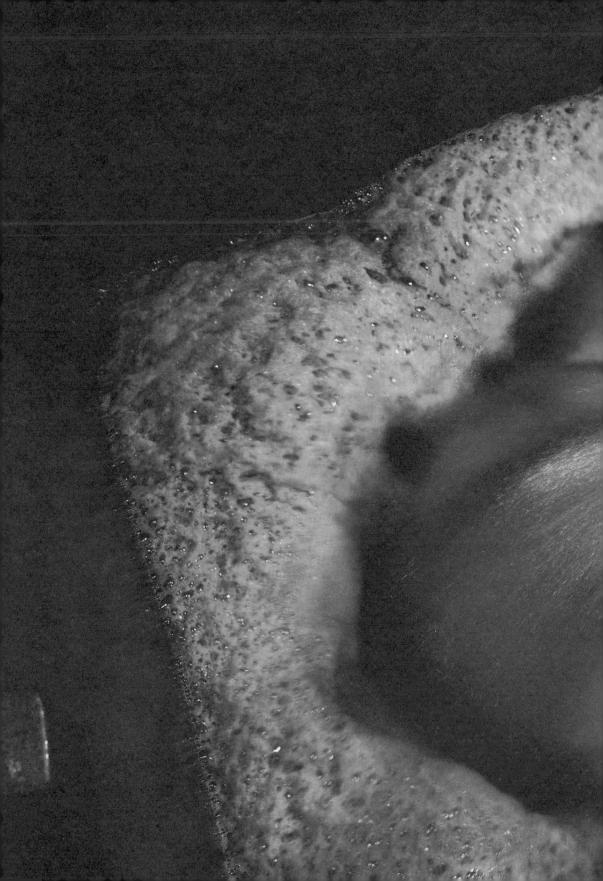

Glossary of Tasting Terms

Aromas and flavours

almond
Secondary aroma and flavour usually associated with maturity. Sometimes present in reds but more readily associated with chardonnay, especially aged bottles.

amontillado
Sherry-like character in wine. Found in some aged reds. One step from oxidation – but it can give an attractive complexity!

animal
Leather, farmyard notes. Can be detracting if overly pungent.

aniseed
A top note in wine related to ripeness of fruit and alcohol.

apricot
A very attractive primary character in whites. An appealing tertiary aroma in reds often associated with freshness and full maturity.

asparagus
A character found in wines from a cooler growing season. It suggests a wine at the cusp of ripeness. It can also be used as a varietal descriptor of sauvignon blanc.

bacon
A smoked meat character, probably derived from oak.

balsamic
A lifted, slightly volatile note that can give freshness and complexity. It can also detract from a wine if too dominant.

barnyard
A character associated with *brettanomyces*. If it's too much it can overwhelm the aromas. A metallic texture on the palate can follow. Winemakers have a tendency to slam it, but if subliminally present it can add lovely complexity to wine.

bay leaf
A lovely herb-like quality that gives freshness and aromatics.

bergamot
An obscure descriptor and fragrant note alluding to fruit complexity. Similar to Turkish Delight or vermouth.

bilgy
An unattractive dirty character in wine, reminiscent of stagnant pond water.

bitter chocolate
A lovely secondary character related to fruit and oak integration.

bitumen
A roasted earth/tarry note often associated with powerful vintages. Not unattractive when balanced with decent fruit.

blackberry
A black fruit character, usually associated with young ripe shiraz (and sometimes cabernet sauvignon) especially from warmer regions including Barossa and McLaren Vale.

black cherry
A black fruit character, associated with very young but well concentrated fruit.

black olive
A secondary note particularly associated with high-quality cabernet sauvignon.

black pepper
A classic secondary note found in shiraz, sometimes described as 'Rhônish' by Australian tasters. It is usually associated with a cool but even growing season. Can be very attractive.

black plum
A black fruit character, associated with young ripe shiraz or cabernet sauvignon.

black truffle
A beautiful aromatic found in great old red wine. Difficult to really describe but it's magical!

blueberry
A youthful, sometimes exuberant, ripe fruit character found in young red wine.

Bordeaux-like
A vague reference to Bordeaux, renowned particularly for its cabernet sauvignon and merlot dominant blends. Often used when wines possess cedar/leafy notes.

bouquet
Nowadays *aroma*, *nose* and *bouquet* mean the same thing. Twenty years ago 'bouquet' meant the bottle-age characters in wine.

boysenberry
An exuberant black fruit with slight musky notes. Always associated with young red wine. Generally evolves into other characters after three or four years.

bramble/brambly
An aroma reminiscent of hedgerows and fresh leaves. The palate will often have a grippy firmness.

brettanomyces/brett
A barnyard or horse blanket–type character that flattens the aromas, tightens up and strips the palate. Winemakers hate it. Wine lovers enjoy it when it is faintly noticeable. Rarely found in Penfolds wines, post-1996 vintages.

brine/briny
A salty, minerally note often found in ageing red wine. Very attractive when balanced with sweet fruit notes.

Burgundian
The Burgundy reference point is generally used as praise for complexity, supple body and texture.

camomile
An alluring white flower aroma that gives fragrance and lift.

camphor
A character close to eucalyptus and mint. Lifts the aromas, but when dominant it can detract.

capsicum
A fragrant note in wine suggesting a cool growing season.

cassis
The classic aroma of cabernet sauvignon. A beautiful pure blackcurrant note that pervades through the wine as it ages.

cedar
A lifted note; often the perfect accompaniment to cassis. It derives from both fruit and oak.

charcoal
A smoky character in wine, verging on graphite.

charcuterie
Dried or cured meats. These are notes derived from both oak and the ageing process.

Chinese medicine shop
A term that has become vogue in Hong Kong and China. An incredible mixture of spices and sometimes mint. Perhaps a descriptor that praises low levels of *brettanomyces* in wine?

chinotto
Sarsaparilla and cola are also similar notes. A very attractive character associated with the ageing of red-fruited wines.

choco-berry
The classic descriptor for a Penfolds red with bottle age. It's a shortening of chocolate and blackberry.

Christmas cake
Rich mixed dark and dried fruits. It is often associated with the very best parcels of shiraz fruit. Panforte is a similar descriptor.

Christmas spice
A vague term that evokes rich, dried spiced fruits.

cigar box
An evocative and desirable character often found in aged cabernet sauvignon; an appealing combination of tobacco and cedar.

cinnamon
A spice note associated with young wine and new oak.

citric
A tangy acid note. Very attractive in young whites, particularly riesling. Suggests freshness and youth.

cocoa
Dark chocolate notes without any sweetness. Associated with oak and early maturation characters.

coffee
Ground coffee and espresso are variations. A nuance often seen in young intense red wines matured in oak.

cola
An evolving note usually found in lighter red-fruited wines.

compote
A similar term to fresh stewed fruit.

confectionary
Some wine writers call this jube or lolly character. It is normally used to describe wines that have a boiled sweet, or one-dimensional ripe fruit character. Not to be confused with pure fruit notes. Often associated with esters in young wine.

confit/confiture
Another word for fruit preservative/ jam/fruit glace, but usually a very positive term suggesting intensity, fruit sweetness and tang.

cranberry
A red fruit character usually found in lighter style or medium-bodied wines.

crème brûlée
A toffee, caramel note that is usually found in aged wines. It can add wonderful complexity or detract from a wine, depending on its intensity.

crème de cassis
A pure fruit variant of cassis, suggesting beautiful ripe notes. Another classic descriptor for cabernet sauvignon aromas.

cumin
A spice note associated with oak and maturation.

dark berry/dark fruit
A catch-all description or abbreviated term for dark fruits including blackberry, blackcurrant, black cherry.

dark chocolate
A classic tasting term for the Penfolds red wine style. A character associated with oak, ripe fruit tannin and maturation. Usually found in wines with intensity and concentration.

dates
A dried fruit character found in evolving ripe shiraz, particularly.

demi-glace
Pan scrapings, pan juices are similar yet less attractive terms. It suggests a fresh savoury, meaty, even salty, complexity in wine.

developed
Evolved or mature in character.

dried fruits
The smell of raisins, prunes and dried apricots, etc. Often found in red wines from warm to hot growing seasons.

dried lavender
A fragrant/lifted aroma similar to herb garden.

dried leaves
An attractive fragrant but secondary aroma often found with ageing wine. *Sous bois* (under tree) is a similar term.

dried out
A term often used to suggest that the wine has lost its fruit aromas and flavours.

dried roses
A very attractive aromatic note often associated with younger wines from a cool growing season.

dusty
A smell derived from oak, tannin or even bottle age rather than fruit.

earth/earthy
A classic tasting note for an aged red wine, matured in oak.

elderberry
A black fruit character used to describe very youthful and intense fruit. Many young Penfolds reds, with substance, possess this character.

espresso
An intense and hugely attractive coffee note found in intense and concentrated Penfolds vintages.

ethereal
An emotive and imprecise term suggesting lifted and enticing aromas.

eucalypt
A lifted, almost volatile, minty/ oily note. It's a character found in some Australian reds, possibly because eucalyptus trees are present near vineyards. If overt, it can be unattractive.

evolved
Mature or developed.

expressive
Lively, compelling and characterful.

fading/faded
The wine has reached its optimum drinking window and now beginning to lose freshness, lustre and balance.

fennel
Similar to aniseed and liquorice.

feral
A strong off note in wine. A sentiment rather than an accurate descriptor that suggests the wine is 'not quite right!'

ferric
A smell of rusty old nails.

floral
Aromas and flavours reminiscent of scented flowers.

forest floor
The smell of undergrowth in a forest. *Sous bois* is the French term but used often by English-speaking tasters. Often associated with graceful middle-aged reds.

forward
The wine is already showing advanced age. Not necessarily bad.

frankincense
A highly perfumed, incense-like aroma.

fresh/freshness
Clean, bright, attractive and natural smelling.

fruit complexity
A myriad fruit aromas exuding a rich, heady and beautiful perfume.

gamey
The smell of wild game. Often used with rich red wines with a touch of development.

garrigue
Beautiful scent of wild herbs including rosemary, lavender, marjoram and verbena.

geranium
A pungent note in wine, suggesting faulty development.

ginger
A lovely character associated with new oak.

graphite
The smell of lead pencil.

green pea
An under-ripe green note in wine.

grilled nut
The gorgeous smell of barrel fermentation.

gun flint
Cordite, struck match notes.

herbaceous
Freshly cut grass, often associated with aromatic wines.

herbal
At the cusp of ripeness; a fragrant but slightly green note.

herb garden
An attractive descriptor. Highly scented and perfumed.

incense
Smoky/musky aromas.

inky
(1) A description of appearance.
(2) A secondary aromatic note that often accompanies black fruit characters. (3) A tactile density descriptor!

iodine
A complex seaweedy/salty aroma.

jammy
Very ripe fruit character.

kirsch
A lifted black cherry aroma.

lead pencil
A character often associated with cabernet franc, but also an oak descriptor.

leafy
A fresh aromatic note, often found in cabernet sauvignon at the cusp of ripeness.

leather
A tertiary/mature note often found in wines of advanced age. Can suggest the wine is at its end point.

liquorice
An attractive lifted note associated with ripeness and alcohol. Prevalent in young, fresh, buoyant wines.

luscious/lush
Rich, generous and voluminous.

mature
The wine has reached cruising altitude. Generally a positive comment suggesting the wine has freed itself from the vivacity of youth.

meat stock
A very attractive secondary character reminiscent of concentrated roasted meats. Similar to pan juices/pan scrapings.

meaty
An attractive character associated with fruit complexity and barrel fermentation.

menthol
A fresh lifted note, similar to eucalypt. When combined with dark chocolate characters, can be very attractive.

mint
A fragrant note, often found in Coonawarra Cabernet Sauvignon.

miso
A secondary character reminiscent of a staple Japanese soup. A combination of saltiness, meaty complexity and sesame. Very moreish and evoking the idea of umami.

mocha
A complex note of chocolate and coffee; associated with new oak and fruit complexity.

molasses
Rich powerful aroma suggesting concentration.

Moroccan spice
Attractive secondary note associated with oak and fruit integration.

mouthfeel
The texture of wine in the mouth.

mulberry
A primary fruit character found in ripe shiraz.

mushroom
A tertiary note found in older wines.

musky
A youthful perfumed note often associated with grenache.

nose
Another word for bouquet.

nougat
A sweet new oak character.

nutmeg
An oak maturation character.

nutty
A character derived from barrel fermentation.

orange peel
Found in both white and red wines. Suggests botrytis in white wine but in red wine it is usually associated with age.

oxidised
The wine has lost its freshness because of over-exposure to air. A fault.

panforte
Rich ripe notes found in young and old wine. A combination of fruit concentration and new oak.

pan juices/pan scrapings
An intense roasted meat character associated with age and barrel fermentation.

pastille
A clear fruit note – especially found in young, perfectly ripe cabernet sauvignon.

pencil shavings
A cedar, lead pencil note derived from oak.

pepper
A lifted note often associated with a cool growing season or region and shiraz.

pine resin
An unattractive, oily note.

plush
Rich, dense, voluminous and concentrated.

polish
A fresh sweet waxy smell.

polished
A stylised, defined or well-finished wine.

porty
Fruit-rich, slightly spirituous aromas and flavours.

pot pourri
Aromas of dried flowers/mixed florals.

powerful
A wine with potent aromas and flavours.

praline
An elegant dark chocolate, sometimes nutty, note connected with shiraz and oak.

prune
Very ripe dried fruit aromas often associated with hot vintages.

pure fruit
Beautiful perfectly ripe, pristine clear fruit aromas.

raisin
Over-ripe or developed note in tawny styles.

raspberry
A lovely fresh red fruit character that is generally found in light- to medium-weighted wines, particularly shiraz.

red cherry
A red fruit note associated with medium-concentrated red wine styles.

redcurrant
A red fruit note found in lighter-weighted Cabernet Sauvignon.

red plum
A red fruit character that is found in medium-concentrated wines.

Rhônish
A wine reminiscent of Côte du Rhône style.

rhubarb
An attractive sweet and herbal aroma.

roasted capsicum/roasted red pepper
A rich sweet, slightly smoky aroma.

roasted chestnut
A character perceivably derived from new oak.

roasted hazelnut
Found in aged chardonnay and barrel-matured reds.

roasted walnut
A maturation character rather than new oak. It may be linked to barrel fermentation.

rose petal
Lovely fragrant note often found in Pinot Noir.

rustic
An imprecise descriptor suggesting earthy or slightly astringent characters in wine.

sage
A gentle lift of mint. A character often seen in Eden Valley fruit, especially Shiraz.

saline
Salty.

sandalwood
Lifted aromas reminiscent of joss sticks.

sappy
A green note in wine often associated with stalks in the vinification process. Sublime when done well.

sarsaparilla
Root beer, chinotto-like aromas.

savoury
Not overly sweet, with developed maturation notes or underlying oak.

sea breezy
A fresh salty aroma.

seasoned oak
Another expression for older oak.

seaweed
A briny, slightly iodine note, usually associated with early secondary development.

sesame seed
A beautiful note similar to umami; derived from maturation and fruit development.

shellac
Varnish.

silage
A sweet tobacco, molasses, slightly stale aroma.

smoke/smoky
A character associated with oak maturation. Sometimes in the fruit.

sous bois
The smell of the forest floor.

soy
A very attractive salty/yeasty/caramelised note usually derived from bottle age.

spearmint
A fresh sweet mint herbal note.

spices
A generic note meaning the wine is evolved with fresh secondary notes reminiscent of mixed spices. Sometimes refers to oak flavour.

stale
Lacks freshness.

star anise
A very attractive lifted, slight aniseed note.

stemmy
Another variation of stalky.

stewy/stewed fruits
Intense sweet, fresh, concentrated fruit aromas.

strawberry
Beautiful ripe red fruit note, often associated with pinot noir and lighter-weighted wines.

sumptuous
Generous, rich and opulent. Usually reserved for great vintages.

sweaty
A very imprecise term. A slightly salty/stale note, possibly a low-level fault.

sweet fruit
The natural sweetness and presence of fruit pervades through the wine. A classic descriptor for Penfolds reds.

Szechuan peppercorn
Hot and spicy.

tar
Intense earthy/bitumen note associated with a warm vintage and age.

tawny-like
Smells like a fortified wine with lifted raisin notes.

tea leaf
An imprecise reference point. A fresh fragrant herb, spring flower aroma like English Breakfast tea.

toasted/toasty
Lovely complex character associated with aged Riesling or barrel maturation in reds.

tobacco
A tertiary character in red wine or a descriptor for Sangiovese.

toffee
A sweet molasses type of aroma. Suggests concentration and development.

tomato leaf
A slightly herbal note often associated with Cabernet Sauvignon.

truffle
A magical, almost indescribable aroma of forest floor, roasted spices and dried florals. One of the most desirable tertiary aromas that can be found in red wine. A rare sensation.

Turkish Delight
A rosewater/bergamot note in wine.

umami
A Japanese term used in wine tasting. A salty, savoury note often found in miso soup.

underbrush
A variation of forest floor and *sous bois*.

unevolved
Young or elemental. No signs of fruit complexity or bottle development.

vanilla
Sweet note associated with new French oak. Sometimes seen in American oak too.

Vegemite
Australian term based on a brand of yeast extract.

vegetal
The smell of under-ripe fruit.

violet
A beautiful aroma that can only result from picking the fruit at the perfect cusp of ripeness.

volatile/volatile acidity/VA
A controversial subject. All wines possess some volatile acidity. Max Schubert used this element in wine to lift the aromas and flavours. At its extreme it can be a fault. In this case it's called vinegar!

white pepper
A subtle, spicy note often found in cool-climate Shiraz.

wood spice
An ambiguous term for oak-matured characters.

wood varnish
A lifted note connected with a strong new oak influence. It can add complexity or it can detract, depending on the mix.

Palate

after-taste
The persistent taste of wine after it has been consumed.

alcohol
This key element, when pronounced, brings a fiery taste to wine. If it's balanced, it brings buoyancy, elevated mouthfeel and vinosity.

al dente
Slightly chewy texture, like perfectly cooked spaghetti.

angular
A rigid tannin structure.

backbone
The tannin spine of the wine.

big structure
Strong flavours, concentration and tannins.

brawny
A muscular presence.

bristle
Pronounced sweeping tannins that brush through the wine.

buoyant/buoyancy
The classic Max Schubert reference for wines with sweet fruit, freshness and generosity. A key characteristic in young Penfolds reds.

chalky
Perfectly ripe powder-dry tannins.

chewy
A slightly firmer tannin structure that gives a sinewy tannin presence.

chocolatey
Ripe plentiful, supple tannins coupled sometimes with a chocolate flavour sensation.

classic/classically proportioned
Relates to the classic Penfolds red winemaking style of rich ripe fruit, barrel ferment complexity and chocolatey tannins.

composed
All the elements are in harmony.

concentration
High, medium and low concentration can relate to the intensity, power and definition of the wine.

dense
Jam-packed, compressed, laden or concentrated with fruit, oak and tannins!

density
Relates to the concentration and texture of the wine.

drink now
Best to drink up soon, within a few years.

drink soon
Probably best to drink within the next five years or so.

drinking window
A very imprecise prediction of a wine's lasting power. Wine fades rather than ends at a particular point.

elegant
A term suggesting a well-balanced wine with understated power.

elemental
All the elements of fruit, oak, tannin and acidity are juxtaposed rather than integrated.

expansive
Full of flavour and generosity.

fade/fading
The wine is losing its freshness, balance and definition. All the elements are unravelling.

fine acidity
A balanced, unobtrusive line of acidity.

fine-grained
A classic descriptor for Cabernet Sauvignon tannins at optimum ripeness and expression.

finesse
All the elements are gracefully balanced.

finish
The way the wine ends at the back palate.

firm
A rigid tannin structure.

fleshy
Ample fruit and density.

fragmented/fragmenting
The wine is beginning to fall apart.

full-bodied
Rich and flavourful.

generous
Full-flavoured and approachable.

glossy
An element of refinement, polish.

graphite
A textural quality with supple ultra-fine tannins, often associated with oak (see lead pencil).

gravel/gravelly
A fair mouthful of chunky tannins.

green
Under-ripe.

grip/grippy
Firm, grasping and dominating tannins.

gritty
A noticeable gravelly texture.

inky
Saturated in flavour and without overt tannin presence.

juicy
Fruity, fresh and youthful.

lacy
Very fine supple, loose-knit tannins.

leafy
Slightly green, often under-ripe notes. Can bring freshness and interest.

length
Relates to how the wine lasts and disappears on the palate. A wine with great length has significant presence and persistence.

lighter style
A lightly weighted or less concentrated style.

line and length
A classic Australian tasting term borrowed from cricketing parlance. It relates to the way acidity, tannin and flavours move across the palate.

long
Another term for length.

loose-knit
The tannins are spread evenly out over the palate.

masculine
Sinewy or muscular in structure.

medium-bodied
Moderately concentrated.

metallic
An unattractive sharp palate structure sometimes associated with *brettanomyces*.

minerally
A controversial term, but used to describe the assorted sensations of acidity on the palate.

muscular
A firmly structured and solid wine.

new oak
American and French oak add flavour and structure to wine. The characters of new French oak can range from vanilla, ginger and nutmeg to roasted chestnut. American oak can bring a malty sweetness or lavish coconut, tropical note. If well seasoned, it can also show vanilla notes. Texturally oak can add to the tannin structure in wine.

old oak
Also called seasoned oak. When wine is aged in older oak, the maturation characters are all derived from interaction of fruit, tannin and acidity. The ingress of oxygen into barrels is important and cannot be similarly obtained through storage in stainless steel.

palimpsest
Multi-layered.

past
The wine is generally past its best drinking.

peak
The optimum drinking period.

Penfolds style
Rich, physiologically ripe fruit, rounded tannins and integrated oak.

pippy
Juicy with slightly sappy tannins.

poised
Balanced, well-defined.

polished (tannins)
Perfectly ripe and finely structured.

powdery
Dusty tannin structure.

residual (oak/sugar, etc.)
The faint remnants/residue.

resolved
The tannins are now perfectly integrated via polymerisation.

restrained
Reserved in nature. Neither overt or forward. Not much to give at this stage.

rich/richness
Concentrated and flavourful.

ripe (tannins and fruit)
Fully mature fruit or tannins.

robust
A vigorous and full-bodied wine.

rusty
The tannins are firm, sinewy and a touch metallic. Like rusty old nails! Ferric is a similar term.

satin
Perfectly smooth tannins.

saturated
Flooded with fruit or flavour.

savoury
Characters derived mainly from oak ageing and bottle age.

seamless
A fashionable term meaning all the elements are in balance.

searing
Intense acidity or tannin.

secondary
The second phase of wine development when all the elements are integrated and bottle age characters are now present.

sheen
(1) A glossy impression on the palate. The flavours and textures glide across the palate. (2) Also refers to lustre of the wine.

silky
Soft and smooth tannin structure.

sinewy
Muscular firm, but unyielding tannin structure.

sinuous
Graceful in structure and flowing in flavour.

slinky
Fine elegant tannins with slightly firm structure.

smooth
Soft and easy to drink.

solid
Firm and dense.

soupy
Densely structured, almost cumbersome.

structure
The basic architecture of the wine: fruit, oak, tannin and acidity.

style
The overall look or genre of the wine.

stylish
Smartly put together.

substantial
Ample in structure; a wine with considerable presence.

supple
Sinuous and smooth.

sweet fruit
The classic Penfolds descriptor. The natural sweetness derived from fruit rather than sugar or alcohol.

tannin plume
Beautiful tail of tannins that follows the fruit at the finish.

tannin slick
The residue of tannins left on the back palate, once the fruit has disappeared.

tertiary notes
The wine is developing into old age. The primary fruit generally has disappeared.

texture
The consistency and the way the wine feels in the mouth.

tight
The acidity and tannins are giving a linear structure. More bones than flesh.

underpowered
The fruit flavours and concentration are not sufficient for the structure of the wine.

velvety
The smooth chocolatey texture/tannin structure of the wine.

vertex
The zenith or peak of development.

vigour/vigorous
The energy and velocity of the wine on the palate.

vinosity
A distinct freshness, presence and attack of wine.

volume
Relates to the body and presence of the wine.

voluminous
Having great volume and fullness of flavour.

well-concentrated
Density and richness is balanced with intensity of fruit/flavour.

youthful
A young wine.

Acknowledgements

I would like to thank all of *The Rewards of Patience* project team for the detailed planning and staging of tastings around the world. Much of the new material was uncovered through the State Library of South Australia and trove.com.au. Veteran Penfolds winemaker John Bird provided me with invaluable advice and superb insights to the winemaking culture at Penfolds during the 1960s and 1970s. John Davoren's children, Mary Ryan, Kath Elton and Rob Davoren, SJ, helped me piece together new insights about their father and the family's connection with Penfolds. The late Ray Beckwith, lucid to the last, talked about his research work during the 1930s and 1940s and how it interconnected with the development of table wine at Penfolds. Murray Marchant, Magill's senior winemaker during the 1950s and 1960s, was at Max Schubert's side during the early development of Grange. He further unravelled the politics of the day and put the competitive rivalries of Max Schubert and John Davoren into perspective. Ex-Chief Winemaker Don Ditter was also helpful on several occasions. His technical contribution to Penfolds is so important: he recalibrated winemaking standards and set the brand on course for great success, after a difficult period of consolidation during the 1970s. John Duval should also be acknowledged for his long and productive Penfolds career.

I visited various sources to cross-check and to find inspiration. Valmai Hankel, South Australia's unofficial wine historian, has unearthed material and written several great narratives about the state's winemakers and vineyards. Philip White's wickedly amusing and politically charged editorials about South Australia's wine landscape are just wonderful, yet highlight the inadequacies of heritage protection. Chris Shanahan and David Farmer are passionate advocates of Penfolds history; they added new and important perspectives about Penfolds, especially Magill Estate. A letter to Peter Gago from Max Lake, published almost in its entirety, is a beautiful vignette of how Max Schubert engaged with the outside world for feedback and support during the 1960s. Huon Hooke's biography of Max Schubert, Max Lake's *Classic Wines of Australia*, various editions of the *Wine Review* by Len Evans, Sandy Coff's observations about her father Max Schubert and various Penfolds texts belonging to Bill and Collene Kalb were again referenced. Andrea Pritzker, Yanni Wu, Fongyee Walker and David Wei translated various tasting reviews for this text.

Anders Josephson, an important Penfolds collector during the 1980s and 1990s, was a hugely influential figure during my formative years as a wine auctioneer. When I write about Penfolds, I often think of him. His contribution to the fame of Penfolds as a collectible is inestimable.

This edition could not have been written without the help of the Penfolds winemaking team and the crucial collaboration of Peter Gago. Behind the scenes Tammy Atlee, Penfolds Global Brand Business Manager, commissioned and managed this significant book project. Clare Coney edited the text meticulously. Her observations, advice and experience have brought an extra precision and flow to the manuscript. Hardie Grant's editorial and design team, headed by project editor Helen Withycombe, publishing director Fran Berry and designer Hamish Freeman have taken the Seventh Edition to a new and exciting level. Richard Humphrys and Milton Wordley contributed stunning photographs. Finally and certainly not least, I would like to thank my wonderful and long-suffering wife, Bobby Caillard, for helping me during the research stages, and the painstaking transcription of my tasting notes and timelines. I would not have been able to write this book without her.

Penfolds is an Australian national treasure and one of the world's greatest wine brands. I feel lucky and privileged to have been given the access, the trust and the friendship of so many people connected to Penfolds.

About the Author

Andrew Caillard, MW, is a veteran Australian wine auctioneer and wine critic. He is the co-founder of Langton's, Australia's leading fine wine auction house and publisher of the influential Langton's Classification of Australian Wine. In addition to his role at Langton's, he works with Australia's largest retail wine group. A recognised authority on ultra-fine Australian wine, Andrew Caillard also works as a freelance writer, tasting panellist, columnist, international wine judge and author. His articles appear in *Australian Gourmet Traveller Wine* and other magazines. For over twenty-five years he has documented the history and secondary market performance of Penfolds through various published editorials, five editions of *Penfolds The Rewards of Patience*, Langton's Classification of Australian Wine and participation in the Penfolds Red Wine Re-corking Clinics. Andrew Caillard is also the associate producer of *Red Obsession*, a feature film documentary about Bordeaux and China, which premiered at the Berlin International Film Festival in February 2013. He is a Roseworthy Graduate in Wine Marketing (1984) and a Master of Wine (1993). In 2011 Andrew Caillard was awarded the prestigious Australian Wine Communicator of the Year, and appointed a 'Baron of the Barossa'.

Penfolds Magill Estate Winery
and Restaurant
78 Penfold Road, Magill SA 5073
Cellar Door telephone:
+61 (0)8 8301 5569
Restaurant telephone:
+61 (0)8 8301 5551

Penfolds Barossa Valley Winery
Tanunda Road, Nuriootpa SA 5355
Cellar Door telephone: +61 (0)8 8568 8408

Penfolds Consumer Relations
Australia: 1300 651 650
New Zealand: 0800 651 650
United States and Canada: 1800 255 9966
United Kingdom: 020 8843 8400

Published in 2013 by Hardie Grant Books

Hardie Grant Books (Australia)
Ground Floor, Building 1
658 Church Street
Richmond, Victoria 3121
www.hardiegrant.com.au

Hardie Grant Books (UK)
Dudley House, North Suite
34–35 Southampton Street
London WC2E 7HF
www.hardiegrant.co.uk

A Cataloguing-in-Publication entry is available from
the catalogue of the National Library of Australia at
www.nla.gov.au
The Rewards of Patience (Seventh Edition)
ISBN 978 1 74270 607 8

Publishing Director: Fran Berry
Project Editor: Helen Withycombe
Editor: Clare Coney
Design Manager: Heather Menzies
Designer: Pfisterer + Freeman
Production: Todd Rechner

Colour reproduction by Splitting Image Colour Studio
Printed and bound in China by 1010 Printing
International Limited